VOLUME ONE HUNDRED AND THIRTY ONE

ADVANCES IN
PROTEIN CHEMISTRY AND STRUCTURAL BIOLOGY

Protein Interaction Networks

VOLUME ONE HUNDRED AND THIRTY ONE

ADVANCES IN PROTEIN CHEMISTRY AND STRUCTURAL BIOLOGY

Protein Interaction Networks

Edited by

ROSSEN DONEV
MicroPharm Ltd,
United Kingdom

ELSEVIER

ACADEMIC PRESS
An imprint of Elsevier

Academic Press is an imprint of Elsevier
50 Hampshire Street, 5th Floor, Cambridge, MA 02139, United States
525 B Street, Suite 1650, San Diego, CA 92101, United States
The Boulevard, Langford Lane, Kidlington, Oxford OX5 1GB, United Kingdom
125 London Wall, London, EC2Y 5AS, United Kingdom

First edition 2022

ISBN: 978-0-323-99231-2
ISSN: 1876-1623

For information on all Academic Press publications
visit our website at https://www.elsevier.com/books-and-journals

Publisher: Zoe Kruze
Acquisitions Editor: Leticia Lima
Developmental Editor: Naiza Ermin Mendoza
Production Project Manager: Abdulla Sait
Cover Designer: Victoria Pearson

Typeset by STRAIVE, India

Contents

Contributors

Abdulaziz S. Alothaim
Department of Biology, College of Science in Zulfi, Majmaah University, Majmaah, Saudi Arabia

V. Anu Preethi
School of Computer Science and Engineering, Vellore Institute of Technology, Vellore, India

Ambritha Balasundaram
Laboratory of Integrative Genomics, Department of Integrative Biology, School of Biosciences and Technology, Vellore Institute of Technology, Vellore, India

Ibrahim Baris
Department of Molecular Biology and Genetics, Koc University, Istanbul, Turkey

Sayoni Chatterjee
Laboratory of Integrative Genomics, Department of Integrative Biology, School of Biosciences and Technology, Vellore Institute of Technology, Vellore, India

Xiaoyin Chen
Department of Computer Science, University of California, Irvine, CA, United States

João Victor Paccini Coutinho
GlycoProteomics Laboratory, Department of Parasitology, ICB; Department of Parasitology, Institute of Biomedical Sciences, University of São Paulo, São Paulo, Brazil

Ankur Datta
Laboratory of Integrative Genomics, Department of Integrative Biology, School of Biosciences and Technology, Vellore Institute of Technology, Vellore, India

Brent J. Frederisy
Department of Computer Science, University of California, Irvine, CA, United States

Milana Frenkel-Morgenstern
Cancer Genomics and BioComputing of Complex Diseases Lab, Azrieli Faculty of Medicine, Bar-Ilan University, Safed, Israel

A.T. Gaspar
Center for Neuroscience and Cell Biology (CNC); Center for Innovative Biomedicine and Biotechnology (CIBB), University of Coimbra, Coimbra, Portugal

C. George Priya Doss
Laboratory of Integrative Genomics, Department of Integrative Biology, School of Biosciences and Technology, Vellore Institute of Technology, Vellore, India

Ramanathan Gnanasambandan
Laboratory of Integrative Genomics, Department of Integrative Biology, School of Biosciences and Technology, Vellore Institute of Technology, Vellore, India

Wayne B. Hayes
Department of Computer Science, University of California, Irvine, CA, United States

R. Hephzibah Cathryn
Laboratory of Integrative Genomics, Department of Integrative Biology, School of Biosciences and Technology, Vellore Institute of Technology, Vellore, India

G.V. Kameshwari Gollakota
Laboratory of Integrative Genomics, Department of Integrative Biology, School of Biosciences and Technology, Vellore Institute of Technology, Vellore, India

Amy D. Kanne
Department of Computer Science, University of California, Irvine, CA, United States

Manoj Kumar Kashyap
Amity Stem Cell Institute, Amity Medical School, Amity University Haryana, Gurugram, India

Ibrahim Halil Kavakli
Department of Chemical and Biological Engineering; Department of Molecular Biology and Genetics, Koc University, Istanbul, Turkey

Pasha Khosravi
Department of Computer Science, University of California, Irvine, CA, United States

Thales Kronenberger
Department of Internal Medicine VIII, University Hospital Tuebingen; Department of Pharmaceutical and Medicinal Chemistry, Institute of Pharmaceutical Sciences, Eberhard-Karls-Universität; Cluster of Excellence iFIT (EXC 2180) "Image-Guided and Functionally Instructed Tumor Therapies", University of Tuebingen; Tuebingen Center for Academic Drug Discovery & Development (TüCAD2), Tuebingen, Germany

Janaina Macedo-da-Silva
GlycoProteomics Laboratory, Department of Parasitology, ICB; Department of Parasitology, Institute of Biomedical Sciences, University of São Paulo, São Paulo, Brazil

Suely Kazue Nagahashi Marie
Cellular and Molecular Biology Laboratory (LIM 15), Neurology Department, Faculdade de Medicina FMUSP, Universidade de Sao Paulo, Sao Paulo, Brazil

Benedict A. Mbakogu
Department of Computer Science, University of California, Irvine, CA, United States

Tamma Medha
Laboratory of Integrative Genomics, Department of Integrative Biology, School of Biosciences and Technology, Vellore Institute of Technology, Vellore, India

I.S. Moreira
Department of Life Sciences, Center for Neuroscience and Cell Biology, Coimbra University; CNC—Center for Neuroscience and Cell Biology, CIBB—Center for Innovative Biomedicine and Biotechnology, Coimbra, Portugal

Sumit Mukherjee
Cancer Genomics and BioComputing of Complex Diseases Lab, Azrieli Faculty of Medicine, Bar-Ilan University, Safed, Israel

Sunanda Biswas Mukherjee
Cancer Genomics and BioComputing of Complex Diseases Lab, Azrieli Faculty of Medicine, Bar-Ilan University, Safed, Israel

Simon Ngao Mule
Department of Parasitology, Institute of Biomedical Sciences, University of São Paulo, São Paulo, Brazil

Nuri Ozturk
Department of Molecular Biology and Genetics, Gebze Technical University, Kocaeli, Turkey

Giuseppe Palmisano
GlycoProteomics Laboratory, Department of Parasitology, ICB; Department of Parasitology, Institute of Biomedical Sciences, University of São Paulo, São Paulo, Brazil; School of Natural Sciences, Macquarie University; Faculty of Science and engineering, Macquarie University, Sydney, NSW, Australia

L.F. Piochi
Center for Neuroscience and Cell Biology (CNC); Center for Innovative Biomedicine and Biotechnology (CIBB), University of Coimbra, Coimbra, Portugal

A.J. Preto
Center for Neuroscience and Cell Biology (CNC); Center for Innovative Biomedicine and Biotechnology (CIBB); Institute for Interdisciplinary Research, University of Coimbra; PhD Programme in Experimental Biology and Biomedicine, Institute for Interdisciplinary Research (IIIUC), University of Coimbra, Casa Costa Alemão, Coimbra, Portugal

Saravanan Ramya
Computer Aided Drug Design and Molecular Modeling Lab, Department of Bioinformatics, Science Block, Alagappa University, Karaikudi, Tamil Nadu, India

Livia Rosa-Fernandes
GlycoProteomics Laboratory, Department of Parasitology, ICB; Department of Parasitology, Institute of Biomedical Sciences, University of São Paulo, São Paulo, Brazil

N. Rosário-Ferreira
Center for Neuroscience and Cell Biology (CNC); Center for Innovative Biomedicine and Biotechnology (CIBB); Department of Chemistry, Coimbra Chemistry Center, University of Coimbra, Coimbra, Portugal

Chandrabose Selvaraj
Computer Aided Drug Design and Molecular Modeling Lab, Department of Bioinformatics, Science Block, Alagappa University, Karaikudi, Tamil Nadu, India

Gurunathan Rubha Shri
Computer Aided Drug Design and Molecular Modeling Lab, Department of Bioinformatics, Science Block, Alagappa University, Karaikudi, Tamil Nadu, India

Sanjeev Kumar Singh
Computer Aided Drug Design and Molecular Modeling Lab, Department of Bioinformatics, Science Block, Alagappa University, Karaikudi, Tamil Nadu, India

Ramamoorthy Siva
Laboratory of Integrative Genomics, Department of Integrative Biology, School of Biosciences and Technology, Vellore Institute of Technology, Vellore, India

S. Udhaya Kumar
Laboratory of Integrative Genomics, Department of Integrative Biology, School of Biosciences and Technology, Vellore Institute of Technology, Vellore, India

Rajendran Vijayakumar
Department of Biology, College of Science in Zulfi, Majmaah University, Majmaah, Saudi Arabia

Siyue Wang
Department of Computer Science, University of California, Irvine, CA, United States

Carsten Wrenger
Department of Parasitology, Institute of Biomedical Sciences, University of São Paulo, São Paulo, Brazil

Salma Younes
Department of Biomedical Sciences, College of Health and Sciences, Qatar University, QU Health, Doha, Qatar

Hatem Zayed
Department of Biomedical Sciences, College of Health and Sciences, Qatar University, QU Health, Doha, Qatar

CHAPTER ONE

On the current failure—but bright future—of topology-driven biological network alignment

Siyue Wang, Xiaoyin Chen, Brent J. Frederisy, Benedict A. Mbakogu, Amy D. Kanne, Pasha Khosravi, and Wayne B. Hayes*
Department of Computer Science, University of California, Irvine, CA, United States
*Corresponding author: e-mail address: whayes@uci.edu

Contents

Advances in Protein Chemistry and Structural Biology, Volume 131
ISSN 1876-1623
https://doi.org/10.1016/bs.apcsb.2022.05.005

Abstract

Since the function of a protein is defined by its interaction partners, and since we expect similar interaction patterns across species, the alignment of protein-protein interaction (PPI) networks between species, based on network topology alone, should uncover functionally related proteins across species. Surprisingly, despite the publication of more than *fifty* algorithms aimed at performing PPI network alignment, few have demonstrated a statistically significant link between network topology and functional similarity, and *none* have demonstrated that orthologs can be recovered using network topology alone. We find that the major contributing factors to this surprising failure are: (i) edge densities in most currently available experimental PPI networks are *demonstrably* too low to expect topological network alignment to succeed; (ii) in the few cases where the edge densities are high enough, some measures of topological similarity easily uncover functionally similar proteins while others do not; and (iii) most network alignment algorithms to date perform poorly at optimizing even their own topological objective functions, hampering their ability to use topology effectively. We demonstrate that SANA—the *Simulated Annealing Network Aligner*—significantly outperforms existing aligners at optimizing their own objective functions, even achieving near-optimal solutions when the optimal solution is known. We offer the first demonstration of global network alignments based on topology alone that align functionally similar proteins with *p*-values in some cases below 10^{-300}. We predict that topological network alignment has a bright future as edge densities increase toward the value where good alignments become possible. We demonstrate that when enough common topology is present at high enough edge densities—for example in the recent, partly synthetic networks of the *Integrated Interaction Database*—topological network alignment *easily* recovers most orthologs, paving the way toward high-throughput functional prediction based on topology-driven network alignment.

1. Introduction

"No protein is an island unto itself."

Physical Biology of the Cell, 2nd Ed. (2013), Phillips, Kondev, Theriot, and Garcia (2013, Figure 4.20).

"A complex network of protein interactions underlies cell function."

Molecular Biology of the Cell, 6th Ed., (2015), Alberts et al. (2015, p. 166).

1.1 Motivation

The statements above—taken from reputable textbooks less than a decade old—arise from one of the canonical assumptions of modern biology: that a protein's function is intimately tied to its set of interaction partners, as well as the broader context of the network in which they are all embedded. Thus, to understand cellular function requires understanding its network of protein-protein interactions (PPIs). Luckily, just as with individual proteins, the PPI network of one species is not an island unto itself: proteins come from genes, and there is significant genetic similarity across species. For example, the genomes of humans and chimps are about 96%–97% identical (Waterson et al., 2005); about 80% of mouse genes have a direct 1-to-1 ortholog in human (Pennacchio, 2003); and life-critical genes can be near-identical even across species as far apart as yeast and human (Pearson, 2013). Given the significant genetic sequence similarity between species, and since proteins arise from genes, we expect that proteins across different species that arise from orthologous genes will perform similar functions—and since function is *defined* by one's interaction partners, orthologous proteins across closely-related species—or even across distantly related species that perform critical ancestral functions—should often share interaction partners. In the parlance of network analysis, *we expect that PPI networks across species will exhibit a significant amount of topological network similarity.*

Most computational prediction of protein function today is sequence based. Sequence data are ubiquitous and cheap to acquire, and there is a robust and sophisticated ecosystem of algorithms for their comparison and alignment. Some methods attempt to predict function directly from sequence (e.g., Kulmanov & Hoehndorf, 2020), while others use longer chains of reasoning including sequence, structure, binding domains, interfaces, and compatible interaction partners (e.g., Zhang, Freddolino, & Zhang, 2017). Existing functional information anywhere along this path may implicate the original protein with a similar function via the "guilt-by-association" principle. However, this road from sequence analysis to function prediction is long, complex, and error prone: there are examples of proteins with no sequence similarity having near-identical function (Furuse, Fujita, Hiiragi, Fujimoto, & Tsukita, 1998; Schlicker, Domingues, Rahnenführer, & Lengauer, 2006); examples of proteins with identical sequence having multiple, completely different functions (Kabsch & Sander, 1984; Morrone et al., 2011); and even examples of structural but not functional similarity (Madsen et al., 1999).

In contrast, PPI interactions can be *directly* measured via yeast two-hybrid and other methods (Walhout & Vidal, 2001), allowing direct detection of interactions and allowing a much shorter path to predictions based on guilt by association. Finally, since functional similarity can exist even in the absence of sequence similarity, network-based methods may be able to predict function in cases where sequence cannot. Furthermore, functions are naturally encoded using a network schema—witness how the word "pathway" is commonly used to describe how a set of entities coordinate their interactions to perform some function.

Given the growing amount of network data, and the expected similarity of PPI networks across species, it is surprising that, despite the existence of over *fifty* published papers attempting to solve the PPI network alignment problem (see reviews such as Clark & Kalita, 2014; Mamano & Hayes, 2017; Milenković, Zhao, & Faisal, 2013), none of them have been able to use topology *alone* to recover cross-species links between proteins with known similar function such as orthologs, and only a few have been able to show a statistically significant link between Gene Ontology (GO)-based function and topology-based network alignment—unless sequence information is also used (Davis, Yaveroğlu, Malod-Dognin, Stojmirovic, & Pržulj, 2015; Faisal, Zhao, & Milenković, 2014; Gaudelet, Malod-Dognin, & Pržulj, 2018; Kuchaiev, Milenković, Memišević, Hayes, & Pržulj, 2010; Kuchaiev & Pržulj, 2011; Malod-Dognin & Pržulj, 2019). As a result, almost all network alignment algorithms today use sequence similarity between protein pairs to guide network alignments. Even more worrisome, when trying to balance the objectives of aligning proteins using network topology vs aligning proteins that have similar sequences, a negative correlation has been widely observed: topology-weighted network alignments are far less able to align functionally similar proteins than sequence-similarity-weighted network alignments. The connection between topological network similarity and functional similarity has been so tenuous for so long that many authors now refer explicitly to a "sequence-topology trade-off," effectively abandoning the promise of independently using the expected common network topology across species to uncover functional similarity (Alberich, Alcala, Llabres, Rossello, & Valiente, 2019; El-Kebir, Heringa, & Klau, 2011; Gligorijević, Malod-Dognin, & Pržulj, 2015; Kalecky & Cho, 2018; Klau, 2009; Malod-Dognin, Ban, & Pržulj, 2017).

While it is clear that the best functional predictions will ultimately come from integrating all relevant sources of data—sequence, structure, and network connections—each source of data being integrated should first pass a stringent "litmus test" demonstrating its ability, alone or in conjunction with other data, to improve the result. The widespread belief in the "sequence-topology trade-off" makes it clear that network topology has not yet passed this litmus test. Our goal in this paper is to explain this surprising failure.

1.2 The sequence-topology "trade-off"

To date, the only success in identifying biologically relevant common subnetworks of interactions across species using topology *alone* have used *graphlets* and their derivatives (Davis et al., 2015; Gaudelet et al., 2018; Malod-Dognin & Pržulj, 2019), which have also seen modest success in detecting functional similarity in network alignments (Faisal et al., 2014; Kuchaiev et al., 2010; Kuchaiev & Pržulj, 2011; Milenković, Ng, Hayes, & Pržulj, 2010). To our knowledge, in order to gain biological relevance, *every* other network alignment algorithm applied to PPI networks has had to guide the alignment with objective functions that include protein-pair sequence similarities, in order to encourage sequence—and thus functionally—similar protein pairs to align to each other. This fact has led to a widespread belief in a so-called "sequence-topology trade-off," (Alberich et al., 2019; Malod-Dognin et al., 2017) so that virtually every alignment algorithm either imposes sequence-based restrictions (El-Kebir et al., 2011; Elmsallati, Msalati, & Kalita, 2018; Ibragimov, Malek, Guo, & Baumbach, 2013; Kalecky & Cho, 2018; Kazemi, Hassani, Grossglauser, & Modarres, 2016; Klau, 2009; Malek, Ibragimov, Albrecht, & Baumbach, 2016; Mohammadi, Gleich, Kolda, & Grama, 2016; Xie et al., 2016; Yasar & Çatalyürek, 2018; Zhu, Li, Liu, Qin, & Yu, 2017), or has a literal trade-off in an objective function of the form $\alpha T(a) + (1 - \alpha) S(a)$, where a is an alignment, $T(a)$ measures the topological similarity exposed by the alignment according to some topological measure, $S(a)$ is an objective based on sequence similarity, and α is a balancing parameter that makes the trade-off explicit (Aladağ & Erten, 2013; Alkan & Erten, 2015; Chindelevitch, Ma, Liao, & Berger, 2013; Clark & Kalita, 2015; Crawford & Milenković, 2015; Gong, Peng, Ma, & Huang, 2015; Hashemifar, Ma, Naveed, Canzar, & Xu, 2016; Hashemifar & Xu, 2014; Kuchaiev & Pržulj, 2011; Malod-Dognin et al., 2017; Malod-Dognin & Pržulj, 2015; Mir, Naghibzadeh, & Saadati, 2017; Neyshabur, Khadem, Hashemifar, & Arab, 2013; Patro & Kingsford, 2012; Saraph & Milenković, 2014; Singh, Xu, & Berger, 2008; Sun, Crawford, Tang, & Milenkovic', 2015; Vijayan, Saraph, & Milenković, 2015). Virtually all methods that use the explicit trade-off have found that functional similarity is positively correlated with the weight given to sequence, and negatively correlated with the weight given to topology (Clark & Kalita, 2014; Hashemifar & Xu, 2014; Kazemi et al., 2016; Malod-Dognin et al., 2017; Patro & Kingsford, 2012).

1.3 Possible reasons for the failure

The failure of purely topology-driven network alignments to uncover functional similarity may be due to several factors. In order of decreasing pessimism, possible reasons include:

R0: Perhaps there is little or no correlation between common network topology and common protein function/orthology across species. In addition to being severely at odds with most of modern biology, in Supplementary material we explicitly demonstrate the astronomically low probability of this case (cf. Table S1 in Supplementary material).

R1: Perhaps there is a strong topology-function relationship in the *true* PPI networks, but our current data are too noisy or incomplete (Ideker & Krogan, 2012), or network coverage too uneven (Rolland et al., 2014) for *any* algorithm to detect the underlying common topology. We address this hypothesis by leveraging Information Theory (cf. Section 3.1).

R2: Perhaps current data *do* exhibit, to some extent, the expected underlying topology-function relationships, but currently available measures of topological network similarity fail to capture this relationship. For example, some objective functions may *saturate* before others (cf. Section 4.2), meaning they are inherently incapable of detecting relevant topological features even if those features exist. We address this hypothesis by measuring the ability of various objective functions to recover known functional relationships (cf. Section 4).

R3: Perhaps current topological measures *are* capable of exposing the expected functional similarities, but existing algorithms do not adequately optimize their topological objective functions due to the difficulty in optimally solving the NP-complete problems that arise in topological network alignment. We address this in Sections 3.2.1 and 3.2.2.

Hypothesis R0 may be dismissed virtually out of hand: it is assumed by modern medicine (Thomas et al., 2012), contradicts our understanding of molecular evolution (Lockhart, Steel, Hendy, & Penny, 1994), is at odds with recent explicit demonstrations of functional identity between orthologs of distantly related species (Kachroo et al., 2015), and is inconsistent with the astronomically significant *p*-values of interologs (common interactions between species—cf. Table S1 in Supplementary material).

1.4 Contribution

In this paper, we attempt to disentangle the latter three hypotheses, but in the order R1, R3, and finally R2. In particular, we walk the reader through the following observations:

Regarding R1: In Section 3.1 we propose a novel hypothesis based on Information Theory that places a lower bound on the amount of information required to uniquely specify one alignment (for example, a "correct" one) out of all possible alignments (cf. Eq. 2). We then suggest ways to estimate the formal information *content* of PPI input networks—essentially, their edge density—to test if there is enough information to satisfy the requirement. Though quantifying network information content is extremely difficult and requires significant theoretical development, we show that a preliminary closed-form estimate agrees with existing literature as well as our own computational experiments to within a few percent in edge density. We note that no pair of existing BioGRID networks have edge densities that satisfy the hypothesis, suggesting that *no* algorithm based on topology alone is capable of producing high-quality global network alignments between existing networks. Finally, we show that the recent, partially synthetic IID networks (Kotlyar, Pastrello, Malik, & Jurisica, 2018) have edge densities well above the hypothesized bound, and that excellent alignments can easily be found using several existing measures of topological similarity such as EC (Kuchaiev et al., 2010) or S^3 from MAGNA (Saraph & Milenković, 2014).

Regarding R3: We demonstrate that nine of the most recent, competitive global network aligners all produce suboptimal alignments according to their own objective functions, by showing that SANA—the *Simulated Annealing Network Aligner*—provides far more optimal values of these objectives than the original aligners do themselves (cf. Fig. 3). We further show that, given enough simulated annealing time (an hour is enough for existing PPI networks), SANA is able to produce alignments with near-optimal values of the objective function in all cases where the optimal value is known.

Regarding R2: Since SANA is a random search algorithm, each alignment is different even though all alignments achieve almost the same (and presumably near-optimal) scores of the desired objective function. This means that SANA can be used to *randomly sample* alignments from the frontier consisting of near-optimal alignments according to any given objective function. This allows us to compare the alignments produced by different topological objectives, confident that the alignments being compared are among the best possible for each chosen topological objective. We compare these objectives for their ability to (a) align proteins that share a statistically significant number of GO terms (Section 4.4), and (b) recover known orthologs between species in current PPI data (Section 4.5). Finally, we argue that no topological measure is *always* best, by demonstrating that different measures of topological similarity are applicable in different circumstances (cf. Section 6.6).

2. Preliminaries

2.1 Pairwise global network alignment (PGNA)

For the purposes of this paper, we define an alignment as a 1-to-1 mapping between the nodes of exactly two networks. More formally, let $G_1 = (V_1, E_1)$ and $G_2 = (V_2, E_2)$ be two graphs (networks) with node sets V_i and edge sets E_i, $i = 1, 2$, and assume without loss of generality that $|V_1| \leq |V_2|$. A *pairwise global network alignment* (PGNA) is a 1-to-1 mapping a from V_1 to V_2. (We show later how to extend it to a many-to-many alignment.) Given a particular alignment a, assume we have some *objective function* $f(a)$ that measures the "quality" of a in some fashion. Our goal is to find an alignment a^* that maximizes the objective:

$$f(a^*) \geq f(a)\ \forall a. \tag{1}$$

Fig. 1 depicts a schematic diagram of a small PGNA, and schematically depicts how the commonly used topological measures EC and S^3 are computed. In Fig. 1, we use colors to depict the original networks: blue for G_1 and red for G_2. Though there are many possible alignments between them, Fig. 1 depicts the visually obvious one where the square 1-2-3-4 of G_1 has simply been moved to the right until it is "on top of" the a-b-c-d square in G_2. The resulting alignment is depicted at the far right, where aligned nodes and edges are depicted in purple (a mix of blue and red), and unaligned nodes and edges retain their original color.

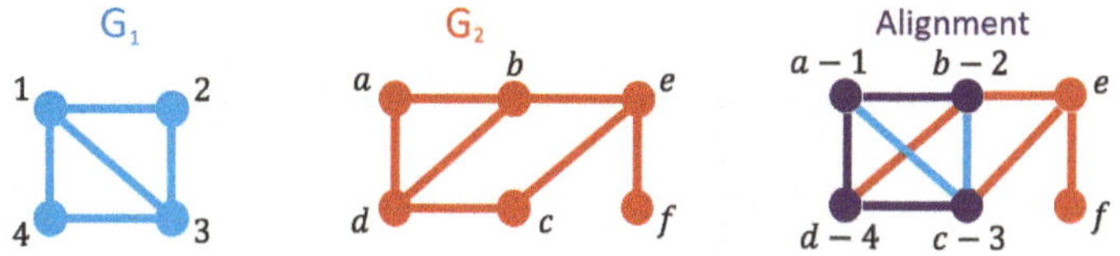

Fig. 1 A schematic depiction of a 1-to-1 pairwise global network alignment (PGNA). The input graphs are G_1 (*blue*, with fewer nodes) and G_2 (*red*, with more nodes). The alignment can be depicted itself as a network: aligned nodes and edges are *purple* (depicting a mix of *red* and *blue*), while unaligned nodes and edges retain their original color. Note there are no *blue nodes* in the alignment because, being a 1-to-1 mapping, every *blue node* must be aligned to a *red* one, producing *purple*. Two commonly used topological measures are EC and S^3, which in our schematic can be computed as $EC = |\text{purple edges in the alignment}|/|\text{all edges of } G_1|$, while $S^3 = |\text{purple edges in the alignment}|/|\text{edges of all colors between purple nodes}|$. Thus, the depicted alignment has EC$=\frac{3}{5}$ and $S^3 = \frac{3}{6}$.

Note that this is not the *best* alignment: rotating G_1 90 degrees in either direction causes its "diagonal" edge to align with the corresponding diagonal edge in G_2; this causes the EC to increase from 3/5 to 4/5, and the S^3 score to increase from 3/6 to 4/5—the denominator of S^3 decreases by one since we gained one purple edge but we *lost* two: one blue, one red.

2.2 Measuring topological similarity

There are many ways to measure topological similarity. In the context of network alignment, similarity is defined as a function of the alignment: just as with sequence similarity, we can only measure topological "similarity" once an alignment is specified, and the goal is to find an alignment which maximizes the observed topological similarity. Table 1 lists (to our knowledge) all measures of topological similarity introduced in the past 15 years in the context of pairwise PPI network alignment, along with the algorithms that introduced and/or used them; these are the measures we will be studying below for their ability to recover functionally similar regions and/or orthologs between species.

2.3 SANA: The simulated annealing network aligner

An alignment algorithm consists of two orthogonal components: first, a *objective function* M that measures the topological similarity $M(A)$ observed in a given alignment $A : V_1 \rightarrow V_2$; and second, a *search algorithm* designed to search the space of all possible alignments looking for ones that score well according to M.

Table 1 Objective functions and alignment algorithms compared.

Objective function	**Used by**	**Brief description**
EC (Kuchaiev et al., 2010)	GRAAL, MAGNA++, OptNetAlign, SANA	The percent of edges in the smaller network that align to edges in the bigger one
Symmetric Substructure Score (S^3) (Saraph & Milenković, 2014)	MAGNA(++), OptNetAlign, SANA	Similar to EC, but symmetric and measured only on the subgraph induced by the node alignment
GDV Signature Sim. (Milenković & Pržulj, 2008)	GRAAL, MAGNA++, SANA	Node-based similarity measure comparing local graphlet counts around a node
L-GRAAL GDV Sim. (Malod-Dognin & Pržulj, 2015)	L-GRAAL, SANA	L-GRAAL's modified GDV similarity, normalized differently and using only graphlets with 4 or fewer nodes
Spectral Signature Sim. (Patro & Kingsford, 2012)	GHOST, SANA	Based on the distribution of eigenvectors of normalized Laplacian matrices of surrounding subgraphs of each node
Edge Graphlet Vector Sim. (Crawford & Milenković, 2015)	GREAT, SANA	Similar to Graphlet similarity, but scores pairs of edges instead of pairs of nodes
Importance (Hashemifar & Xu, 2014)	HubAlign, ModuleAlign, SANA	Ranks nodes recursively based on their degree and the Importance of adjacent nodes; inspired by Google's PageRank (Page, Brin, Motwani, & Winograd, 1999)
Greedy EC + Degree Diff.	PROPER, INDEX, SANA	Greedily built "local EC," ties broken by degree difference
GDV-Weighted EC (WEC) (Sun et al., 2015)	WAVE, SANA	Overlapping edges receive a score based on the average of the graphlet degree difference scores of their endpoints

Many existing publications convolve the search algorithm and objective function, making it difficult to discern which is more responsible for the quality (good or bad) of a given alignment. Our algorithm, SANA (Mamano & Hayes, 2017), the *Simulated Annealing Network Aligner*, clearly separates the two and uses simulated annealing (SA, Kirkpatrick, Gelatt, & Vecchi, 1983) as the search algorithm. (SANA is available on GitHub at https://github.com/waynebhayes/SANA.) SA has a rich history of successful application to NP-complete problems across a wide array of application domains (Aguiar e Oliveira Junior, Ingber, Petraglia, Rembold Petraglia, & Augusta Soares Machado, 2012; Dowsland & Thompson, 2012; Meise, 1998; Mitra, Romeo, & Sangiovanni-Vincentelli, 1985; Park & Kim, 1998; Romeo, Sangiovanni, & Huang, 1986; Strens, 2003; Suman & Kumar, 2006; Szu & Hartley, 1987; Zhan, Lin, Zhang, & Zhong, 2016). One important aspect of SA is the choice of *temperature schedule*; SANA automatically determines effective temperature limits using an algorithm detailed elsewhere (Hayes & Mamano, 2022). The only significant unknown is the amount of CPU time one should spend traversing the temperature range; for now we determine this empirically, by performing longer-and-longer runs, each from scratch, until the final score stops increasing. For the networks in this paper, we have empirically determined that 1 hour is sufficient on the machines used in this study (a cluster of 100 identical machines, each with 96 GB of RAM and a 3.33-GHz, 24-core Intel X5680 CPU).

3. Testing hypotheses R1 and R3

Recall briefly our hypotheses for the current failure of network alignment: **R1** is that networks contain too little topological information for good alignments to be found; **R2** is that current topological measures of similarity are incapable of finding *biologically* relevant alignments because they are somehow measuring the "wrong" type of topology; and **R3** hypothesizes that current alignment algorithms, even if optimizing biologically relevant measures of topology, are producing alignments that are far from their chosen measure's optimal value, artificially degrading the objective function's ability to find relevant biology.

3.1 Addressing R1 using information theory and edge density

Let $\mathscr{A}$ be the set of all possible alignments—i.e., $\mathscr{A}$ defines the *search space*. If G_1 and G_2 have $n_1 \leq n_2$ nodes, respectively, the total number of possible 1-to-1 alignments is

$$|\mathscr{A}| = \binom{n_2}{n_1} n_1! = \frac{n_2!}{(n_2 - n_1)!}. \quad (2)$$

If we give each element of $\mathscr{A}$ a unique integer identifier between 0 and $|\mathscr{A}| - 1$, then elementary Information Theory (see Section 6.1) tells us that the minimum amount of information required to uniquely specify one alignment, measured in bits (Shannon, 1948), is

$$\begin{aligned}\text{Netbits}(n_1, n_2) &= \log_2(|\mathscr{A}|) = \log_2\left(\frac{n_2!}{(n_2 - n_1)!}\right) \\ &= \log_2\left(\prod_{k=n_2-n_1+1}^{n_2} k\right) = \sum_{k=n_2-n_1+1}^{n_2} \log_2 k.\end{aligned} \tag{3}$$

For typical PPI networks, the log of the values of k in the sum are easily computed, and so the exact value of the information requirement is easily computed (see Eq. 4).

For particular values of n_1 and n_2, we refer to Eq. (3) as Netbits(n_1, n_2); it specifies the absolute minimum amount of input information required to uniquely identify an alignment of n_1 nodes to n_2 nodes—for example, one that we may deem as the "correct" alignment. In a topological alignment of two networks, the only input information we have is the two networks, which can be viewed as two lists of edges. The way we *leverage* that information is encoded in (i) the topological objective function that measures the similarity exposed between the two networks by any given alignment and (ii) the algorithm used to build the alignment, guided at least in part by the topological objective function. We will discuss the algorithms and objectives below, but for now we are tasked with determining how much raw information exists in two networks—and whether it is enough information to satisfy Eq. (3).

While specifying our information *requirement* in Eq. (3) is fairly straightforward, quantifying the information *content* in our pair of networks—formally called a "message" in the parlance of Information Theory—is quite nontrivial; entire modern research areas are built around quantifying information, including the areas of data compression (Johnson, Harris, & Hankerson, 2003), cryptography (Adamek, 2011), and virtually all forms of digital communication (Sakrison, 1968). For now, we conjecture that the amount of information encoded in an *arbitrary* network, without any restrictions on its structure, is approximately equal to its number of edges. This hypothesis arises from the fact that the adjacency matrix for a network consists of bits, with each edge corresponding to a single 1-bit in the matrix. Recent work has shown that networks roughly the size of our PPI networks can be encoded with as few as 1.8 bits per edge, on average, with the authors Dhulipala et al. (2016) also claiming that there is "much room for improvement." According to Shannon Information Theory

(Shannon, 1948), no object—including a graph—can be encoded using fewer bits than its information content, so the *actual* storage requirement for a particular scheme to encode G (in this case, about 1.8 bits per edge) is an *upper* bound on G's information content. The above results suggest that we are justified in claiming that each edge in a network provides approximately one bit of information that can be leveraged toward narrowing the search space of a network alignment. This in turn suggests that, when aligning two networks, they will require $\approx$Netbits(n_1, n_2) edges if we are to isolate a specific alignment a^*—and if there is some idea of a "correct" alignment, then this likely represents the approximate number of *common* edges between the node pairs deemed as the correct alignment.

While the previous paragraph is far from rigorous and will require significant development before details can be made more precise, the premise can easily be tested empirically. Namely, to empirically test Eq. (3), we need realistic-looking PPI networks with (i) enough similarity to expect good alignments to exist, and (ii) enough edges to satisfy the Netbits threshold. To that end, we leverage IID—the *Integrated Interaction Database* (Kotlyar et al., 2018). Each nonhuman mammalian PPI network in IID has been augmented by transferring high-confidence experimental interactions from the human PPI network to the corresponding interologs connecting 1-to-1 orthologs in the nonhuman mammal. As a result, all the mammalian PPI networks in IID are highly similar and have more than enough edges to satisfy the Netbits criterion (Eq. 3). In particular, the IID networks of rat and mouse contain 13,510 1-to-1 orthologs between them with degree at least 1 in both networks. Taking the subgraphs induced on these 13,510 nodes gives two highly similar PPI networks, which we call R and M, each with 13,510 nodes, with the 1-to-1 orthologs providing a "correct" mapping between them; R has 233,289 edges, while M has 237,380—both well above the Netbits value of 165,898 for $n_1 = n_2 = 13,510$. Starting with R and M, we choose a fraction $F \in [0, 1]$ and—independently for each network—randomly remove edges until only a fraction F of the original edges remain. This gives networks R_F and M_F, which we align with SANA for 1 hour, optimizing S^3. (We choose S^3, the *Symmetric Substructure Score*, since the input IID networks are highly symmetric, by design.) Then, looking at the *Common Connected Subgraph* from the alignment, we extract the set of aligned node pairs that are orthologs—i.e., the set of aligned node pairs that were *correctly* aligned. This set is called C, while the complement set in the alignment is U, the set of aligned node pairs that are not orthologs. Fig. 2 plots the number of edges induced on C or U as a function of the cardinality of the respective set. The purple curve is Netbits(n)—where

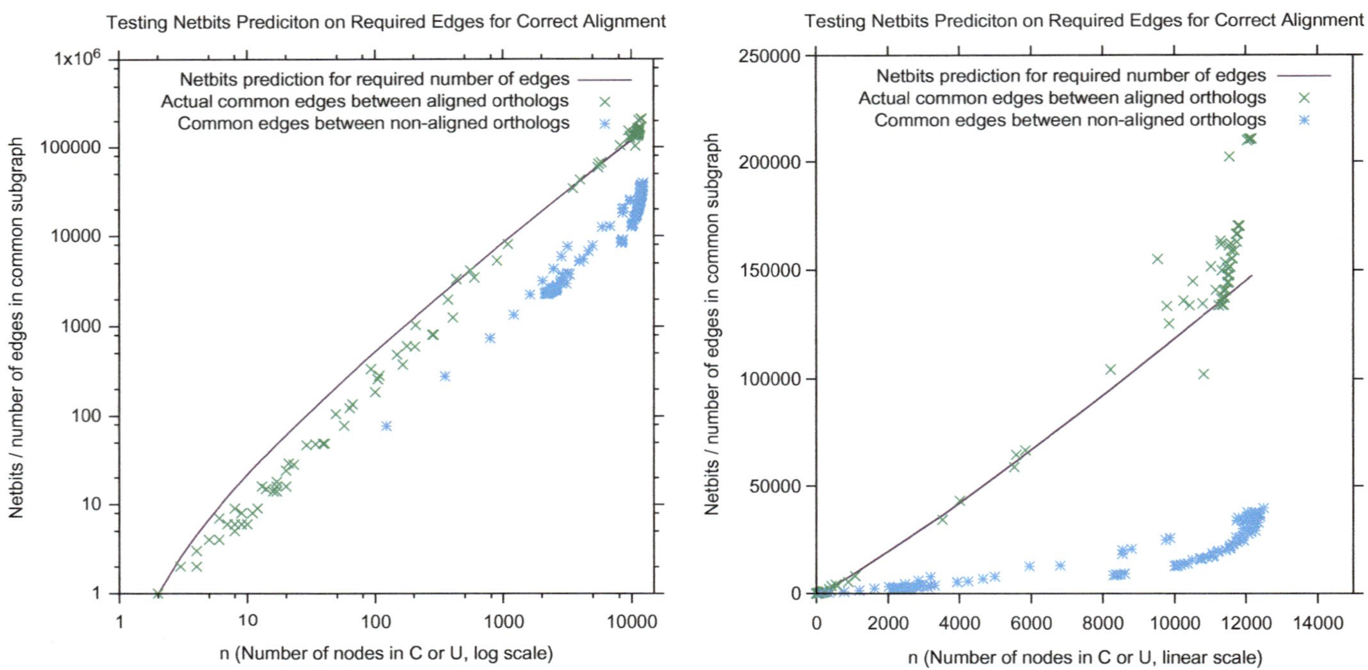

Fig. 2 Testing Information Theory "Netbits" Prediction by aligning rat-mouse orthologs from the *Integrated Interaction Database* (IID, Kotlyar et al., 2018). We take the PPI networks of rat (*R. norvegicus*) and mouse (*M. musculus*) from IID, induced on the 1-to-1 orthologs between them. Then, for various values of $F \in [0, 1]$, we independently choose a fraction F of edges from each network and align them with SANA for 1 hour optimizing S^3. Then, from the *Common Connected Subgraph*, we extract the set of 1-to-1 orthologs, C, that were correctly aligned, and U, the set that were incorrectly aligned. We then plot Netbits($|C|$) (*purple curve*), the actual number of common edges induced on C (*green* x's), and the actual number of common edges induced on U (*blue* x's), as a function of either $|C|$ or $|U|$, as appropriate. (Note the *left* and *right figures* plot the same data: log scale on the *left*, linear on the *right*.) We see that the Netbits curve closely follows the number of common edges between correctly aligned orthologs, whereas incorrectly aligned orthologs have far fewer common edges than the Netbits value.

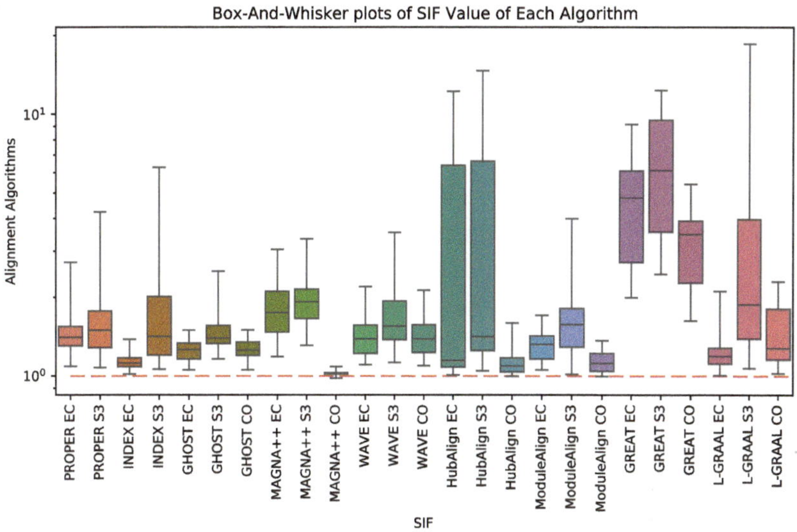

Fig. 3 SANA universally outperforms other aligners at optimizing topological objectives. For each competing search algorithm, we plot the *SANA Improvement Factor*, or *SIF*, which is the ratio of SANA's score to that of the other aligner, for the measures EC, S^3, and the competitor's "custom" objective ("CO") if one exists. Note that the vertical axis is logarithmic. The *dotted red line* represents a SIF ratio of 1—i.e., SANA and the other algorithm achieved the same score. The height of each box marks the lower and upper quartile of SIF across the 28 pairs of species (in some cases fewer—see text), while the line inside the box marks the median SIF. The whiskers mark the minimum and maximum SIF values. Note that even the *bottom* whiskers rarely dip below 1, meaning even SANA's *worst* performance is usually better than that of the competing aligners.

n is $|C|$ or $|U|$ as appropriate—i.e., the predicted number of bits required to correctly align n nodes. The green x's plot the actual number of common edges induced on C—that is, the number of common edges between correctly aligned orthologs (i.e., correctly recovered *interologs* according to IID). The blue x's depict the number of edges induced on U—that is, between *incorrectly* aligned orthologs in the same alignments. (Note that for each alignment, we have the identity $|U| + |C| = 13,510$, so that each alignment has one green x and one blue x, but they are at different horizontal locations.) As we can see in the left figure (using a log scale), when the number of correctly aligned orthologs is a small fraction of the total (i.e., green x's with $n \lesssim 1000$), we can get correct alignments with slightly fewer edges than the Netbits prediction; but when a large fraction of the orthologs are correctly aligned (best observed in the right figure that uses a linear scale), alignments with fewer common edges than predicted by Netbits are rare—i.e., most alignments with a large fraction of correctly aligned ortholog pairs have a number of common edges at or above the Netbits threshold. In contrast, the total number of edges induced between incorrectly aligned orthologs is *always* well below the Netbits threshold.

Incidentally, the green x's below the purple curve in Fig. 2 are not terribly surprising: while the general idea is to suggest that each edge provides "approximately" 1 bit of information, the general structural properties of the graph can impart significant additional information. For example, consider a tree consisting of a root node and three linear "arms" extending outward, each with a different, arbitrary length. Such a three-armed tree can have an arbitrary number of nodes, each of which is the sole occupant of its automorphism orbit—meaning that when aligning the tree to itself, there is only one alignment that achieves an S^3 score of 1, and that alignment correctly aligns every node to itself. Thus, the S^3 measure in this case is able to isolate the single, unique *correct* alignment in an arbitrarily large search space with n nodes but only $n - 1$ edges—*far* fewer edges than Netbits(n, n) for large n. Thus, Eq. (3) is clearly only an approximation on the number of required edges for a correct alignment, though Fig. 2 suggests it tends to become statistically more likely to be a good approximation to a lower bound as n increases.

To conclude, we have offered a theoretical argument, supported by quantitative empirical evidence in good agreement with the theory (cf. Fig. 2), for the observed widespread failure of topology-only global network alignments: it is due, at least in part, to insufficient edge densities in existing experimental PPI networks. In the absence of artificially inflated edge densities in databases such as IID, these low edge densities make it impossible for *any* algorithm based on topology alone to reliably produce global network alignments with robust biological value.

3.2 Addressing R3: Inadequate optimization of chosen topological objective functions

Our ultimate goal is to produce the best possible topological alignments according to the topological measures of Table 1, in order to measure how well each of them is able to uncover functional similarity. To ensure a fair comparison between the objective functions of Table 1, we need to show that SANA provides a level playing field. To do this, we demonstrate in Section 3.2.1 that SANA provides *near-optimal* solutions—that is, alignments with near-optimal values of the objective function—in all cases the optimal value of the objective is known. One could not ask for a more level playing field than allowing each objective function to reach its (near-) optimal value. Second, to demonstrate that no other published algorithms are capable of providing this level playing field, in Section 3.2.2 we demonstrate that SANA's near-optimal alignments dramatically outscore each of the algorithms of Table 1 at simultaneously optimizing EC, S^3, and their own topological objectives.

3.2.1 SANA achieves near-optimal solutions when the optimal solution is known

In general, given an appropriate temperature schedule and enough CPU time, simulated annealing converges to an optimal solution with probability one (Ingber, 1989). To provide evidence that SANA is performing at this level, we offer two lines of evidence. First, Table 2 demonstrates that SANA produces alignments that score extremely close to optimal ones computed by the Hungarian algorithm (Milenković et al., 2010), while taking only a fraction of the time on large networks. For edge-based NP-complete cost functions such as EC and S^3, Figs. S1 and S2 in Supplementary material show that SANA easily aligns any existing PPI network with a subgraph of itself—effectively providing quick solutions to the subgraph isomorphism problem—a well-known NP-complete problem (Garey & Johnson, 1979).

3.2.2 SANA outscores other aligners even at optimizing their own objectives

We compare SANA's ability to simultaneously optimize EC, S^3, and the objectives of Table 1 to that of the aligners listed in the same table. (See Section 6.3 for details.) Since SANA almost universally outperforms all other search algorithms in all measures, we introduce the metric *SANA Improvement Factor*, or *SIF*, which is the ratio of SANA's value of any objective f vs the value of f achieved by the other alignment algorithm, when both are applied to the same network pair:

$$SIF(f, G_1, G_2) = \frac{\text{SANA's score on objective function } f \text{ aligning } G_1, G_2}{\text{Other Search Algorithm's score on } f \text{ aligning } G_1, G_2}.$$

Fig. 3 depicts SIF scores against the aligners of Table 1 across all 28 pairs of BioGRID networks (cf. Section 6.3). As we can see, SANA almost universally outperforms other aligners at simultaneously optimizing EC, S^3, and the competing aligner's own custom objective. In particular, out of a total of 188 tests, SANA was only marginally beaten in the following handful of cases: HubAlign beat SANA by 0.3% in Importance in 1 out of 28 BioGRID

Table 2 Achieving provably near-optimal solutions: We compare SANA's ability to optimize the local (node-pairwise) orbit degree vector similarity (Milenković & Pržulj, 2008) to that of the Hungarian algorithm as implemented by H-GRAAL (Milenković et al., 2010), which provides provably optimal solutions when the objective consists only of node-pair similarities.

G_1	Nodes	G_2	Nodes	$T_{H-GRAAL}$	T_{SANA}	$Score_{H-GRAAL}$	$Score_{SANA}$	SANA/optimal
syeast0	1004	syeast05	1004	2.87	20	0.946163	0.946139	0.99997
syeast0	1004	Yeast	2390	53.43	60	0.86285	0.86276	0.9999
Yeast	2390	Human	9194	5567.85	120	0.883	0.882	0.998
RNorvegicus	1657	HSapiens	13276	566.03	120	0.918	0.917	0.998
CElegans	3134	AThaliana	5897	5516.58	120	0.959	0.956	0.996

T_X is the run-time in minutes for $X = H - GRAAL, SANA$; similarly for $score_X$. Time does not include the precomputation of the orbit degree similarity matrix, which is identical for both programs. The far right column is SANA's score as a fraction of the optimal score produced by H-GRAAL Milenković et al. (2010). Networks are from the H-GRAAL suite, which include "synthetic yeast" (syeast) from Krogan et al. (2006a), yeast2 from Collins et al. (2007), human1 from Radivojac et al. (2008), and BioGRID networks from 2013. (We do not attempt more recent BioGRID networks since H-GRAAL would require months, rather than days, on modern BioGRID networks.)

pairs (RN-CE)—but on the same pair, SANA outperformed HubAlign's EC and S^3 scores by 0.9% and 158%, respectively; for its own objective function, ModuleAlign outperformed SANA in 2 out of 28 network pairs: by 0.3% for RN-HS and 0.2% for CE-DM—while on the same pairs SANA outperformed ModuleAlign's scores by 23% in EC and 160% in S^3 for RN-HS, and by 12% in EC and 50% in S^3 for CE-DM; MAGNA++ outperformed SANA by 2% in 1 out of 28 pairs (RN-MM), while on the same pairs SANA outperformed MAGNA++'s EC and S^3 scores by 103% and 105%, respectively.

3.2.3 Summary: SANA provides a near-optimal level playing field for objective function comparison

Together, Table 2 and Figs. 3, S1, and S2 in Supplementary material give us confidence that SANA provides a level playing field for comparing objective functions because it produces alignments with near-optimal scores of the topological objective. More explicitly,

- Table 2 demonstrates that for node-based objective functions for which the Hungarian algorithm can provide exactly optimal solutions, SANA comes within a fraction of a percent of the optimal value in a reasonable amount of CPU time.
- Figs. S1 and S2 in Supplementary material demonstrate that for edge-based objective functions which are NP-complete, SANA correctly aligns each of the depicted PPI networks to itself in a few minutes.
- Fig. 3 tells us that SANA almost universally outperforms all other aligners at optimizing their own objective functions. In the handful of cases that the other aligner performs better, it does so by only a minuscule amount in *one* of the three objectives being optimized, while SANA soundly beats it in the other two.

4. Addressing R2: Measuring functional relevance of topological objective functions

4.1 Functional relevance

Given a network alignment a, let $f : a \rightarrow \mathcal{R}$ be some measure of topological similarity exposed by a, and let $g : a \rightarrow \mathcal{R}$ be some measure of functional similarity across nodes aligned by a. We say that an alignment a is "functionally relevant" if it scores higher in g than some user-defined threshold.

Let f^* be the highest possible topological score across all alignments and let A^* be the set of alignments that achieve that score. Similarly, let g^* be the highest possible functional score, and let B^* be the set of alignments that achieve it. Since the goal of topological network alignment is to uncover functionally similar nodes across networks, we want f to have the following properties: (i) there should be a high correlation between the values of $f(a)$ and $g(a)$ so that increasing f also tends to increase g (a high Spearman correlation would be best, though a high Pearson correlation will suffice); (ii) there should be substantial overlap between the sets A^* and B^*—or, at the very least, between alignments that score close to f^* and those that score close to g^*; and (iii) the fraction of alignments in A^* that are also in B^*—formally, the ratio $|A^* \cap B^*|/|A^*|$, or its "near-optimal" equivalent—should not be vanishingly small. If these three properties are satisfied, then f is functionally relevant measure of topological similarity, because optimizing f will move us toward optimizing g, and (near-)optimal alignments under f will have a good chance of exposing high functional similarity among aligned nodes.

Conversely, if any one of the above properties fails, then f is not a functionally relevant measure of topological similarity. Respectively: (i) if f does not correlate with g, then increasing f will not increase g—the latter being necessary to expose alignments with high functional similarity; (ii) even when f and g are *correlated*, if topologically optimal alignments are not close to functionally optimal ones, then there is no benefit to aggressively pushing f toward its optimal value; (iii) if A^* contains functionally (near-) optimal alignments but they are greatly outnumbered by alignments with low functional score, then a randomly chosen alignment from A^* has only a small chance of having high functional similarity. In any of these cases, f fails the test of being a "functionally relevant" topological objective function.

4.2 Objective function saturation

It is the failure of point (iii) above that we refer to as "objective function saturation": assume that (1) the value of g^* is known, (2) alignments that score g^* also have optimal topological score f^*, but (3) there exist many alignments with optimal topological score f^* that nonetheless score significantly below g^* in functional similarity. Then, we say that f *saturates*: like a sponge that is far too small to soak up a spill, it reaches its full capacity long before the job is done.

Saturation can be caused by at least two fundamental issues: (a) not enough topological input information, resulting in the inability of *any* topology-based method to isolate "good" individual alignments; and (b) the topological objective function having insufficient discriminative power even when enough information is available. The first case is addressed by our appeal to Information Theory; we address the second in our detailed comparison of objectives in the following sections. (Section 1 in Supplementary material provides detailed schematic examples of saturation.)

4.3 Evaluating the functional relevance of topological measures

We now compare the measures of topological similarity proposed in Table 1—taken from a wide range of publications—for their ability to produce alignments that highlight functionally similar proteins across species. To that end, ideally we would like to test each of them for the three properties in Section 4.1.

As has been noted by others (Kuchaiev et al., 2010; Milenković et al., 2010) and discussed in several places above, there may exist a large (but difficult to estimate) number of alignments with optimal or near-optimal topological scores. We have provided evidence (cf. Figs. 2 and S2 in Supplementary material; Table 2; Sections 3.2.1 and 4.2) that SANA tends to provide alignments with near-optimal topological scores when given enough CPU time (an hour seems enough on existing PPI networks). Since simulated annealing is a random search algorithm, SANA may produce a different alignment each time it is run. Since it produces near-*optimal* alignments, it can thus be viewed as providing a ***random sample of alignments with near-optimal scores*** according to the objective function used. (We cannot claim the sample is unbiased since we currently know nothing about the distribution. That is, even if we achieve an alignment in A^*, we cannot currently claim we are sampling A^* in an unbiased manner.) Thus, SANA effectively allows us to simultaneously test all three properties of functional relevance, viz.: (i) since SANA searches the alignment space randomly, if the value of f and g are uncorrelated, then increasing f will not tend to increase g; (ii) if alignments that are near-optimal in g do not have scores that are near-optimal in f, then SANA's ability to push the value of f to near-optimality will have little functional benefit; and (iii) even if near-optimal scores in g have near-optimal scores in f but f saturates, then a randomly chosen near-optimal alignment according to f is unlikely to uncover one of the relatively few alignments that are well scoring in g.

In the following sections, we evaluate a large set of alignments, across objective functions from Table 1 and BioGRID pairs from Table 3. Each of SANA's output alignments consists of pairs of proteins providing a 1-to-1 mapping from proteins in G_1 to those in G_2. For each pair of BioGRID networks from Table 3, and for each objective in Table 1, we ran SANA 100 times for 1 hour each. Across all cases, the objective function values agreed to less than 1%, 3%, and 7% in 51%, 90%, and 100% of cases, respectively, suggesting that 50% of SANA's alignments achieved a topological score within 1% of optimal, a further 40% were within 3% of optimal, and the remaining 10% were within 7% of optimal.

4.4 Recovery of common Gene Ontology terms

The Gene Ontology (The Gene Ontology Consortium, 2008) describes our understanding of genes and proteins. It consists of thousands of descriptive annotations called *GO terms* which are arranged into a hierarchical description of molecular functions (MF), biological processes (BP), and cellular components (CC). Genes and gene products, such as proteins, are "annotated" with various GO terms as we learn what those genes/proteins do. GO terms high in the hierarchy can be vague and can annotate thousands of proteins, while more specific GO terms lower in the hierarchy tend to annotate fewer proteins. Well-understood genes and proteins can have many GO annotations, while lesser understood proteins may have few or no GO annotations, or be annotated with only high-level, vague GO terms.

To evaluate the functional relevance of each topological similarity measure f from Table 1, we performed 100 independent runs of SANA optimizing f for 1 hour, across all 28 network pairs from Table 3. Given an alignment, for each GO term g we counted the number, k_g, of aligned protein pairs that share g, and then evaluated the statistical significance of k_g by computing the probability that a random alignment would have k_g or more protein pairs sharing g. This provides a p-value for each single GO term, in each single alignment (cf. Section 6.5). Since we performed 100 alignments for each of the 28 network pairs for each of topological objectives, we have 100 samples of each GO term's statistical significance in each pair of networks, for each objective. Fig. 4 presents the cumulative distributions of the p-values of 7 measures, for the network pairs yeast–human (left) and mouse–human (right) of BioGRID 3.4.164. (We choose yeast–human because these are by far the most complete PPI networks; and mouse–human because mouse is the most complete mammalian PPI network after human.)

Table 3 Major networks from BioGRID (v3.4.164, August/September 2018), sorted by mean degree.

Nodes	Edges	Eq. (3)	Ratio	Degree	Density	Name	Abbr.	Species
5984	104962	66455	1.58	35.08	0.00586	Baker's yeast	SC	*Saccharomyces cerevisiae*
17200	282181	217200	1.30	32.81	0.00191	Human	HS	*Homo sapiens*
8728	46364	101678	0.46	10.62	0.00122	Fruit fly	DM	*Drosophila melanogaster*
9364	34725	110037	0.32	7.42	0.00079	Cress	AT	*Arabidopsis thaliana*
2811	8931	28156	0.32	6.35	0.00226	Fission yeast	SP	*Schizosaccharomyces pombe*
6777	18108	76477	0.24	5.34	0.00079	Mouse	MM	*Mus musculus*
3194	5572	32581	0.17	3.49	0.00109	Worm	CE	*Caenorhabditis elegans*
2391	3554	23392	0.15	2.97	0.00124	Rat	RN	*Rattus norvegicus*

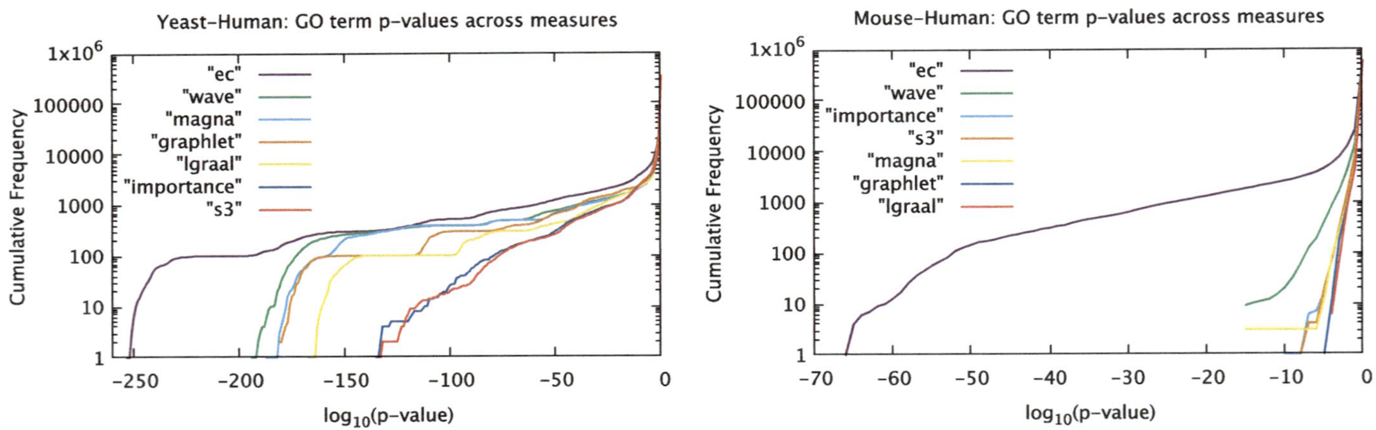

Fig. 4 Cumulative distribution of *p*-values of individual GO terms across 100 alignments of yeast–human (*left*) and mouse–human (*right*), for alignments optimizing each objective function that was computable for these two network pairs. For each GO term *g*, we use the tail of the hypergeometric distribution to compute the *p*-value that the observed number of aligned protein pairs would share *g* in a random alignment (see Section 6.5). Note that all scales are logarithmic. The legend of each figure lists the objectives best-to-worst for that pair. We observe that there are vast differences in the functional similarity uncovered by optimizing the different objective functions. In particular, EC is by far the best measure for uncovering functional similarity between aligned proteins in BioGRID 3.4.156. Its yeast–human alignments contain over 3800 GO terms with *p*-values below 10^{-10}, over 500 with *p*-values below 10^{-100}, and over 100 below 10^{-200}. The second most effective objective is WAVE's WEC (a graphlet-weighted version of EC), with over 2100 GO terms with *p*-values below 10^{-10}, almost 400 with *p*-values below 10^{-100}, but none below 10^{-200}. The mouse–human alignments have the same top two measures in the same order (EC and WAVE), though in this case EC hugely dominates all measures, with WAVE as a distant second. The other five measures (Importance, S^3, MAGNA, graphlet, and L-GRAAL) show no solid ordering between them here, nor in other pairs of networks (cf. Supplementary material).

We observe that the topological measures vary quite substantially in their ability to align a statistically significant number of protein pairs that share GO terms: EC seems by far the best, with WAVE's "WEC" (a graphlet-based weighted EC) also performing significantly better than other measures.

While Fig. 4 displays the distribution of p-values of individual GO terms, computing the p-value of an entire alignment based on GO terms is non-trivial since most GO terms in the hierarchy are not independent of each other. While in principle it should be possible analytically to derive the interrelationships between GO terms, we have instead employed the *Empirical Brown's Method* (Poole et al., 2016) (see Section 6.5) to compute p-values for each alignment, accounting for the nonindependence across GO terms. Fig. 5 is similar to Fig. 4, but for entire alignments rather than individual GO terms. As we can see, the EC measure again dominates, and the strongest p-values are for the most dense networks, with the best EC-driven alignments between yeast and human having p-values of 10^{-2000}, while even the *worst* yeast–human alignments by this measure have p-values of about 10^{-1200}. As can be seen in Supplementary material Info, these two network pairs are not atypical: fully *half* of all network pairs have strong p-values when alignment is driven by EC.

4.5 Topology-based recovery of thousands of orthologs between major BioGRID species

Aligning a large set of proteins annotated with a particular GO term to another large set annotated by the same GO term provides plenty of statistical evidence that topology is able to align functionally similar *regions* to each other. However, while aligning a whole region of the PPI network in one species to a similar region in the other species is encouraging, it provides little *predictive* power about individual proteins. In addition, correctly aligning *one specific pair* of proteins is far more stringent than aligning one set to another set but allowing arbitrary permutations among members of each set. Thus, in this section we highlight the ability of network topology to recover the "correct" alignment between specific pairs of orthologous proteins between species.

Consider Table 4: collectively, across all 28 pairs of species and all objectives, we recovered almost 3000 known orthologous protein pairs; as seen in the bottom row of Table 4, this recovery rate has a cumulative p-value of about 10^{-144}. (The objectives for GHOST and GREAT were too expensive to compute for some of the largest network pairs—cf. the blank entries in Table 4). Observing the different columns of Table 4, we see that different

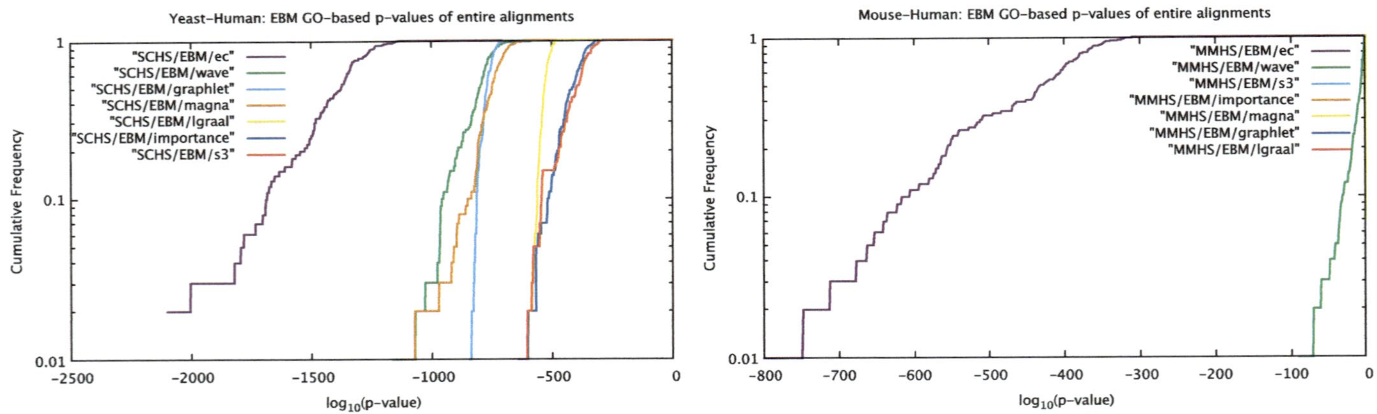

Fig. 5 Similar to Fig. 4, except all the *p*-values for individual GO terms have been combined using the *Empirical Brown's Method* (Poole, Gibbs, Shmulevich, Bernard, & Knijnenburg, 2016), producing a single, holistic *p*-value for each pairwise alignment. There are 100 alignments for each network pair and for each objective function. We see again that EC is the dominant measure, producing individual alignments with GO-term-based *p*-values universally below 10^{-1000} for yeast–human and 10^{-300} for mouse–human. In the case of yeast–human, all other measures have *p*-values that, though quite significant, are hundreds of orders of magnitude less so than EC. In the case of mouse–human, all objectives except EC and WAVE have so many GO terms with a *p*-value of 1 (i.e., totally insignificant due to zero proteins sharing that GO term despite there being a nonzero number that *could* share it) that their whole-alignment *p*-values are so close to 1 that they are invisible against the border on the *right side* of the plot. (More network pairs are listed in Supplementary material.)

Table 4 Recovered orthologs.

	Obj.													
Nets	**Orth. pairs**	**EC**	**S^3**	**Imp.**	**Wave**	**Grt**	**Ghst**	**Mag**	**Gr**	**Lg**	**Sum**	**Uniq**	**Rate**	**P**
SC-HS	1245	**72**	52	64	18	–	–	19	8	5	238	172	13.8%	10^{-44}
MM-HS	6286	**101**	37	34	77	–	–	26	5	7	287	282	4.5%	10^{-17}
SP-SC	1242	**71**	38	36	65	–	–	33	1	3	247	227	18.3%	10^{-16}
SP-HS	945	**34**	5	2	7	–	–	5	0	0	53	53	5.6%	10^{-15}
SC-DM	955	**43**	36	38	23	–	–	7	1	1	149	126	13.2%	10^{-12}
DM-HS	3511	**52**	28	31	46	–	–	14	1	1	173	161	4.6%	10^{-08}
RN-HS	2223	**38**	15	14	5	–	–	6	1	1	80	79	3.6%	10^{-07}
RN-MM	2267	**68**	43	39	28	39	37	29	13	5	301	276	12.2%	10^{-06}
CE-HS	1137	**24**	7	5	8	–	–	1	0	0	45	45	4.0%	10^{-06}
SP-DM	806	**29**	29	29	9	–	13	8	0	0	117	102	12.7%	10^{-06}
RN-DM	865	**27**	12	16	17	15	21	5	2	4	119	110	12.7%	10^{-05}
RN-AT	534	10	**19**	13	13	14	8	7	4	3	91	82	15.4%	10^{-04}
CE-DM	999	**27**	20	12	15	19	15	7	1	1	117	113	11.3%	10^{-03}
MM-AT	1233	25	**29**	22	19	–	11	16	4	6	132	126	10.2%	10^{-03}
MM-DM	2108	**40**	26	31	31	–	19	28	7	8	190	175	8.3%	0.02
RN-SC	357	**14**	6	7	4	–	–	5	1	1	38	38	10.6%	0.03
SP-AT	844	11	**18**	14	15	–	7	7	2	3	77	69	8.2%	0.05

Continued

Table 4 Recovered orthologs.—cont'd

Nets	Orth. pairs	Obj.											
		EC	S^3	Imp.	Wave	Grt	Ghst	Mag	Gr	Lg	Sum	Uniq	Rate	*P*
CE-AT	562	**11**	11	5	5	7	9	7	0	0	55	53	9.4%	0.4
AT-HS	1864	**17**	9	6	6	–	–	4	3	0	45	41	2.2%	0.5
RN-SP	371	**19**	11	11	6	10	9	7	1	1	75	65	17.5%	0.8
SP-MM	812	**17**	6	10	10	–	4	6	1	1	55	49	6.0%	1
SP-CE	403	15	13	17	12	**20**	8	8	2	1	96	84	20.8%	1
CE-SC	362	**9**	5	5	9	–	–	3	1	1	33	33	9.1%	1
SC-MM	745	**14**	13	8	8	–	–	4	1	3	51	44	5.9%	1
DM-AT	1477	**19**	14	16	16	–	13	9	4	3	94	92	6.2%	1
CE-MM	1136	**19**	5	14	9	–	11	11	1	0	70	68	6.0%	1
RN-CE	726	**25**	11	16	16	13	15	10	0	4	110	104	14.3%	1
SC-AT	977	8	5	6	7	–	–	**9**	7	2	44	43	4.4%	1
total	36992	**859**	523	521	504	137	200	301	72	65	3182	2912	7.87%	5×10^{-144}

Each row is one of 28 species pairs from BioGRID 3.4.164, using the 2-letter abbreviations of Table 3. Orth.pairs=number of orthologs between the BioGRID network pair according to NCBI Homologene. Columns EC through lg report the number of unique orthologs recovered in 100 alignments of SANA optimizing said measure (abbreviations: *grt*, GREAT; *ghst*, GHOST; *mag*, MAGNA; *gr*, graphlets; *lg*, L-GRAAL); the best performing measure is in boldface. A dash "–" indicates the algorithm failed to produce similarities for SANA to optimize. Objectives are sorted left-to-right by total recovery (last row, with the totals for GREAT and GHOST adjusted for failures). "sum" is total orthologs recovered across all nine measures, including duplicates recovered by more than one measure; "uniq"=sum with duplicates removed; rate=uniq/orth.pairs; *p*-value=probability of the boldfaced number of orthologs being properly paired at least once across 100 random alignments (see Section 6.4). Rows are sorted by *p*-value. Finally, all *p*-values in the last column have been increased by a factor of nine to account for the multiple hypothesis testing across the nine objective functions.

topological objectives result in different ortholog recovery rates; the columns are sorted by the last row (total recovered orthologs across all species). Furthermore, comparing the "uniq" column with "sum" demonstrates that there is not much overlap between the orthologs recovered by different objectives so that each objective appears to have value in that they recover near-orthogonal sets of orthologs. As discussed in Section 6.6, this suggests that no single topological measure is "best" at recovering orthologs. Interestingly, this conclusion is quite different from the GO-term-based p-values discussed above (cf. Section 6.5), where EC was dominant. One clear difference is that GO-term-based evaluations are more global, while ortholog recovery is more local. Clearly more research is needed before we have a full understanding of these results.

Interestingly, yeast–human has the top p-value and is a very close second to mouse–human in raw number of recovered orthologs. Observing Table 3, we see that mouse is about an order of magnitude less complete than human, and significantly less complete than yeast—and yet mouse–human does *better* than yeast–human at raw number of recovered orthologs. Likely this is because mouse and human are more closely related than yeast and human. We interpret these observations to bode extremely well for future network analyses: viz., if mouse–human does so well with such incomplete network coverage, and yeast–human does so well despite the vast taxonomic distance between them, then topology-driven network alignments are likely to become even more effective as the volume of high-quality network data continues to grow.

5. Discussion

5.1 Statistical significance

To our knowledge, the p-values displayed in this paper are orders of magnitude more significant than published p-values derived from any other topology-only methods of PPI network analysis. We believe our results provide—by far—the strongest evidence to date demonstrating the relationship between global PPI network topology and functional similarity between regions of PPI networks, as well as for individual proteins. While we have several lines of evidence to soundly reject R0, we have demonstrated that the widespread failure of existing topological methods are likely a combination of all three of R1, R2, and R3. Namely,

- Information theory suggests that most existing PPI networks have too low an edge density for topology-only methods to produce robust, well-defined global alignments. However, repeated sampling of high-scoring global alignments may expose local regions that are robustly aligned; our companion paper (Wang, Atkinson, & Hayes, 2022) demonstrates that robust GO-term predictions are possible even using decade-old BioGRID and GO releases.

- Though many topological objective functions have been proposed in the literature, the algorithms published to optimize those objectives almost universally fail to optimize those objectives to their maximum values. This failure to produce near-optimal solutions has hampered the ability of previous aligners to produce statistically significant topology-function relationships.
- When optimized more fully, the various objective functions show a wide disparity in their ability to align functionally similar proteins—at least on current data. (This may change as more data become available.)
- The most effective objective function on current data appears to be EC, which is able to align functionally similar regions of the denser PPI networks with *p*-values hundreds, or in some cases thousands, of orders of magnitude below published *p*-values in the literature. However, the dominance of EC may be temporary—an artifact of its ability to align networks of highly disparate of completeness; we expect more symmetric measures, such as S^3, to become more prominent as the densities of PPI networks become more comparable.

As we can see from Table 4, *all* of the topological measures tested are able to recover a statistically significant number of orthologs between species, especially when the network density is high enough in both networks. However, the EC measure recovers the greatest number of orthologs for almost all species pairs from BioGRID. Additionally, EC is usually the best measure for recovering common GO terms, and far outperforms all other measures on the most complete PPI networks. As we explain in Section 6.6, we believe this is likely due simply to the wide disparity in network completeness across the BioGRID networks; if the 2018 IID networks (Kotlyar et al., 2018) are any indication of the amount of similar topology that *truly* exists, then more symmetric measures such as the S^3 score are likely to come to the fore as PPI networks become more complete. At the moment, however, we note that using EC alone is not a good idea: the "uniq" column of Table 4 shows that there is little overlap in the orthologs recovered by the various measures, suggesting that all of them have merit and that more information may be gleaned by using all of them rather than any one of them, including EC.

Fig. 2 (as well as Fig. S5 in Supplementary material) demonstrates that orthologs that were recovered in Table 4 typically had a degree in their own networks of 3–5× the degree of orthologs that were not recovered. Not surprisingly, this clearly indicates that degree—essentially a surrogate for local edge density—plays a large role in the ability of network topology to recover biologically relevant alignments. We believe the discussion around Hypothesis R1 in Section 3.1, especially the number of bits discussed in relation to the size of the search space (Eq. 2), provides the beginnings of an understanding of *how much* topology is required before "good" topological network alignments can be expected to exist.

The number of BioGRID edges for human has been growing steadily by about 30% per year over the past decade. This strong growth of PPI network data suggests that the information theory requirements will be met within the next few years, suggesting a bright future on the horizon for topology-based analysis of PPI networks.

6. Methods

6.1 Information theory in the context of network alignment

Information theory, first introduced by Claud Shannon in 1948 (Shannon, 1948), is a very mature and well-developed field with many texts available (e.g., Klir, 2005; Kullback, 1997). One question it can help answer is, "What is the bare minimum amount of information required to answer a well-posed question?" Information is often quantified in bits. For example, consider the domain of a set with exactly two elements, $\{a, b\}$. Uniquely identifying one of the elements requires exactly 1 bit of information: the answer to the question "is it *a*?" (1 for yes, 0 for no, the latter directly implying the chosen member is *b*). If there are 4 items $\{a, b, c, d\}$, then two bits are required: the first bit splits the set in two (eg $\{a, b\}$ and $\{c, d\}$), and the second bit chooses between the first and second element of the appropriate pair. In general, if there are $n = 2^k$ elements in a set, one absolutely requires at least k bits to uniquely identify one element in that set. Taking the base-2 log of both sides, we see that $\log_2(n) = k$ bits are required to uniquely identify one element among n. Thus, from the base-2 log of Eq. (2), we can compute the absolute bare minimum amount of information, in bits, required to uniquely identify one alignment. If we would be satisfied—or are only able—to identify any member of a set of N "equivalent" alignments, the information requirement is reduced by log_2N bits.

We emphasize that this bound is a fundamental limit. It does not depend on the objective function used, or the search algorithm. It is a statement that, if not enough information is on hand, then in general it will be impossible for *any* algorithm to isolate a relevant alignment.

Thus, the flip side of this question is, "how much information, in bits, do we have at hand, given two networks we are trying to align?" If we can quantify the information content of the two networks, we can determine if they contain *enough* information to identify relevant alignments. The problem is that precisely measuring the amount of information in a "message" (e.g., a network) is very hard. This is why, for example, compression algorithms are constantly being improved: how much information, in bits, is in an English sentence? It is clearly fewer than the bits required to represent all its characters, b/c fr xampl we cn remv most vwls and stll ndrstnd th sntnce. (Note: the spelling mistakes are *intentional*, to demonstrate our point.)

So, the information content of a network is hard to estimate—and can depend on context. For example, if we first specify the domain of networks to be cliques and nothing else, then we need only specify the number of nodes in the clique—even a 1000-node clique requires only enough bits to specify the number 1000 (10 bits), which is far less than the number of edges. However, we do not know enough about the structure of PPI networks to constrain them to any particular subset of networks (a PPI network may not even be connected). So for now, given that the number of "1" bits in the adjacency matrix of an undirected network is exactly the number of edges, we hypothesize that the number of edges provides an approximate lower bound on the amount of information contained in a network, and leave a better estimate to further work.

As Fig. 2 demonstrates, the gap between our estimate of the information content of networks and the number of bits required to uniquely specify an alignment is only a few tens of percent—though this still amounts to thousands of bits. Closing this gap further will require significant advances toward a theory of network alignment and significant development in the formal topological information content of PPI networks.

6.2 Exactly computing the logarithm of large integers

Eq. (2) represents the size of the search space when aligning two networks; given that n_1 and n_2 are typically in the thousands, that number is enormous. However, we can easily compute its base-2 logarithm *exactly*, as follows:

$$\log_2\left(\frac{n_2!}{(n_2-n_1)!}\right) = \log_2[n_2 \times (n_2-1) \times (n_2-2) \times (n_2-3) \times \cdots \times (n_2-n_1+1)] = \log_2 \prod_{k=n_2-n_1+1}^{n_2} k = \sum_{k=n_2-n_1+1}^{n_2} \log_2(k). \tag{4}$$

Since k has values ranging "only" into the thousands, we can easily compute all the required base-2 logarithms and sum them. These are the values listed in the third column of Table 3, for $n_1 = n_2 = n$ for an n-node network.

6.3 Comparing SANA's alignments to those of competing aligners

In the cases of GRAAL, MAGNA++, OptNetAlign, L-GRAAL, HubAlign, ModuleAlign, WAVE, PROPER, and INDEX, we natively programmed their custom objective functions into SANA. In the case of GHOST, we modified GHOST to output its similarities for all node pairs, and then read that similarity file into SANA. In the case of GREAT, we used GREAT to generate required edge-based graphlet degree vectors (EGDVs) and their pairwise similarities, and then read in the edge-pair matrix that evaluates edge-to-edge similarity. (Note that this is a *large* matrix of size $m_1 \times m_2$, where m_i is the number of *edges* in network G_i.) Since we are interested in alignments based *only* on topology, we omit comparison against any algorithm that *requires* sequence—i.e., any algorithm that cannot perform topology-only network alignment. For example, via personal communication with the authors, we have verified that NATALIE (Klau, 2009), NATALIE 2.0 (El-Kebir et al., 2011), and PrimAlign (Kalecky & Cho, 2018) all *require* the use of pairwise sequence similarities. Several published alignment papers (Alberich et al., 2019; Gligorijević et al., 2015; Malod-Dognin et al., 2017) explicitly state that they regard the use of sequence similarities as necessary, and in any case their topological objectives are usually some minor variant of EC and not worth considering independent of EC.

We ran each of these algorithms on all 28 combinations of the 8 largest PPI networks from BioGRID version 3.4.156 from January 2018.

6.4 Computing the *p*-value of recovered orthologs

Given a pair of networks G_1, G_2 with $n_1 \leq n_2$ nodes, assume h is the number of known orthologs between them. To compute the p-value of having recovered r orthologs, we need to know the expected number of aligned orthologs in an alignment chosen uniformly at random from the search space. (Note that this has nothing to do with running SANA; we are estimating the ortholog recovery rate in *random* alignments, not good ones.) We describe two methods for computing this expected value: first, a simpler method that works well when $h/n_1 \ll 1$, and then a more accurate one that, although still approximate, appears to work well for any value of h/n_1.

Given a particular pair of orthologs (u_1, u_2), the probability that u_1 is aligned to u_2 by chance in a random alignment is $1/n_2$; they are misaligned with probability $(1 - 1/n_2)$. Given two orthologous pairs (u_1, u_2) and (v_1, v_2), the probability that both are misaligned is well approximated by $(1-1/n_2)^2$; this value is not exact because, since SANA produces only 1-to-1 alignments, u_1 and v_1 cannot simultaneously be aligned to (for example) u_2. This results in a very weak but difficult to disentangle dependence, which—in this first, simpler method—we shall ignore. Thus, the probability that all h orthologs are misaligned can be approximated by $(1-1/n_2)^h \approx (1 - h/n_2)$ since $1/n_2$ is small. Thus, we see that the probability that *at least one* of any of the h orthologs is correctly aligned is $\approx 1 - (1 - h/n_2)$, which is simply h/n_2. This value makes intuitive sense if we consider that each of the h orthologs contributes an equal "chance" to being aligned, each with probability $1/n_2$.

The above approximation can break down when h/n_2 is not small. The second method involves using the hypergeometric distribution, though still only approximately. We start by noting that when $n_1 < n_2$, then only n_1 out of the n_2 nodes of G_2 are aligned. In a random alignment, we therefore expect that, on average, only $h' = \frac{n_1}{n_2} h$ of G_2's orthologs are included among its aligned nodes. (For simplicity, assume the value of h' is rounded to the nearest integer.) This in turn directly implies that there are only h' orthologs *available* for alignment with G_1, even though G_1's nodes include all h of its orthologs. Recall the hypergeometric distribution $\mathcal{H}(i, m, M, N)$ describes the probability of i successes in m draws from a finite population N in which M objects have the desired property. To estimate the probability of 1 or more aligned orthologs, we should compute $1 - \mathcal{H}(0, n_1, h', n_1^2)$, since there are n_1 aligned pairs (the number of draws), and h' available "successes" out of n_1^2 pairs—the first n_1 coming from G_1, the second from the subset of G_1's nodes that are aligned. This would be exact except for the complication that, though the hypergeometric distribution already accounts for non-replacement, we have the additional constraint imposed by the 1-to-1 property of our alignments: after choosing an aligned pair (u, v) from the population of N, we must remove not only that pair, but *all other* pairs that contain *either* u or v. We call this the *1-to-1 constraint*, and it violates the model of the hypergeometric distribution in which the population is *constant* and that only *one member* of the population is removed with each draw. To approximately account for this, we must estimate the size of a surrogate population that provides good estimates of the desired probabilities. We have found empirically that using n_1 throughout gives probabilities that are far too low, but that good results are obtained if we substitute n_1 with the approximate number of orthologs across "valid" populations—i.e., populations that satisfy the 1-to-1 constraint. Thus, our surrogate population, of constant size N' designed to mimic the effect of a dynamically decreasing population, is

$$N' = n_1^2 - \sum_{k=1}^{h''} k = n_1^2 - \frac{h''(h''+1)}{2}, \tag{5}$$

where h'' is the arithmetic mean of h and h'. We then use N' instead of N as the constant population size for the hypergeometric distribution above, arriving finally at the expression $\lambda_{n_1,n_2,h} = 1 - \mathcal{H}(0, n_1, h', N')$ which estimates the mean recovery rate of aligned orthologs in random alignments.

We generated 10,000 random alignments of each of the 28 species pairs, in order to empirically estimate the mean ortholog recovery rate of random alignments, and compared the result to both analytical estimates. The result, depicted in Fig. 6, shows good agreement.

With the per-alignment rate of recovering ortholog pairs in a single random alignment, we now finally come to the task of estimating the p-values in Table 4. We interpret each random alignment as a Bernoulli trial with "success" defined as "at least one pair of correctly aligned orthologs," with probability computed via the second method above. The result is that the number of recovered orthologs per random alignment follows a Poisson distribution with expected recovery rate $\lambda_{n_1,n_2,h}$. Since each alignment is independent, and the expected total rate among T independent Poisson distributions is the sum of their individual rates, we see that in T random alignments, the expected number of recovered orthologs is $T\lambda_{n_1,n_2,h}$. The p-value listed in the last column of Table 4 is computed taking i to be the number of actual recovered orthologs for the "winning" measure

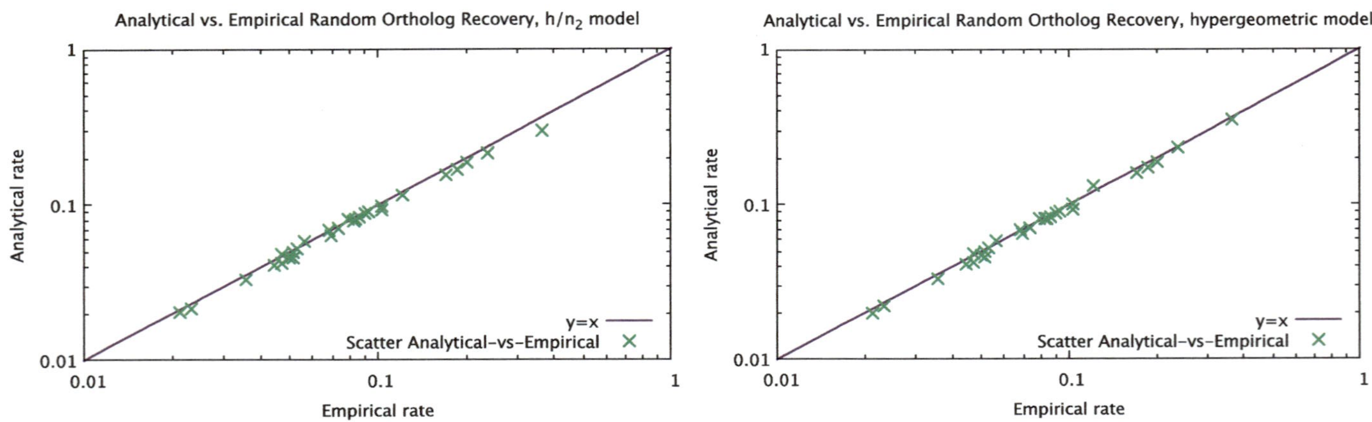

Fig. 6 Estimating rate of ortholog pair recovery in *random* alignments. We compare the empirically determined mean number of paired orthologs recovered in random alignments across 10,000 randomly generated alignments (horizontal axis) vs the analytical version using the simpler model (*left*), and the more complex hypergeometric model of Eq. (5) (*right*). They diverge slightly toward the upper right, where the estimates from the simpler h/n_2 model are slightly less accurate than the hypergeometric-based one.

(e.g., $i = 72$ for the EC measure in the SC-HS row), and then summing the tail of the Poisson distribution from i upward toward infinity until new terms are too small to change the floating-point sum.

Note that this analysis works even in the case of non-1-to-1 orthologs, so long as we count every possible pair among non-1-to-1 orthologs. So, for example, if y_1, y_2, y_3, y_4 are orthologs in yeast that are each orthologous to all of h_1, h_2, h_3 in human, then there are $4 \times 3 = 12$ totals pairs. The number in the "orthologs" column of Table 4 accounts for all such pairs.

6.5 Computing the *p*-value of shared GO terms in an alignment

We use GO terms downloaded from the Gene Ontology website in September 2018, and eliminate all sequence-based GO terms. In particular, we only allow GO terms with evidence codes EXP, HDA, HEP, HGI, HMP, IC, IDA, IEP, IGI, IKR, IMR, IPI, IRD, NAS, ND, and TAS. Similar to the p-value computation of recovered orthologs detailed above, we use the hypergeometric distribution $\mathcal{H}(i, m, M, N)$, where i is the observed number of protein pairs sharing GO term g, m is the number of protein pairs in the alignment, N is the number of protein pairs in the full Cartesian product of $V_1 \times V_2$ (i.e., all possible protein pairings between G_1 and G_2), and M is the number of such pairs that share g.

6.6 On the importance of choosing the right measure of topological similarity

To demonstrate that no one topological measure is universally best, we will compare two commonly used measures of topological similarities: EC (Kuchaiev et al., 2010) and S^3—the *Symmetric Substructure Score* (Saraph & Milenković, 2014). Both are a ratio of edge counts. In both cases, the numerator is the number of "purple" edges in the alignment (cf. Fig. 1). Clearly, more purple is better since purple edges highlight exactly what we mean by "common topology" between networks. The difference between EC and S^3 is in the denominator: EC simply uses the number of edges in G_1, since that is clearly an upper bound on the number of purple edges. However, as pointed out by Saraph and Milenković (2014), EC is asymmetric with respect to edge "color": if we swap G_1 and G_2, then the value of EC changes. Even more worrisome is that unaligned "red" edges are not penalized at all by EC: an alignment can have many unaligned red edges so long as blue edges are minimized. To rectify this, the *Symmetric Substructure Score* S^3 was created (Saraph & Milenković, 2014):

the numerator is still the number of purple edges, but the denominator counts *all edges of all colors* in the induced subgraph on the purple nodes. Thus, if G_1 can be embedded with the same numerator in two different places in G_2, then S^3 will prefer the embedding that has fewer unaligned *red* edges, even though both embeddings have the same EC score.

The S^3 score clearly is a better measure when we know—or expect—that two networks genuinely contain regions of high similarity—meaning there exist large regions with highly similar *induced* structure. In that case, a "good" alignment will exist in which there is a large number of purple edges, and very few red or blue unaligned edges. Such is the case, for example, with any alignment between the mammals of the 2018 release of the IID networks (Kotlyar et al., 2018). The mammal PPI networks in IID-2018 were created by taking the experimentally determined edges for that mammal, and adding in all edges between proteins that had 1-to-1 orthologs with human. The result is that most of the edges in these networks are transferred directly from human experimental PPIs, which makes them all *highly* similar to the human PPI network—possibly overly so. For example, the next most complete mammal PPI network after human in BioGRID 3.4.164 is mouse, but it contains only about 20,000 edges on 7000 nodes, compared to human's 280,000 edges on 17,000 nodes (cf. Table 3). In other words, the mouse PPI network is likely less than 11% complete (20,000/280,000≈0.11). Observing Table S2 in Supplementary material, we see that when IID synthetically augments mammalian PPI networks using interologs from human, then using SANA to optimize S^3 for 1 hour between human and any of these networks recovers about 12,500 out of 16,000 orthologous proteins—essentially a 75% rate of "node correctness" between species. Conversely, optimizing EC recovers only about 10,200 orthologs, a decrease of about 20%. Clearly, S^3 is the better measure in cases like this where the PPI networks contain large regions that are almost identical.

Now consider a very different scenario: starting with some complete network G_2 that has more than enough edges to self-align according to Eq. (3), and let G_1 be a subnetwork where 95% of G_2's edges have been removed at random. In the parlance of Fig. 1, the "correct" alignment would have an EC score of exactly 1 (since every blue edge from G_1 exists in G_2), but the S^3 score is only 0.05, since 95% of red edges between purple nodes remain unaligned. In this case, S^3 may prefer to align the nodes of G_1 over the "least dense" regions of G_2, in order to minimize the number of unaligned red edges. This may result in sacrificing purple edges in the numerator in order to reduce red edges in the denominator, resulting in an alignment far from the "correct" one. Thus, we see that when edge densities are unbalanced, EC may be a more appropriate measure than S^3, precisely *because* it is an imbalanced measure, and this imbalance matches the imbalance in edge densities.

This latter case—of EC being a better measure when edge densities are grossly imbalanced—is exactly what we see when we use SANA to align virtually any pair of BioGRID networks: optimizing EC results in alignments that are much better able to align regions of high functional similarity as measured in our companion paper by ability to successfully predict GO terms (Wang et al., 2022). This is presumably because the BioGRID networks have highly variable network coverage so that given any two of them, one is likely to have far more edges than the other simply due to different levels of completeness.

7. Data availability

The BioGRID networks we used are available from TheBioGRID.org; only physical interactions between nodes in the stated species were included. SANA is available in source code form at https://github.com/waynebhayes/SANA.

Acknowledgments

We thank Annie Raichev for running NAUTY (Mckay, 2010) on the BioGRID Human PPI network to count the number of automorphisms.

Author contributions

ADK produced the initial SIF results that launched the project, while BJF updated the SIF results on a wider range of networks and aligners. PK and BAM performed all locking + saturation experiments. XC performed all H-GRAAL and most information theory experiments. SW and BJF together performed all runs and initial analyses for Table 4. WBH conceived and guided the entire project and performed some of the analyses related to information theory, frequency, GO terms, and *p*-values, and created the final manuscript based on drafts of the respective sections written by each author. All authors were given the opportunity to review the final manuscript.

References

Adamek, J. (2011). *Foundations of coding: Theory and applications of error-correcting codes with an introduction to cryptography and information theory*. John Wiley & Sons.

Aguiar e Oliveira Junior, H., Ingber, L., Petraglia, A., Rembold Petraglia, M., & Augusta Soares Machado, M. (2012). Adaptive simulated annealing. In *Stochastic global optimization and its applications with fuzzy adaptive simulated annealing* (pp. 33–62). Berlin, Heidelberg: Springer Berlin Heidelberg. https://doi.org/10.1007/978-3-642-27479-4_4.

Aladağ, A. E., & Erten, C. (2013). SPINAL: Scalable protein interaction network alignment. *Bioinformatics*, *29*(7), 917–924. https://doi.org/10.1093/bioinformatics/btt071. http://bioinformatics.oxfordjournals.org/content/29/7/917.abstract.

Alberich, R., Alcala, A., Llabres, M., Rossello, F., & Valiente, G. (2019). Alignet: Alignment of protein-protein interaction networks. *arXiv:1902.07107*.

Alberts, B., Johnson, A., Lewis, J., Raff, M., Roberts, K., & Walter, P. (2015). *Molecular biology of the cell. The extracellular matrix of animals*. New York, NY: Garland Science. Available from http://www.ncbi.nlm.nih.

Alkan, F., & Erten, C. (2015). SiPAN: Simultaneous prediction and alignment of protein–protein interaction networks. *Bioinformatics*, *31*(14), 2356–2363.

Chindelevitch, L., Ma, C.-Y., Liao, C.-S., & Berger, B. (2013). Optimizing a global alignment of protein interaction networks. *Bioinformatics*, *29*(21), 2765–2773. https://doi.org/10.1093/bioinformatics/btt486. http://bioinformatics.oxfordjournals.org/content/29/21/2765.abstract.

Clark, C., & Kalita, J. (2014). A comparison of algorithms for the pairwise alignment of biological networks. *Bioinformatics*, *30*(16), 2351–2359.

Clark, C., & Kalita, J. (2015). A multiobjective memetic algorithm for PPI network alignment. *Bioinformatics*, *31*(12), 1988–1998. https://doi.org/10.1093/bioinformatics/btv063. http://bioinformatics.oxfordjournals.org/content/31/12/1988.abstract.

Collins, S. R., Kemmeren, P., Zhao, X.-C., Greenblatt, J. F., Spencer, F., Holstege, F. C. P., … Krogan, N. J. (2007). Toward a comprehensive atlas of the physical interactome of saccharomyces cerevisiae. *Molecular and Cellular Proteomics*, *6*(3), 439–450. https://doi.org/10.1074/mcp.M600381-MCP200. http://www.mcponline.org/content/6/3/439.abstract.

Crawford, J., & Milenković, T. (2015). Great: Graphlet edge-based network alignment. In *2015 IEEE International conference on bioinformatics and biomedicine (BIBM)* (pp. 220–227).

Davis, D., Yaveroğlu, Ö. N., Malod-Dognin, N., Stojmirovic, A., & Pržulj, N. (2015). Topology-function conservation in protein–protein interaction networks. *Bioinformatics*, *31*(10), 1632–1639. https://doi.org/10.1093/bioinformatics/btv026.

Dhulipala, L., Kabiljo, I., Karrer, B., Ottaviano, G., Pupyrev, S., & Shalita, A. (2016). Compressing graphs and indexes with recursive graph bisection. In *Proceedings of the 22nd ACM SIGKDD international conference on knowledge discovery and data mining* (pp. 1535–1544).

Dowsland, K. A., & Thompson, J. M. (2012). Simulated annealing. In *Handbook of natural computing* (pp. 1623–1655). Springer.

El-Kebir, M., Heringa, J., & Klau, G. W. (2011). Lagrangian relaxation applied to sparse global network alignment. In *IAPR International conference on pattern recognition in bioinformatics* (pp. 225–236). Springer.

Elmsallati, A., Msalati, A., & Kalita, J. (2018). Index-based network aligner of protein-protein interaction networks. *IEEE/ACM Transactions on Computational Biology and Bioinformatics (TCBB)*, *15*(1), 330–336.

Faisal, F. E., Zhao, H., & Milenković, T. (2014). Global network alignment in the context of aging. *IEEE/ACM Transactions on Computational Biology and Bioinformatics*, *12*(1), 40–52.

Furuse, M., Fujita, K., Hiiragi, T., Fujimoto, K., & Tsukita, S. (1998). Claudin-1 and-2: Novel integral membrane proteins localizing at tight junctions with no sequence similarity to occludin. *The Journal of Cell Biology*, *141*(7), 1539–1550.

Garey, M. R., & Johnson, D. S. (1979). *Computers and intractability–A guide to the theory of NP-completeness*. New York: W. H. Freeman and Company.

Gaudelet, T., Malod-Dognin, N., & Pržulj, N. (2018). Higher-order molecular organization as a source of biological function. *Bioinformatics*, *34*(17), i944–i953.

Gligorijević, V., Malod-Dognin, N., & Pržulj, N. (2015). *Bioinformatics*, *32*(8), btv731. FUSE: Multiple network alignment via data fusion.

Gong, M., Peng, Z., Ma, L., & Huang, J. (2015). Global biological network alignment by using efficient memetic algorithm. *IEEE/ACM Transactions on Computational Biology and Bioinformatics*, *13*(6), 1117–1129.

Hashemifar, S., Ma, J., Naveed, H., Canzar, S., & Xu, J. (2016). Modulealign: Module-based global alignment of protein–protein interaction networks. *Bioinformatics*, *32*(17), i658–i664.

Hashemifar, S., & Xu, J. (2014). HubAlign: An accurate and efficient method for global alignment of protein-protein interaction networks. *Bioinformatics*, *30*(17), i438–i444. https://doi.org/10.1093/bioinformatics/btu450.

Hayes, W., & Mamano, N. (2022). *Automatic temperature endpoints for simulated annealing*. (Submitted for publication).

Ibragimov, R., Malek, M., Guo, J., & Baumbach, J. (2013). GEDEVO: An evolutionary graph edit distance algorithm for biological network alignment. In *OASIcs-OpenAccess series in informatics: Vol. 34*. Schloss Dagstuhl-Leibniz-Zentrum fuer Informatik.

Ideker, T., & Krogan, N. J. (2012). Differential network biology. *Molecular Systems Biology*, *8*(1), 565.

Ingber, L. (1989). Very fast simulated re-annealing. *Mathematical and Computer Modelling*, *12*(8), 967–973.

Johnson, P. D., Jr, Harris, G. A., & Hankerson, D. C. (2003). *Introduction to information theory and data compression*. Chapman and Hall/CRC.

Kabsch, W., & Sander, C. (1984). On the use of sequence homologies to predict protein structure: Identical pentapeptides can have completely different conformations. *Proceedings of the National Academy of Sciences of the United States of America*, *81*(4), 1075–1078.

Kachroo, A. H., Laurent, J. M., Yellman, C. M., Meyer, A. G., Wilke, C. O., & Marcotte, E. M. (2015). Systematic humanization of yeast genes reveals conserved functions and genetic modularity. *Science*, *348*(6237), 921–925.

Kalecky, K., & Cho, Y.-R. (2018). PrimAlign: PageRank-inspired Markovian alignment for large biological networks. *Bioinformatics*, *34*(13), i537–i546.

Kazemi, E., Hassani, H., Grossglauser, M., & Modarres, H. P. (2016). PROPER: Global protein interaction network alignment through percolation matching. *BMC Bioinformatics*, *17*(1), 527.

Kirkpatrick, S., Gelatt, C. D., & Vecchi, M. P. (1983). Optimization by simulated annealing. *Science*, *220*(4598), 671–680.

Klau, G. (2009). A new graph-based method for pairwise global network alignment. *BMC Bioinformatics*, *10*(Suppl. 1), S59. https://doi.org/10.1186/1471-2105-10-S1-S59.

Klir, G. J. (2005). *Uncertainty and information: Foundations of generalized information theory*. John Wiley & Sons.

Kotlyar, M., Pastrello, C., Malik, Z., & Jurisica, I. (2018). IID 2018 update: Context-specific physical protein–protein interactions in human, model organisms and domesticated species. *Nucleic Acids Research*, *47*(D1), D581–D589.

Krogan, N. J., Cagney, G., Yu, H., Zhong, G., Guo, X., Ignatchenko, A., … Greenblatt, J. F. (2006a). Global landscape of protein complexes in the yeast Saccharomyces cerevisiae. *Nature*, *440*, 637–643.

Kuchaiev, O., Milenković, T., Memišević, V., Hayes, W., & Pržulj, N. (2010). Topological network alignment uncovers biological function and phylogeny. *Journal of The Royal Society Interface*, 7(50), 1341–1354. https://doi.org/10.1098/rsif.2010.0063.
Kuchaiev, O., & Pržulj, N. (2011). Integrative network alignment reveals large regions of global network similarity in yeast and human. *Bioinformatics*, *27*, 1390–1396. https://doi.org/bioinformatics/btr127.
Kullback, S. (1997). *Information theory and statistics*. Courier Corporation.
Kulmanov, M., & Hoehndorf, R. (2020). DeepGOPlus: Improved protein function prediction from sequence. *Bioinformatics*, *36*(2), 422–429.
Lockhart, P. J., Steel, M. A., Hendy, M. D., & Penny, D. (1994). Recovering evolutionary trees under a more realistic model of sequence evolution. *Molecular Biology and Evolution*, *11*(4), 605–612.
Madsen, P., Celis, J. E., Rasmussen, H. H., Vorum, H., Anant, S., Gromov, P., … Celis, J. E. (1999). Psoriasis upregulated phorbolin-1 shares structural but not functional similarity to the mRNA-editing protein apobec-1. *Journal of Investigative Dermatology*, *113*(2), 162–169.
Malek, M., Ibragimov, R., Albrecht, M., & Baumbach, J. (2016). CytoGEDEVO-global alignment of biological networks with cytoscape. *Bioinformatics*, *32*(8), 1259–1261.
Malod-Dognin, N., Ban, K., & Pržulj, N. (2017). Unified alignment of protein-protein interaction networks. *Scientific Reports*, 7(1), 953.
Malod-Dognin, N., & Pržulj, N. (2015). *Bioinformatics*, *31*, 2182–2189. https://doi.org/10.1093/bioinformatics/btv130. L-GRAAL: Lagrangian graphlet-based network aligner. http://bioinformatics.oxfordjournals.org/content/early/2015/02/28/bioinformatics.btv130.abstract.
Malod-Dognin, N., & Pržulj, N. (2019). *Bioinformatics*, *35*, 3727–3734. Functional geometry of protein interactomes.
Mamano, N., & Hayes, W. B. (2017). SANA: Simulated annealing far outperforms many other search algorithms for biological network alignment. *Bioinformatics (Oxford, England)*, *33*, 2156–2164.
Mckay, B. D. (2010). *Nauty*. http://users.cecs.anu.edu.au/bdm/nauty.
Meise, C. (1998). On the convergence of parallel simulated annealing. *Stochastic Processes and Their Applications*, *76*(1), 99–115.
Milenković, T., Ng, W. L., Hayes, W., & Pržulj, N. (2010). Optimal network alignment with graphlet degree vectors. *Cancer Informatics*, *9*, 121.
Milenković, T., & Pržulj, N. (2008). Uncovering biological network function via graphlet degree signatures. *Cancer Informatics*, *6*, 257–273.
Milenković, T., & Pržulj, N. (2008). Uncovering biological network function via graphlet degree signatures. *Cancer Informatics*, *6*. CIN–S680.
Milenković, T., Zhao, H., & Faisal, F. E. (2013). Global network alignment in the context of aging. In *Proceedings of the international conference on bioinformatics, computational biology and biomedical informatics, BCB'13*. (pp. 23:23–23:32). New York, NY, USA: ACM. https://doi.org/10.1145/2506583.2508968.
Mir, A., Naghibzadeh, M., & Saadati, N. (2017). INDEX: Incremental depth extension approach for protein–protein interaction networks alignment. *Biosystems*, *162*, 24–34.
Mitra, D., Romeo, F., & Sangiovanni-Vincentelli, A. (1985). Convergence and finite-time behavior of simulated annealing. In *1985 24th IEEE Conference on decision and control* (pp. 761–767).
Mohammadi, S., Gleich, D. F., Kolda, T. G., & Grama, A. (2016). Triangular alignment (TAME): A tensor-based approach for higher-order network alignment. *IEEE/ACM Transactions on Computational Biology and Bioinformatics*, *14*(6), 1446–1458.
Morrone, A., McCully, M. E., Bryan, P. N., Brunori, M., Daggett, V., Gianni, S., & Travaglini-Allocatelli, C. (2011). The denatured state dictates the topology of two

proteins with almost identical sequence but different native structure and function. *Journal of Biological Chemistry*, *286*(5), 3863–3872.

Neyshabur, B., Khadem, A., Hashemifar, S., & Arab, S. S. (2013). NETAL: A new graph-based method for global alignment of protein-protein interaction networks. *Bioinformatics*, *29*(13), 1654–1662. https://doi.org/10.1093/bioinformatics/btt202. http://bioinformatics.oxfordjournals.org/content/29/13/1654.abstract.

Page, L., Brin, S., Motwani, R., & Winograd, T. (1999). *The pagerank citation ranking: Bringing order to the web*. (Tech. Rep.). Stanford InfoLab.

Park, M.-W., & Kim, Y.-D. (1998). A systematic procedure for setting parameters in simulated annealing algorithms. *Computers and Operations Research*, *25*(3), 207–217. https://doi.org/10.1016/S0305-0548(97)00054-3.

Patro, R., & Kingsford, C. (2012). Global network alignment using multiscale spectral signatures. *Bioinformatics*, *28*(23), 3105–3114. https://doi.org/10.1093/bioinformatics/bts592. http://bioinformatics.oxfordjournals.org/content/28/23/3105.abstract.

Pearson, W. R. (2013). An introduction to sequence similarity ("homology") searching. *Current Protocols in Bioinformatics*, *42*(1). 3–1.

Pennacchio, L. A. (2003). Insights from human/mouse genome comparisons. *Mammalian Genome*, *14*(7), 429–436.

Phillips, R., Kondev, J., Theriot, J., & Garcia, H. G. (2013). *Physical biology of the cell*. New York, NY: Garland Science. Available from http://www.ncbi.nlm.nih.

Poole, W., Gibbs, D. L., Shmulevich, I., Bernard, B., & Knijnenburg, T. A. (2016). Combining dependent P-values with an empirical adaptation of Brown's method. *Bioinformatics*, *32*(17), i430–i436.

Radivojac, P., Peng, K., Clark, W. T., Peters, B. J., Mohan, A., Boyle, S. M., & Mooney, S. D. (2008). An integrated approach to inferring gene-disease associations in humans. *Proteins*, *72*(3), 1030–1037. http://www.ncbi.nlm.nih.gov/pubmed/18300252.

Rolland, T., Taşan, M., Charloteaux, B., Pevzner, S. J., Zhong, Q., Sahni, N., … Vidal, M. (2014). A proteome-scale map of the human interactome network. *Cell*, *159*(5), 1212–1226.

Romeo, F., Sangiovanni, V. A., & Huang, M. (1986). An efficient general cooling schedule for simulated annealing. In *Proceeding of IEEE international conference on computer aided design*.

Sakrison, D. J. (1968). *Vol. 968. Communication theory: Transmission of waveforms and digital information*. New York: Wiley.

Saraph, V., & Milenković, T. (2014). MAGNA: Maximizing accuracy in global network alignment. *Bioinformatics*, *30*(20), 2931–2940.

Schlicker, A., Domingues, F. S., Rahnenführer, J., & Lengauer, T. (2006). A new measure for functional similarity of gene products based on gene ontology. *BMC Bioinformatics*, 7(1), 302.

Shannon, C. E. (1948). A mathematical theory of communication. *Bell System Technical Journal*, *27*(3), 379–423.

Singh, R., Xu, J., & Berger, B. (2008). Global alignment of multiple protein interaction networks with application to functional orthology detection. *Proceedings of the National Academy of Sciences of the United States of America*, *105*(35), 12763–12768. https://doi.org/10.1073/pnas.0806627105. http://www.pnas.org/content/105/35/12763.abstract.

Strens, M. J. A. (2003). Evolutionary MCMC sampling and optimization in discrete spaces. In *ICML'03: Proceedings of the twentieth international conference on machine learning*. http://citeseerx.ist.psu.edu/viewdoc/summary?doi=10.1.1.3.637.

Suman, B., & Kumar, P. (2006). A survey of simulated annealing as a tool for single and multiobjective optimization. *Journal of the Operational Research Society*, *57*(10), 1143–1160.

Sun, Y., Crawford, J., Tang, J., & Milenkovic', T. (2015). Simultaneous optimization of both node and edge conservation in network alignment via WAVE. In M. Pop, & H. Touzet (Eds.), *Vol. 9289. Algorithms in bioinformatics. Lecture notes in computer science* (pp. 16–39). Germany: Springer Berlin Heidelberg. https://doi.org/10.1007/978-3-662-48221-6_2.

Szu, H., & Hartley, R. (1987). Fast simulated annealing. *Physics Letters A, 122*(3-4), 157–162.

The Gene Ontology Consortium. (2008). The gene ontology project in 2008. *Nucleic Acids Research, 36*(Suppl. 1), D440–D444. https://doi.org/10.1093/nar/gkm883. http://nar.oxfordjournals.org/content/36/suppl_1/D440.abstract.

Thomas, P. D., Wood, V., Mungall, C. J., Lewis, S. E., Blake, J. A., & Gene Ontology Consortium. (2012). On the use of gene ontology annotations to assess functional similarity among orthologs and paralogs: A short report. *PLOS Computational Biology, 8*(2), e1002386.

Vijayan, V., Saraph, V., & Milenković, T. (2015). MAGNA++: Maximizing accuracy in global network alignment via both node and edge conservation. *Bioinformatics, 31*(14), 2409–2411.

Walhout, A. J., & Vidal, M. (2001). High-throughput yeast two-hybrid assays for large-scale protein interaction mapping. *Methods, 24*(3), 297–306.

Wang, S., Atkinson, G. R. S., & Hayes, W. B. (2022). SANA: Cross-species prediction of gene ontology GO annotations via topological network alignment. *npj Systems Biology*. accepted.

Waterson, R. H., Lander, E. S., Wilson, R. K., Chimpanzee, T., & Sequencing, A. C. (2005). Initial sequence of the chimpanzee genome and comparison with the human genome. *Nature, 437*(7055), 69.

Xie, J., Xiang, C., Ma, J., Tan, J., Wen, T., Lei, J., & Nie, Q. (2016). An adaptive hybrid algorithm for global network alignment. *IEEE/ACM Transactions on Computational Biology and Bioinformatics (TCBB), 13*(3), 483–493.

Yasar, A., & Çatalyürek, Ü. V. (2018). An iterative global structure-assisted labeled network aligner. In *Proceedings of the 24th ACM SIGKDD international conference on knowledge discovery & data mining* (pp. 2614–2623). ACM.

Zhan, S.-h., Lin, J., Zhang, Z.-j., & Zhong, Y.-w. (2016). List-based simulated annealing algorithm for traveling salesman problem. *Computational Intelligence and Neuroscience, 2016*, 8.

Zhang, C., Freddolino, P. L., & Zhang, Y. (2017). COFACTOR: Improved protein function prediction by combining structure, sequence and protein–protein interaction information. *Nucleic Acids Research, 45*(W1), W291–W299.

Zhu, Y., Li, Y., Liu, J., Qin, L., & Yu, J. X. (2017). GMAlign: A new network aligner for revealing large conserved functional components. In *2017 IEEE international conference on bioinformatics and biomedicine (BIBM)* (pp. 120–127). IEEE.

CHAPTER TWO

From single-omics to interactomics: How can ligand-induced perturbations modulate single-cell phenotypes?

L.F. Piochi[a,b,†], A.T. Gaspar[a,b,†], N. Rosário-Ferreira[a,b,c], A.J. Preto[a,b,d,e], and I.S. Moreira[f,g,*]

[a]Center for Neuroscience and Cell Biology (CNC), University of Coimbra, Coimbra, Portugal
[b]Center for Innovative Biomedicine and Biotechnology (CIBB), University of Coimbra, Coimbra, Portugal
[c]Department of Chemistry, Coimbra Chemistry Center, University of Coimbra, Coimbra, Portugal
[d]PhD Programme in Experimental Biology and Biomedicine, Institute for Interdisciplinary Research (IIIUC), University of Coimbra, Casa Costa Alemão, Coimbra, Portugal
[e]Institute for Interdisciplinary Research, University of Coimbra, Coimbra, Portugal
[f]Department of Life Sciences, Center for Neuroscience and Cell Biology, Coimbra University, Coimbra, Portugal
[g]CNC—Center for Neuroscience and Cell Biology, CIBB—Center for Innovative Biomedicine and Biotechnology, Coimbra, Portugal
*Corresponding author: e-mail address: irina.moreira@cnc.uc.pt

Contents

† First Co-authors.

Advances in Protein Chemistry and Structural Biology, Volume 131
ISSN 1876-1623
https://doi.org/10.1016/bs.apcsb.2022.05.006

Abstract

Cells suffer from perturbations by different stimuli, which, consequently, rise to individual alterations in their profile and function that may end up affecting the tissue as a whole. This is no different if we consider the effect of a therapeutic agent on a biological system. As cells are exposed to external ligands their profile can change at different single-omics levels. Detecting how these changes take place through different sequencing technologies is key to a better understanding of the effects of therapeutic agents.

Single-cell RNA-sequencing stands out as one of the most common approaches for cell profiling and perturbation analysis. As a result, single-cell transcriptomics data can be integrated with other omics data sources, such as proteomics and epigenomics data, to clarify the perturbation effects and mechanism at the cell level. Appropriate computational tools are key to process and integrate the available information.

This chapter focuses on the recent advances on ligand-induced perturbation and single-cell omics computational tools and algorithms, their current limitations, and how the deluge of data can be used to improve the current process of drug research and development.

1. Introduction

1.1 Fundamentals of the drug discovery process

The drug discovery concept encloses the methodologies for both *de novo* drug design and drug repurposing. *De novo* drug design pipelines, via the traditional, experimental approach, take more than a decade and reportedly cost between one to three billion US dollars per drug (Nosengo, 2016). Since 2011, this value is increasing (Hughes, Rees, Kalindjian, & Philpott, 2011), meaning that the identification of novel drugs and the improvement of known ones is becoming increasingly expensive (Akhoon, Tiwari, & Nargotra, 2019; Nosengo, 2016). Producing new drugs is a low rate success process, often associated with the poor clinical efficiency achieved, the unpredictable toxicity of the drug and with the difficulty in producing synthesizable drug-like compounds suitable for commercialization (Sun, Gao, Hu, & Zhou, 2022). Hence, the industry is shifting toward drug repurposing approaches. In drug repurposing, the research objective is the identification of novel targets for an already marketed drug. Nosengo et al. estimated that the repurposing process involves an average investment of 300M US dollars in 6.5 years, which is 30–40% cheaper in comparison with *de novo* design (Akhoon et al., 2019; Nosengo, 2016).

The task of Research and Development (R&D) of new drugs involves four phases, starting with the identification of a therapeutic target, potential compounds to target it, followed by preclinical, and clinical tests. In the preclinical phase, lead candidates are submitted to *in vitro* followed by *in vivo* assays to assess Absorption, Distribution, Metabolism, Excretion and Toxicity (ADMET) properties. In the clinical phase, the pharmacokinetic properties of the drug are refined. This is a critical phase where 90% of the drugs fail (Sun et al., 2022). To circumvent the question of compounds failing after showing high *in vitro* potency, scientists sought to perform a deeper description of the intrinsic properties of a molecule that would better showcase its ability as a therapeutic compound (Di, Kerns, & Carter, 2009). As the gathering and processing of high-throughput data is expensive and delays drug development, an increasing interest for Computer-Aided Drug Design (CADD) methods for identification of potential candidates and targets using pharmacodynamic and pharmacokinetic data arose (Prieto-Martínez, López-López, Eurídice Juárez-Mercado, & Medina-Franco, 2019). Once the drug is approved by the Food and Drug Administration (FDA), in America, or European Medicines Agency (EMA), in Europe, it starts to be commercialized and monitored by the responsible entity (Hughes et al., 2011; Prieto-Martínez et al., 2019).

The main challenges, and opportunities, at the R&D of new drugs, stand with the mining of large amounts of information, associated with single-cell and cell population. Just as the gene expression differs from cell to cell, variations in metabolites concentration and cellular dynamics also differ. These observations made single-cell omics research to focus on the analysis of variations among cell populations and their impact on their biological responses (Nassar, Raddassi, & Wu, 2021). Therefore, the drug discovery R&D is evolving toward single-cell data analysis, to clarify the disturbances caused by a compound in the biological system, providing information from the genotype to the phenotype (Nassar et al., 2021). Understanding the dynamics of a population of cells from single-cell analysis means to recognize the alterations induced at the signaling network-level, which, compared with the proteomic-level, provide a broader insight on how the components are affected and their influence regarding the biological mechanism or pathway they are involved in.

Single-cell studies have been relevant not only for immunophenotype analysis of the drug candidate, through evaluation of the drug candidate's immunotoxicology on population of cells, but also for understanding the alterations that occur in the cell, through pharmacodynamic markers (Nassar et al., 2021). Taken together, this knowledge highlights the importance of the relationship between biological information through different omics in the drug discovery process. In particular, we cannot forget that understanding the triad sequence-three-dimensional (3D) structure-function of proteins is fundamental in any quest for new drugs or new drug targets. Merging this information with the one retrieved from transcriptomics data is capturing scientists' attention, as it can lead to a better understanding of the overall biological mechanisms involved (Boodhun, 2018; Li, Cao, Westhof, & Miao, 2020). In this chapter, we focus our attention on the role and interrelation of protein and single-cell, and how they can contribute to the design of better therapeutic approaches.

1.2 Overwhelming data quantity and complexity in biology

Advances in high-throughput technologies in the last decade paved the way for a variety of omics studies, generating large amounts of data used for a number of contexts, including drug discovery (Paananen & Fortino, 2020), biomarker identification (Matthews, Hanison, & Nirmalan, 2016), among others (Kato, Takahashi, & Saito, 2011; Van Emon, 2016; Zhang, Xia, Wang, Yang, & Baird, 2018). Recently, the data surge has become even more valuable due to the emergence of new ways of looking into it, such as spatial omics and multi-omics studies, which potentiates new breakthroughs in the analysis of biological systems, including at a single-cell level (Stuart & Satija, 2019). However, massive amounts of data can only solve problems when adequate processing is available. Thus, this "biological big data" accelerated the need for the development of computational tools able to process and integrate the generated information (Argelaguet, Cuomo, Stegle, & Marioni, 2021; Forcato, Romano, & Bicciato, 2021). By adopting mathematical models to analyze high-throughput data, scientists reach a better understanding on a variety of subjects including the structure of biological networks, whose comprehension is essential for the development of novel therapeutic agents. In its turn, characterizing the architecture of biological networks can facilitate the understanding of the functional intricacies of cells and tissues (Barabási, Gulbahce, & Loscalzo, 2011).

In spite of sample annotation and experimental variation (Sturm, List, & Zhang, 2021), biological tissues can be highly heterogeneous, being composed of a variety of cellular subpopulations with individual omics profiles and exhibiting distinct responses to stimuli (Cha & Lee, 2020). While bulk sequencing methods can provide a general perspective on the effects of perturbations on a tissue, responses generated by larger cell subpopulations may hide the effects on other groups of cells (Srivatsan et al., 2020; Trapnell, 2015). Similarly, bulk analysis of tissues can make the identification of disease microenvironments harder, concealing potential targets for intervention in therapy (Wu et al., 2021).

Despite this deluge of data, there is still a lack of specific data regarding subpopulations that make up the regulatory network within a tissue limits our knowledge of the interplay between the components of a regulatory network. Consequently, the improvement of existing treatments to a plethora of conditions is highly hindered. Advances in microfluidics and sequencing technologies propelled many studies to address this data gap by conducting omics analyses on a single-cell level, allowing the dissection of networks in biological systems in a scalable fashion (Svensson, Vento-Tormo, & Teichmann, 2018). Accordingly, the number of cells per single-cell RNA sequencing (scRNA-seq) experiment has grown substantially in the last decade, and tens of thousands of cells can now be analyzed in a single study (Svensson et al., 2018).

scRNA-seq is the most popular approach in the context of single-cell omics and plays a role in the characterization of several tissues and their response to external inputs. Additionally, scRNA-seq has been combined with other methods to profile different features of single-cells including, but not limited to, chromatin accessibility (Buenrostro et al., 2015; Lake et al., 2018), epigenetics (Luo et al., 2017; Mulqueen et al., 2018; Rotem et al., 2015), intracellular and cell-surface proteins (Darmanis et al., 2016; Peterson et al., 2017; Stoeckius et al., 2017), genetic perturbation (Dixit et al., 2016), and spatial information (Chen, Boettiger, Moffitt, Wang and Zhuang, 2015; Eng et al., 2019; Wang et al., 2018). This single-cell resolution data in combination with adequate computational methods has been a game changer in the determination of biological networks and new models are constantly being developed to integrate information and predict the outcomes of perturbations (Zappia & Theis, 2021).

There is still much room for improvement as experiment costs decrease and the amount of publicly available information increases. Performing multimodal, systematic analyses of drug-induced perturbations can be a huge step toward better treatment outcomes in most disease scenarios, from neurodegeneration (Xu et al., 2021; Zhou et al., 2020) to cancer (Aissa et al., 2021; McFarland et al., 2020). By interpreting how external molecules can affect the gene regulatory networks and different cell populations, it becomes easier to narrow down the main drivers behind the effects of compounds and identify opportunities to improve their mechanism of action.

Likewise, this data can also be used to develop computational tools to predict how a given cell type, tissue or patient will react to a specific therapy (Chen et al., 2021; Lotfollahi, Susmelj, Donnoi, et al., 2021; Mitra et al., 2016). These insights highlight the potential of multimodal experiments at a single-cell level for drug development, which generate large amounts of systematic data through high-throughput drug screens, and therefore pave the way for interactomics evaluation of compound-induced perturbations.

2. Single-cell omics: From unimodal to multimodal analysis

Technological advances in the last decade have led to a paradigm shift in omics research, as scientists can now obtain single-cell resolution data to tackle previous shortcomings of bulk analysis. With a wide array of applications, these methodologies have already been found helpful in the characterization of tumor microenvironments (Olalekan, Xie, Back, Eckart, & Basu, 2021; Wu et al., 2021), autoimmune diseases and pathways understanding (Gangaev et al., 2021; Gearty et al., 2022), and ligand and genetic screens (Aissa et al., 2021; Hart et al., 2015; McFarland et al., 2020; Rubin et al., 2019). In fact, despite generating valuable information, constraining omics assays to a unimodal analysis limits the scientific potential of these experiments. Nevertheless, scRNA-seq data and its conjugations (i.e., contexts where transcriptomics data is analyzed along with other cellular characteristics) deserve particular attention given its current popularity. There are two main aspects to be considered in respect to multimodal, interactomics studies at the single-cell level: (i) the technology for cell isolation and molecule identification, and (ii) the methods to process and integrate multi-omics data. Both aspects and the research in the single and multi-omics have been evolving substantially in the last few years, and novel approaches are regularly developed, increasing the scale and quality of available datasets as well as the need for new tools to analyze them (Fig. 1).

2.1 Is multimodal single-cell analysis the way to go?

Despite providing a wide perspective on the biological information of a tissue and accounting for the heterogeneity of such, unimodal single-cell omics are often not enough to fully understand how cells behave under certain conditions. By using only, a single type of data for analysis, insights on the multifaceted relationship between different components of biological

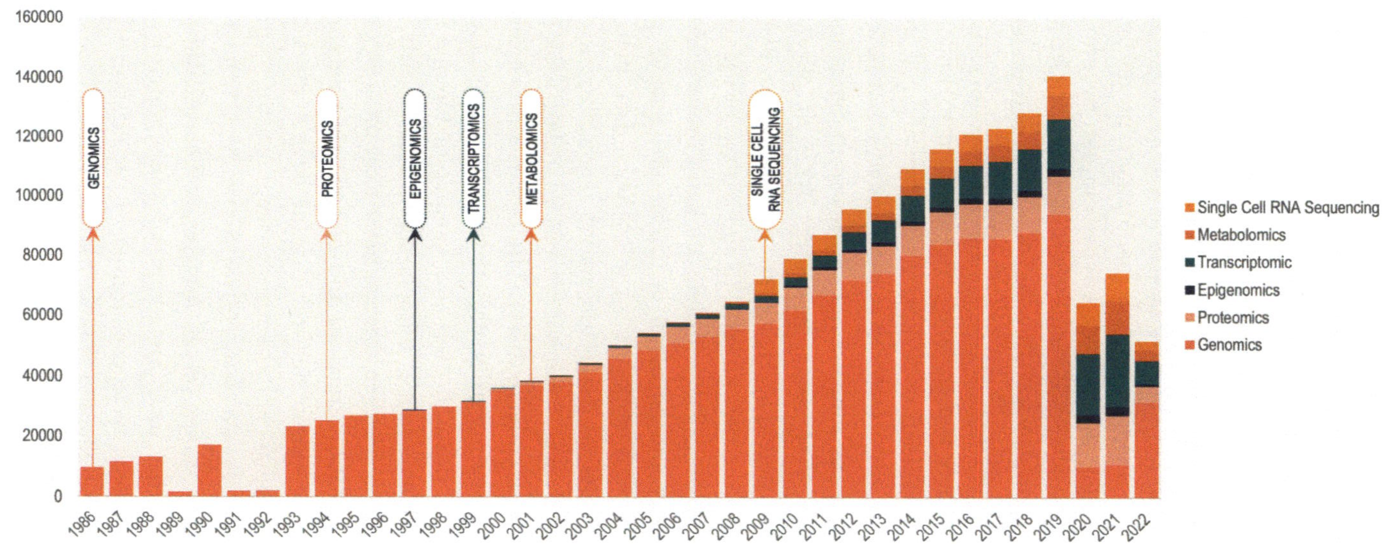

Fig. 1 Scientific publications retrieved from PubMed on April 14, 2022. A cumulative bar chart representing the yearly number of papers of several single omics from 1986 to 2022. PubMed queries: "single cell RNA sequencing," "Metabolomics," "Transcriptomics," "Epigenomics," "Proteomics," and "Genomics." Also identified in rounded text boxes are the first time the terms queried were coined, and date from which the retrieval of PubMed results was considered.

networks are abandoned. For example, while genomics and transcriptomics can provide interesting information, without epigenomics, proteomics or metabolomics data, it is harder to capture the real cell effect. The array of interactions between different components of a biological system can be described, from a systems biology standpoint, as an interactome. Conjugating unimodal data with other data types can thus aid in constructing the interactome of cell subpopulations and be of great assistance in identifying new biological patterns and targets in disturbed cells. Integrative analyses using different data types can also be critical in the discovery of novel cell subpopulations or regulatory networks (Lee, Hyeon, & Hwang, 2020). However, researchers must first understand the relationships between different data types and how to integrate this data much like having frames and making sense of them to understand the movie's plot. Moreover, even with more recent experimental approaches, the cell throughput of these methods is relatively low and new strategies are needed. As such, many studies combine data from different unimodal assays (Rautenstrauch, Vlot, Saran, & Ohler, 2022), an alternative that tackles the scalability problem. However, the use of nonidentical cell populations difficulties the later integration and analysis of the data.

Most models of single-cell data analysis are based on scRNA-seq only, even though they may also be useful in multimodal analyses depending on the protocol. A typical workflow for scRNA-seq data often follows a similar pattern, where the raw data is first used to generate a gene expression matrix that subsequently serves as input for specific analysis tools (Andrews, Kiselev, McCarthy, & Hemberg, 2021). The raw data is first subjected to pre-processing, where it is normally subjected to quality control (e.g., background RNA level subtraction, molecular barcode or Unique Molecular Identifier (UMI) identification, mitochondrial RNA-based filtering), identification of multiplets (when there is more than one cell per analysis unit), normalization, data correction, and feature selection (Luecken & Theis, 2019). There are specific methods that can be used for each of these steps, with the most popular choices for pre-processing being Seurat (Butler, Hoffman, Smibert, Papalexi, & Satija, 2018), Scater (McCarthy, Campbell, Lun, & Wills, 2017) or Scanpy (Wolf, Angerer, & Theis, 2018). These are often coupled with other tools for additional corrections or covariates, including batch effects, cell cycle, pseudotime, or visualization. The output of the pre-processing can be used as input for downstream analyses including dimensionality reduction and visualization, unsupervised clustering and annotation, or differential expression (Andrews et al., 2021; Luecken & Theis, 2019). Most models used for a specific purpose, such as subpopulation classification or inference of network or cell trajectories, tend to have an analogous functional structure but differ mainly in their modeling assumptions and algorithms (Lee, Hyeon, et al., 2020; Raimundo, Meng-Papaxanthos, Vallot, & Vert, 2021).

2.2 Single-cell data integration challenges

The integration of different data types can be done in different ways with respect to the origin and type of the data, as well as the coverage of each modality. Based on recent reviews (Lee, Hyeon, et al., 2020; Miao, Humphreys, McMahon, & Kim, 2021; Stuart & Satija, 2019), the most common integration strategies could be grouped into four classes: (i) correlation or consensus analysis between unimodal data, (ii) integration of different modalities on previously analyzed and calibrated data, (iii) statistical modeling of different data types, (iv) integration and joint analysis of different data types. The first approach (i) is adequate when working with data derived from the same cells, assuming all extracted features to have statistically similar attributes. The second approach (ii) is based on the initial identification of low-dimensional descriptions of raw high-dimensional data of a higher coverage dataset (often scRNA-seq) followed by the integration of features extracted from other modalities. The third approach (iii) aims to build a statistical function that describes the best fit between sets of features between modalities, often requiring previous calibration using other datasets. Finally, the fourth approach (iv) often uses data of multiple modalities with similar coverage, assuming a common latent space between all measurements, irrespective of modality, and clustering cells together to identify the shared variation. All these taken together highlight the potential, but also the challenges to effectively accomplish a true multimodal analysis.

Current methods for the simultaneous analysis of different data types from the same cells have several limitations, mainly in terms of scalability, which must be addressed to bring single-cell analysis to its full potential. New methods are needed for the multimodal characterization of other types of data, including metabolomics and spatial omics, which have been emerging in single-cell research (Longo, Guo, Ji, & Khavari, 2021; Rappez et al., 2021). To mitigate the impact of these shortcomings, computational and statistical tools for integration of multimodal data should be highlighted as viable alternatives. Several tools have been developed for the analysis using specific data types, ranging from classification and regression models to prevision algorithms to identify cell populations or obtain functional insights, particularly using scRNA-seq (Ji, Lotfollahi, Wolf, & Theis, 2021).

The integration of single-cell data is just as essential as the data analysis itself (Fig. 2). Depending on the circumstances, the integration of different data types can be performed either before or after the analysis of one or more datasets. Even when using data that does not originate from the same cell population or from the same experimental protocol, integrating data can allow the identification of components that would not be possible using only one type of data (Rautenstrauch et al., 2022; Stuart & Satija, 2019). The strategy of choice should be carefully considered, and all possible covariates or batch effects must be addressed and corrected to ensure the validity of the analysis. Moreover, integration strategies also encompass their own set of shortcomings. For example, some approaches, particularly of the first type (i), which considers the correlation or consensus analysis between unimodal data, may disregard the differences in functional importance of distinct features by assuming they all have equal connectivity in a network, thus limiting the biological relevance of this kind of analysis (Miao et al., 2021). Moreover, the influence of experimental (e.g., the equipment, reagents, coverage, etc.) and biological (e.g., cell cycle, sex, circadian rhythm, etc.) variations on the data must not be discarded, independently of whether the datasets come from single or multi-omics assays (Luecken et al., 2022).

Due to the intrinsic nature of single-cell methods, the amount of working material per cell is limited and, as a result, observations are often affected by a certain level of noise (due to technical or biological variation) and uncertainty (Miao et al., 2021). This noise can be reduced by ensuring the reproducibility of experiments and applying correction approaches to the data (Xi & Li, 2021), but it is in general difficult to avoid. Multimodal data integration approaches have also been proposed to facilitate analysis by compensating for the loss of information of another modality by introducing additional data. However, this brings its own set of problems that may complicate the perception of noise vs variability between cells (Tini, Marchetti, Priami, & Scott-Boyer, 2018). First, not all approaches exhibit similar coverage, which can introduce additional biases. Second, noise can be amplified by bringing in technical variations of each additional modality, especially if the combined data come from different experiments (Lähnemann et al., 2020). Even though there are popular frameworks for

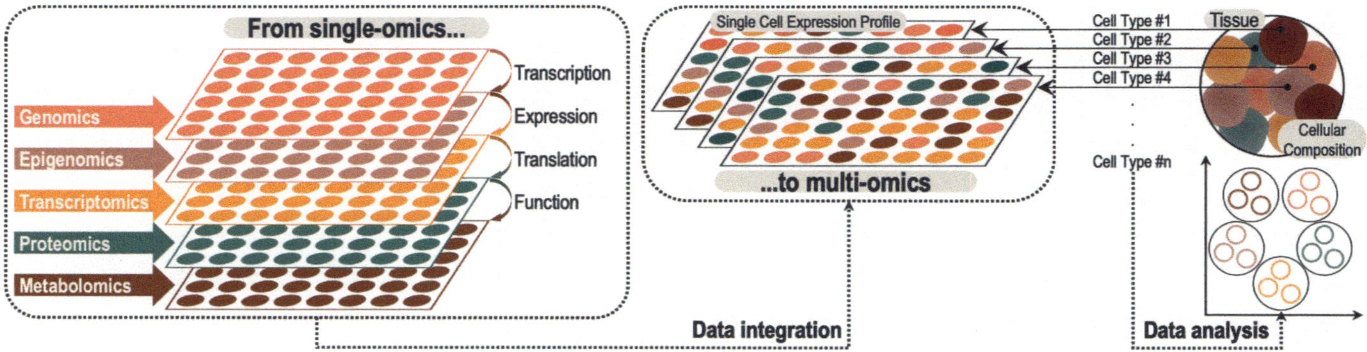

Fig. 2 Multidimensional techniques to extensively analyze underlying biology are required for transformative discoveries in science. Cell sequencing improves the characterization of cell types and states, allowing researchers to gather new knowledge from cells to tissues by combining many single-omics methodologies into a cohesive multi-omics picture.

batch correction of unimodal data (Andrews et al., 2021), pipelines for multimodal data are still needed. Third, matching variables between different modalities into a common format can sometimes be complicated and create more noise (Stuart et al., 2019). Thus, noise can be one of the major limitations of single-cell analysis, particularly when using multimodal data. Nevertheless, different integration approaches can sometimes be adequate depending on the modalities and experimental conditions, reducing the influence of noise on observations (Miao et al., 2021).

This problem becomes even more pronounced when amplification is used in single-cell sequencing approaches as well as in high-throughput experiments with higher resolution, increasing the uncertainty of the data (Lähnemann et al., 2020). That is because single-cell experiments will occasionally sacrifice sensitivity for throughput and lose parts of the data, resulting in data sparsity. Missing values, also known as "dropouts," and multiplets in single-cell sequencing datasets are frequent and must be carefully dealt with. For example, it can be at times problematic to determine if missing values in an expression matrix are the result of genuine lack of expression or no detection during sequencing, especially when using unpaired data (Lee, Hyeon, et al., 2020). Due to this sparsity, different strategies such as imputation or data reconstruction are required to prevent uncertainty during downstream data analysis (Lähnemann et al., 2020). As datasets become larger, dimensionality becomes a problem that makes the development of scalable methods for single-cell analysis more complicated. Furthermore, with the increase in the number of dimensions, not only the computational power required rises but it also becomes harder to infer relationships between network components and to visualize the data without errors (Miao et al., 2021). Some approaches in single-cell sequencing have addressed the scalability question, but the problems have not been entirely solved and other modalities such as single-cell proteomics still need better alternatives (Stuart & Satija, 2019).

At the most, studies integrate multimodal structural data at the same resolution (atomic, molecular or complex-level), or, more recently, integrate multimodal single-cell data from single-cell technologies (Batool, Ahmad, & Choi, 2019; Hashmi & Abbas, 2019). Therefore, current challenges lay on the construction of models able to integrate the highest amount of information as possible to incorporate different data scales and, thereby, create a thorough simulation able to describe a given event from a holistic perspective. Future models should aim to represent the whole instead of the sum of its parts.

For the development of computational methodologies, scientists assume that a given question is the sum of smaller and simpler questions. Thus, organism homeostasis could be interpreted as the sum of each part (Cohen & Harel, 2007). This thinking led computational methodologies to be focused on finding answers for a single part of the whole. That is, approaches so far provide a thorough understanding of very specific events, like atom trajectories in a molecule or how a ligand binds its target, which means that the majority of works do not integrate data originated from smaller and broader events (Chen et al., 2016).

3. Bridging structure and cell data as drivers to understand ligand-induced perturbations

Perturbation in biological contexts could be defined as the changes in basal cellular conditions that take place as a response to external stimuli. These alterations arise from the modulation or disruption of one or more components within a network, be it a gene, a protein, or other biomolecules (Barabási et al., 2011). Given the high complexity and interconnectivity of biological networks, changes in a single component trigger cascade effects many others, and understanding the nature of its interacting factors relationship is pivotal to accelerate the development of more accurate therapies (Barabási et al., 2011). Perturbation experiments often consist of the controlled regulation of specific cell components using perturbagens, followed by the analysis of downstream transcriptomics effects. Genetic perturbation screens using RNA interference or CRISPR approaches have contributed a lot to the understanding of biological networks in recent years (Hart et al., 2015; Rubin et al., 2019). Similarly, pharmacological screens have been paramount in understanding drug's mechanisms-of-action, identifying potential biological targets, and determining drug sensitivity, particularly in cancer cells (Aissa et al., 2021; McFarland et al., 2020). While perturbation experiments are remarkable for research, empirically screening all possible combinations of perturbations is not yet viable. Therefore, several computational models to predict perturbation effects or facilitate the transcriptomics profiling of cells have been recently developed (Burkhardt et al., 2021; Fustero-Torre et al., 2021; Lotfollahi, Susmelj, Donno, et al., 2021; Lotfollahi, Wolf, & Theis, 2019; Squair, Skinnider, Gautier, Foster, & Courtine, 2021).

Many studies have already analyzed the alterations promoted in the transcriptome of various cell types by individual or combined administration of drugs (Aissa et al., 2021; Ben-David et al., 2018; McFarland et al., 2020;

Srivatsan et al., 2020). Likewise, other experiments profiled the proteomics of ligand-induced perturbations, particularly regarding Post-Translational Modifications (PTMs) (Dele-Oni et al., 2021; Nyman et al., 2020), or taken the post drug administration metabolome of cells into account (Dahabiyeh et al., 2020; Dubuis, Ortmayr, & Zampieri, 2018; García-Cañaveras, Castell, Donato, & Lahoz, 2016). Large libraries of ligand-induced perturbation datasets have been built, such as Connectivity Map (Lamb et al., 2006), the Library of Integrated Network-based Cellular Signatures (LINCS) (Subramanian et al., 2017), the Cancer Cell Line Encyclopaedia (CCLE) (Barretina et al., 2012), and the Cancer Therapeutics Response Portal (CTRP) (Basu et al., 2013).

Multi-omics analyses combining different types of ligand perturbation data and different types of omics such as, for e.g., transcriptomics and proteomics (Lee, Song, et al., 2020), or proteomics and genomics (Schmitt et al., 2021) have recently become more popular. However, few studies take additional variables into account, such as combined drug administration, different doses and timepoints, particularly at the single-cell level or with multi-omics analysis, given the cost and substantial number of possible combinations and experimental conditions. It is essential to address these shortcomings to extend the applications of this data in *de novo* drug development and repurposing.

Perturbational modeling can have different focuses, including drug repurposing or design, prediction of drug adverse reactions, synergy effects, drug resistance, drug inactivation, drug response, and drug sensitivity (Wu et al., 2020). Drug resistance, inactivation and response can be defined based on common metrics for well-characterized ligands. Predicting ligand targets and the pathways activated by a ligand is also a common way to study ligand-induced perturbations at a network-level, which often involves the use of classification algorithms (Ji et al., 2021; Nassar et al., 2021). This knowledge can be further applied for prediction of adverse reactions or drug repurposing. Regarding the prediction of synergistic effects between pairs of genes or compounds, classification models or regression models can be employed for categorization of an effect as synergistic or antagonist, and prediction of a synergistic score, respectively (Ji et al., 2021; Wu et al., 2020). Understanding ligand synergy is critical in the development of combination therapies, as these can be key in avoiding drug resistance through additive or multiplicative effects (Preto, Matos-Filipe, Mourão, & Moreira, 2021). Additionally, determining the chemical effects during perturbations in cell-types is the first step toward a biochemical perspective of these events.

For example, *de novo* drug design using transcriptomics profiles and pharmacophore modeling has been used for this purpose, avoiding laborious *in vitro* experiments (Ji et al., 2021; Prieto-Martínez et al., 2019). Moreover, the prediction of genetic sequences using omics profiles to study binding motifs and k-mers further enables the computational prediction of genetic perturbations (Ji et al., 2021).

To match the broad nature of necessary biological questions, a variety of methods have already been applied to ligand-induced perturbations prediction. Among computational approaches, Artificial Intelligence (AI) is widely used for prediction tasks as these techniques are able to readily perform accurate predictions based on a high number of adequately characterized samples (Ji et al., 2021; Vamathevan et al., 2019). As such, the increasing quantity of single-cell data available started to be combined with Machine Learning (ML) algorithms, a branch of AI, for cellular responses for perturbation prediction. Despite most predictors being AI-based, simpler mathematical concepts like factor decomposition (for comprehension of the components in an expression), matrix factorization (to deal with bulk matrices), and distribution modeling (to infer population trends) are widely applied (Ji et al., 2021). Regarding the use of AI techniques, classification and regression algorithms are both applied to interaction and perturbation prediction. Metrics such as K_i (dissociation constant—binding affinity between the inhibitor and the enzyme) or IC_{50} (concentration of inhibitor required to reduce the enzymatic activity to half) are commonly used in these prediction tasks (Batool et al., 2019; Ji et al., 2021). The power of Deep Learning (DL) algorithms, another AI subfield, to deal with big and unstructured data led to its application in studies involving more complex relationships, making them highly useful for integration of multimodal data (Pal, 2020; Wu et al., 2020). For now, perturbational models focus on smaller tasks like quantification and modulation of cellular responses, and prediction of drug targets, interactions, and chemical properties (Ji et al., 2021; Vamathevan et al., 2019).

3.1 Three-dimensional structure role in ligand-induced perturbations

Usually, drug's action is based on inhibition or activation of a target, commonly G Protein-Coupled Receptors (GPCRs), ion channels and kinases families (Santos et al., 2017). Upon target's structure determination, Structure-Based Drug Design (SBDD) methods can be applied for ligand identification, comprehension of the binding mechanisms, recognition of the binding site, lead design and optimization through the identification and manipulation of the physicochemical properties responsible for the binding (Blundell et al., 2006; Yu & MacKerell, 2017).

Since proteins are the most common type of drug targets, it is well known that Protein-Ligand Interactions (PLIs) are associated with molecular recognition events, essential in regulatory networks (Motlagh, Wrabl, Li, & Hilser, 2014). The new interaction occurring at the binding site of a protein upon the binding of either a drug or a biological small molecule induces a chemical shift and structural conformational change in its 3D structure. In turn, these may lead to alterations in the stability of the target that, subsequently, affect its downstream signaling (Kumar & Jernigan, 2021; Williamson, 2013). The need to understand how interactions implicate protein's function is thus evident. In fact, the determination of this relationship was the reason behind the development of methods for tracing structural fluctuations associated with binding events (Kumar & Jernigan, 2021). Binding affinities, which quantitatively estimate the strength of a binding (Nguyen et al., 2021; Shim, Hong, Sohn, & Hwang, 2021), are commonly determined after Virtual Screening (VS) of a family of compounds to a target of interest. They are also vital for the prediction of drug-target interactions (Shim et al., 2021). However, induced perturbations can also happen at remote target sites, which modulates drug's affinity, and consequently influence its efficacy. So, the dynamic and kinetic behavior of the system is also fundamental in drug's efficiency prediction (Ferraro et al., 2021). For this understanding, standard Molecular Dynamic (MD) simulations are highly applied to depict the binding mechanisms involved in PLIs (Jagger, Kochanek, Haldar, Amaro, & Mulholland, 2020) and to predict favorable conformations of protein-ligand systems (Gentile et al., 2022; Zhang, 2011).

AI methods have also been showing their potential to accelerate the study of drug-target interactions since they can analyze large amounts of data and tackle the existent "big data" sources (Vamathevan et al., 2019). Works at the molecular level that use AI algorithms, revealed themselves as a key approach to predict drug response, mainly in cancer patient data, identify new targets and quantify drug resistance (Adam et al., 2020; Bueschbell et al., 2022; Güvenç Paltun, Mamitsuka, & Kaski, 2019; Xia et al., 2022; Yu, Zhou, Gao, & Zha, 2021).

As molecular perturbations induce a variety of effects that propagate up until the organism-level, drug-related studies are starting to address how ligand-induced perturbations diffuse into a network. These can involve protein-DNA, protein-protein interactions, PLIs and so on, leading novel approaches to seek the comprehension of the effects caused at the cell's metabolism after ligand binding (Ji et al., 2021; Rosário-Ferreira, Bonvin, & Moreira, 2022; Zhao, Cao, & Zhang, 2020). Consideration of structural and scRNA-seq data have been shown relevant to address this problem (Rampášek, Hidru, Smirnov, Haibe-Kains, & Goldenberg, 2019). Understanding these outcomes allow the prediction of possible diseases and fragilities to which the organism can be subjected as well as drug's adverse reactions, associated with a given treatment (Zhao et al., 2020). As such, a perspective regarding conformational alterations mechanisms is needed to scale for a broader knowledge (Zhao et al., 2020).

3.2 Single-cell role in ligand-induced perturbations

As highlighted so far, scRNA-seq is the most popular approach in the context of single-cell omics and, accordingly, is the most common when analyzing the effects of perturbations. Many models for different types of analyses have been created for the analysis of scRNA-seq data in recent years, many of which can be applied in the context of ligand-induced perturbations (Ji et al., 2021). Most statistical and computational models aim to model perturbation data or to predict the perturbation profile of cells following treatment with compounds. These can make use of clustering, annotation, and/or inference approaches to characterize cellular identities following perturbations or to predict how cells will respond to perturbation by other agents. Given the previously mentioned limitations of single-cell data, most methods contain challenges that should be addressed in future tools. However, as the data availability increases and these drawbacks are addressed, the quality of available models should also become more fit to deal with this data.

Different methods are used for the visualization, clustering, and profiling of cell populations subjected to perturbations. Dimensionality reduction is often used to counter the high dimensionality-related issues by embedding the expression data into a low-dimensional space, containing the relevant features to interpret or predict the response to perturbation, and thus being an important step for the downstream analysis and visualization of scRNA-seq data. This is usually preceded by a feature selection step, in which the dataset is filtered so that only genes with higher variance-to-mean ratio are kept (Luecken & Theis, 2019). There are different techniques for dimensionality reduction, but the most common are the linear method

Principal Component Analysis (PCA), or the nonlinear methods t-Distributed Stochastic Neighbor Embedding (t-SNE) and Uniform Manifold Approximation and Projection (UMAP). Using orthogonal linear combinations of the variables, PCA preserves the variance and, thus, the Euclidean distances between cells if no principal components are discarded. Given the complexity of scRNA-seq datasets and its architecture, PCA is not able to fully capture the data structure. Nevertheless, as it can still effectively preserve global distances, PCA is often employed as a part of the pre-processing step before proceeding with downstream analysis. As nonlinear approaches, both t-SNE (van der Maaten & Hinton, 2008) and UMAP (McInnes, Healy, Saul, & Großberger, 2018) sacrifice the global structure for local similarity, allowing the visualization of the heterogeneity but do not ensure the preservation of distances. Therefore, despite the limitations of PCA, these algorithms can be used for visualization but should not be used directly for downstream analysis (Andrews et al., 2021). UMAP was previously believed to preserve large-scale structures better than t-SNE, but this notion was recently countered and UMAP was shown to exhibit similar performance to t-SNE as long as both are correctly trained (Kobak & Linderman, 2021). Accordingly, both methods require initialization (e.g., using PCA) and tuning of hyperparameters for optimal performance and preservation of data structures (Luecken & Theis, 2019).

Unsupervised clustering algorithms are often part of a scRNA-seq analysis workflow and are used to group cells with similarities in their expression in response to perturbation. The similarity between cells is detected based on distance matrices using low-dimensional data as input (Luecken & Theis, 2019). These methods use algorithms such as k-means (Hartigan & Wong, 2012), the Louvain algorithm (Blondel, Guillaume, Lambiotte, & Lefebvre, 2008), K-Nearest Neighbor (KNN) graphs, and the methods based on these, such as SC3 (k-means-based) (Kiselev et al., 2017) and PhenoGraph (Louvain-based) (Levine et al., 2015). All these algorithms have their own parameters that should be tuned to optimize the analysis, thus making it sensitive to user-defined settings. After clustering, it is important to annotate the clusters for characterization of gene signatures to obtain biologically relevant insights. The annotation of the data relies on external sources such as the popular Human Cell Atlas (Regev et al., 2017) or the literature, and the identification is done by comparing the differential expression between clusters (Luecken & Theis, 2019). However, if no specific reference datasets are available, the annotation based on marker genes should be carefully performed as these may differ between datasets due to biological variations (e.g., cell cycle, sex, or tissue). Once annotated, the clusters can be further grouped into cell types or subtypes to analyze how the cellular composition is affected following treatment, if sufficient cell numbers are available. Similarly, the clustered data can be used in a network perspective to infer gene-regulatory networks based on collective expression patterns between cells or subpopulations (Cha & Lee, 2020).

Despite the clear advantages of clustering analysis, it is not always adequate to employ them. This is the case when dealing with datasets of cells analyzed under time-dependent treatments such as different timepoints of drug treatment or developmental processes. That is, there are cases in which cells should be analyzed from a continuous perspective instead of the frequently discrete profile of clustering analysis. In general, trajectory inference models gather data from neighboring cells using an inferred "pseudotime" variable to assign the position of cells and reconstruct the continuous process (Van den Berge et al., 2020). Despite its name, this variable can also describe trajectories that are not temporally defined, such as spatial positions (Andrews et al., 2021). Trajectory inference analysis requires the addition of covariates such as timepoints or drug concentration into the analysis. Cell density information is a particularly important aspect to track when performing trajectory analysis and can be a challenge with certain datasets despite some methods trying to tackle this issue (Chen et al., 2019). Trajectory inference analysis often takes place following dimensionality reduction or unsupervised clustering, and the choice of method should consider the characteristics of the dataset that is being analyzed.

3.3 How can we integrate multimodal biological data? Can structure lead to better network analysis while potentiating the single cell contribution?

The fundamentals and scientific relevance of integrative single cell analyses as well as its limitations have already been highlighted. Moreover, how the study of ligand-induced perturbations can shape the field of drug development has also been referenced. But how does all this fit together? It is evident that the field of single-cell omics is continuously evolving, and that new methodologies and models are emerging to address current challenges (Marx, 2022). Therefore, raising questions about opportunities that are currently being set aside can promote the development of novel approaches to widen applications of current and future technologies.

The integration of multiple data types for the analysis of ligand-induced perturbations would undeniably be beneficial for the development and repurposing of drugs, generating important information of biological networks at different levels. As demonstrated by current challenges, this is no easy nor cheap task; nevertheless, it is still possible to highlight which data types and analyses could be more valuable when evaluating the perturbational effects of different compounds. First and foremost, as highlighted throughout this chapter, the combination of transcriptomics and proteomics data is a powerful tool to answer major questions about the mechanism and function of selected ligands:

- How is gene expression altered following treatment with these compounds?
- Which cell types are being affected the most?
- How are the levels of potential target proteins affected by this drug?
- Are these protein targets being post-transcriptionally modified?
- Which are the potential side effects that could result from these alterations?

These and many other questions can be addressed if the right tools are used for processing and analysis. Moreover, the inclusion of metabolomics data, despite being one of the current challenges in integrative single-cell analysis, could provide an even more powerful perspective on the physiological changes that are occurring. Instead of being limited to observing the shifts in gene and protein profiles, one could truly reflect on the downstream effects of such alterations, which could be particularly relevant for, but not limited to, the discovery of compounds targeting metabolic components. Integration of metabolomics data can be done through the quantification of compounds which can be used to infer the activation/inactivation within target regulatory networks, along with off-target information (Alarcon-Barrera, Kostidis, Ondo-Mendez, & Giera, 2022). Likewise, the integration of epigenetics data, can also be paramount to incorporate in the predictive models (Hamamoto, Komatsu, Takasawa, Asada, & Kaneko, 2019).

Similar insights can be drawn from the integration of other data types. But how can structural and biophysical data be integrated into all this? Although in its infancy, there are a few studies that already integrate the effects of molecular alterations at the network level to structural data, a practice that should be expanded for a better understanding of all the different aspects of binding and drug perturbation (Ozturk & Carter, 2021; Zhou et al., 2021).

It is known that a variety of factors at the organism-level (e.g., diet, weight, and diseases) and at the molecular-level can affect ligand-induced perturbations (McCallum, Lip, & Padmanabhan, 2014). Thus, in the drug discovery field, especially in personalized medicine, single-cell omics can be essential to target patient individualities, as well as to propose new therapeutic agents (Jehan, 2019). Particularly, the integration of multimodal data in current and future models may provide a more complete description of a number of events, leading to the generation of more accurate and precise algorithms which otherwise would not be possible to achieve (Miao et al., 2021).

3.4 Datasets and prevision models for drug-induced perturbations

In the last few years, many scRNA-seq datasets of ligand-induced perturbations in various cell types have emerged, along with computational models to analyze such data. While most of the studies that gave rise to these datasets regretfully do not assess the effects of different concentrations or timepoints of several ligands, they still represent a good source of information. These datasets have been fundamental to train or benchmark perturbation prediction tools, further highlighting both the usefulness and the need for this kind of data. In fact, due to the small number of available recent datasets, other models have made use of bulk RNA-seq datasets in combination with scRNA-seq data to complement their training with approaches such as Deep Transfer Learning (DTL) (Wu et al., 2020). Therefore, it becomes relevant to discuss the most recent datasets and models that were made available and discuss their applications and limitations to suggest where single-cell perturbation research is headed. Many general models and perturbation datasets have been compiled recently (Ji et al., 2021), but here we will highlight only those that are related to drug-induced perturbations.

While most currently available scRNA-seq perturbational datasets are genetic, Clustered Regularly Interspaced Short Palindromic Repeats (CRISPR) screens, there are also studies analyzing the transcriptomics effects of small molecules in different cell types. Some have characterized the effects of more than a few sets of drugs and cell lines, highlighting the scarcity of data, which partially explains the limitations of current models. Starting with experiments that tested the largest number of compounds, the dataset generated by Srivatsan et al. (2020) and the one by Ben-David et al. (2018) stands out. The former is possibly the most complete currently available dataset, as different doses of 188 compounds at different timepoints in three different cancer cell lines were assessed. The latter also tested several drugs, using 321 anti-cancer compounds in multiple strains of the

MCF7 breast cancer cell line, but did not evaluate different timepoints nor dosages. Other notable datasets include the ones by Aissa et al. (2021), McFarland et al. (2020), Zhao et al. (2021), Wang, Alnabati, et al. (2021), and Marquina-Sanchez et al. (2020). Despite not reporting a high number of drug combinations, these datasets were selected because different cell types were exposed to more than one compound. Other available datasets were also included in Table 1.

Regardless of data availability, several models have emerged to model drug perturbation data and/or predict effects following treatment. Most models are based on ML or DL and were built using either the Python or R programming languages. Few have been structured specifically with ligand-induced perturbations in mind and, in fact, many have not even been benchmarked or validated using drug perturbation data. Moreover, even fewer accept multimodal data as input, underscoring the need for new tools to analyze multi-omics. Beyondcell accepts scRNA-seq data as input and aims to identify cancer cell subpopulations with differential drug responses and indicate possible treatments (Fustero-Torre et al., 2021). While the drug signature data used in the model is obtained from a bulk RNA-seq dataset, it was validated with data from different scRNA-seq experiments. scDEAL has a similar objective as Beyondcell, aiming to predict cancer drug responses, but accomplishes that in a different way (Chen et al., 2021). That is because it uses a DTL approach to translate drug signatures from bulk RNA-seq datasets to a single-cell level. The model was validated with several datasets, but the accuracy of its prediction varied, and the drug signature data was retrieved from a single bulk RNA-seq dataset. The Compositional Perturbation Autoencoder (CPA) aims to analyze and predict scRNA-seq responses to perturbations (Lotfollahi, Susmelj, Donno, et al., 2021). It was not built specifically for drug-induced perturbations and can only accept expression data as input but was validated using the dataset by Srivatsan et al. (2020). Next, scTenifoldNet compares two scRNA-seq samples and tries to identify which gene networks are differentially affected (Osorio, Zhong, Li, Huang, & Cai, 2020). Despite being validated mainly using genetic perturbation datasets, it was also validated using a multimodal scRNA-seq and ATAC-seq dataset. However, this model requires the combination with other tools for clustering analysis and profiling, as it assumes all cells in the input datasets are of the same type. There are other models that, despite not being directly tested with drug perturbation data, could also be used, or modified for that purpose. These were also included in Table 2.

Table 1 Publicly available ligand-induced perturbation scRNA-seq datasets

Reference	Organism	Tissue	Cell origin	# Cell lines	# Cells	# Drugs	# Transcripts	# Genes	Method	Platforms	Data
Zhao et al. (2021)	Human	Brain, Tumor	Glioblastoma patients	1	16,863	2 (etoposide and panobinostat)	–	–	SCRB-seq (microwell)	Illumina NextSeq 500, Illumina NovaSeq 6000	GSE148842
Tirier et al. (2021)	Human	Bone marrow, Tumor	Myeloma patients	20	212,404	Not specified	–	–	Chromium	Illumina HiSeq 4000	GSE161801
Aissa et al. (2021)	Human, Mouse	Culture, Tumor	In vitro, Mice	2	10,000	4	28,669	8787	Chromium & Drop-seq	Illumina NextSeq500	GSE149383
Wang, Frederick, et al. (2021)	Human	Culture	In vitro	1	372	2 (paclitaxel and celecoxib)	–	9600	Fluidigm C1 + Smart-seq2	Illumina NextSeq 500	GSE162256
McFarland et al. (2020)	Human	Culture	In vitro	24-99	7317 (13 drugs) 13,713 (trametinib)	1 (trametinib) or 13	–	–	MIXseq	Illumina HiSeq X, Illumina NovaSeq 6000	https://figshare.com/s/139f64b495dea9d88c70
Marquina-Sanchez et al. (2020)	Human, Mouse	Pancreas	Ex vivo pancreatic islets (3 humans/3 mice)	2	107,192 (82,463 human)	1 drug (artemether)	142,165	–	Chromium	Illumina HiSeq 4000	GSE147203
Schnepp et al. (2020)	Human	Culture	In vitro	2	324	1 (Docetaxel)	–	12,862	Fluidigm C1 + SMART-seq v4 (C1)	Illumina HiSeq 2500, Illumina HiSeq 4000	GSE140440

Continued

Table 1 Publicly available ligand-induced perturbation scRNA-seq datasets—cont'd

Reference	Organism	Tissue	Cell origin	# Cell lines	# Cells	# Drugs	# Transcripts	# Genes	Method	Platforms	Data
Stewart et al. (2020)	Human, Mouse	Lung, Xenograft, Tumor	Patient-derived mouse xenografts	8	Varied with sample: at least 2000 per sample	1 (cisplatin)	–	50–2153	Chromium	Illumina HiSeq 2000	GSE138267
Kagohara et al. (2020)	Human	Culture, HNSCC	In vitro	3	5000	1 (cetuximab)	–	–	Chromium	Illumina HiSeq 2500	GSE137524
Srivatsan et al. (2020)	Human	Culture	In vitro	3	650,000	188	–	6,238	sciPlex	Illumina NextSeq 550, Illumina NovaSeq 6000	GSE139944
Bell et al. (2019)	Mouse	Culture	In vitro	1	–	1 (BET inhibitor)	–	–	Cel-seq2 (adapted)	Illumina NextSeq 500	GSE122843
Sharma et al. (2018)	Human	Oral, Tumor	Oral cancer patients	20	1302	1 (cisplatin)	–	–	SMARTer (C1)	Illumina HiSeq 2500	GSE117872
Ben-David et al. (2018)	Human	Culture	In vitro	1	26,465	321	–	10–1574	Chromium	Illumina NextSeq 500	GSE114462
Ho et al. (2018)	Human	Culture	In vitro	2	~400	1 (BRAF inhibitors)	–	5000+	Fluidigm C1 & Chromium	Illumina NextSeq 500	GSE108383, GSE108394
Jerby-Arnon et al. (2018)	Human	Culture	In vitro	3	7186	1 (abemaciclib)	–	–	Chromium & Smart-seq2	Illumina NextSeq 500	https://singlecell.broadinstitute.org/single_cell/study/SCP109/melanoma-immunotherapy-resistance

Note: BET: Bromodomain and Extra-Terminal motif; HNSCC: Head and Neck Squamous Cell Carcinoma; SCRB-seq: Single-Cell RNA Barcoding and sequencing.

Table 2 Available models for drug perturbation response prediction

Perturbation type	Model	Datasets	Year	Aim	Method	Computational techniques used	Performance evaluation	Model advantages	Model limitations	Programming language	Reference	Comments
Drugs	BeyondCell	Ben-David et al. (2018), Ho et al. (2018), Jerby-Arnon et al. (2018), Kinker, Greenwald, Tal, Orlova, and Cuoco (2020), Stewart et al. (2020)	2021	Propose specialized cancer treatments through identification of cell subpopulations with distinct drug response	Construction of therapeutic clusters using ranks of differential sensitivity	KNN and QR Decomposition for correlation remotion after normalization; UMAP for dimensionality reduction	BeyondCell was tested using (Ben-David et al., 2018) and further validated in 5 different datasets	• Addresses the tumor heterogeneity challenge • Ranks drug sensitivity • Useful for development of specialized treatments	• Only works with TFs or drugs with known TF targets • Drug signature data is obtained from bulk RNA-seq platforms, and not single-cell experiments • Requires a scRNA-seq matrix and a drug signature collection or a GMT ranked matrix	R	Fustero-Torre et al. (2021)	Effectively selected therapeutic clusters in different datasets and experiments and proposed drugs
Drugs/ Genetic	CPA	Gehring, Hwee Park, Chen, Thomsom, and Pachter (2019), Hagai et al. (2018), Norman, Horlbeck, Replogle, Ge, and Xu (2019), Srivatsan et al. (2020)	2021	Prediction of scRNA-seq responses considering different types of perturbations (e.g., dosage, time, drug, and gene knock-out)	Description of gene expression using autoencoders	Deep Learning Neural Networks	Accuracy and R^2 were used to evaluate classification	• Prediction of gene expression given unknown combinations of perturbations • Able to deal with multiple covariates like different species and cells-type and various doses	• Model's performance was lower for unseen drug combinations • It is only able to deal predict alterations at the gene expression level	Python	Lotfollahi, Susmelj, De Donno, et al. (2021)	The key feature of the CPA vs a normal autoencoder is its latent space disentanglement and the induced interpretability of the perturbations in the context of cell states and covariates

Continued

Table 2 Available models for drug perturbation response prediction—cont'd

Perturbation type	Model	Datasets	Year	Aim	Method	Computational techniques used	Performance evaluation	Model advantages	Model limitations	Programming language	Reference	Comments
Drugs	scDEAL	Aissa et al. (2021), Bell et al. (2019), Kong et al. (2019), Schnepp et al. (2020), Sharma et al. (2018). Bulk RNA data obtained from GDSC	2021	Prediction of cancer drug response at single-cell level by integrating large-scale bulk cell line data	Applies DTL to bulk and single-cell data to predict single-cell drug response	Deep Transfer Learning	F-Score, AUROC and AP are the metrics used for performance evaluation. scDEAL achieved high metric scores for most datasets. Random repetition and stratified sampling were used to test reproducibility	• Predict responses at the single-cell level using bulk data • Reveals potential drug resistance mechanisms using a gradient feature interpretation	• The accuracy of the prediction results in scDEAL varied depending on the collection of bulk gene expression of cell lines • Only retrieves bulk data from a single database	Python	Chen et al. (2021)	The true innovation is to translate drug responses at bulk-level and transfer to predict drug responses in scRNA-seq Another innovation is the integrated gradient feature interpretation, which provides a set of important genes contributing to the prediction of drug responses

General (tested with drug-induced perturbations)	scTenifoldNet	SERGIO-generated synthetic dataset (Avey et al., 2018; Hagai et al., 2018; Kagohara et al., 2020; Little et al., 2019; Zhou et al., 2020)	2020	Identification of genomic elements associated with multiple phenotypes in CRISPR functional screening that uses scRNA-seq as readout	Unsupervised ML algorithm for ranking genes by its differential regulation level	PC Regression, low-rank tensor approximation and manifold alignment for construction of scGRNs	Benchmarked against other algorithms using BEELINE and a reference dataset (GSD). AUROC, AUPRC and time of computing were used to compare performance. scTenifoldNet outperformed other algorithms and exhibited faster computing time	• It can be applied to understand functional analysis, detect the enriched molecular functions and subsequently, the biological processes affected by a perturbation • It can detect differences in cell's population states between 2 samples • Provides a novel comparison tool	• Only accepts a pair of expression matrices as input • Does not perform any clustering analysis for cells; and does not assign cells to cell types—requires other tools • Assumes all cells in both input matrices are of the same type	Python	Osorio et al. (2020)	Able to detect differences in the cell population's state between two samples in a sensitive and scalable manner using both simulated and real data
General (tested with drug-induced perturbations)	trVAE	Kang et al. (2018)	2020	Prediction of perturbation effects on differential gene expression	Uses a Maximum Mean Discrepancy regularized conditional VAE for prediction of gene expression after IFN-β stimulation	ML for prediction of gene expression, UMAP for visualization	Benchmarked vs scGen, sCVI, CycleGAN, and other methods. Performance evaluation was measured by R^2	• trVAE was shown to provide better estimates than other models • It can analyze and predict multiple perturbations	• The uncertainty associated with the probabilistic model needs further quantification	Python	Lotfollahi, Naghipourfar, Theis, and Alexander Wolf (2020)	The resulting model (trVAE) outperforms all other models on benchmark and real datasets. trVAE may be applied to experiments with many biological conditions

Continued

Table 2 Available models for drug perturbation response prediction—cont'd

Perturbation type	Model	Datasets	Year	Aim	Method	Computational techniques used	Performance evaluation	Model advantages	Model limitations	Programming language	Reference	Comments
Drugs	SCATTome	Self-generated experimental data on multiple myeloma	2016	Prediction of drug response of individual cells based on targeted transcriptome using an assortment of ML methods	An ensemble ML algorithm is used for prediction of single-cell drug response	RF, SVM with Gaussian Kernel and SVM with Sigmoid Kernel techniques are used	Model was evaluated using ROC curves and AUROC scores	• High AUROC scores were achieved • It can be applied for identification of subclonal resistance in patient tumor samples and thus, for personalized medicine	• Only trained using qRT-PCR (targeted) seq data • No information regarding scalability	R	Mitra et al. (2016)	SCATTome computes, classifies, predicts, and quantifies drug-resistant subpopulations within MM tumors using the single-cell targeted gene expression data

Note: AP: average precision; AUPRC: area under the precision-recall curve; AUROC: Area Under the ROC curve; KNN: K-Nearest Neighbors; MM: Multiple Myeloma; PBMC: Peripheral Blood Mononuclear Cells; PC: Principal Component; R^2: coefficient of determination; RF: Random Forest; scGRNs: Single-Cell Gene Regulatory Networks; SVM: Support Vector Machine; TF: Transcription Factor; VAE: Variational AutoEncoder.

4. Conclusion

While the aforementioned challenges are being tackled by advances in single-cell methodologies along with statistical and computational approaches, they still limit the range of application and analysis of current data. Therefore, even though new challenges will arise as these technologies evolve, they represent major opportunities to maximize the impact of single-cell data in biomedical research. Moreover, data availability is exponentially increasing over the years and the type of data existent is vast, covering several biological events that can be described from a molecular to a cellular perspective. Following the breadcrumb trail that tracks the path from single-omics to interactomics promptly turns into an ever-extending process that unfolds in both possibilities and problems associated. Although it is possible to pinpoint technologies that exhibit promising features and results, such as scRNA based data and novel AI-based approaches for protein-ligand interaction prediction and characterization, the integration of these poses a deeper conundrum. Without surpassing this barrier, the field of possibilities will become increasingly limited, perhaps to an alt.

The greatest contemporary challenges faced at the edge of biomedical research such as patient oriented drug design would benefit—perhaps even require—effective integration of omics data into purposeful solutions. It is vital to follow a holistic perspective to tackle the whole tissue, organ, or body problems without, in the process, raising more (or simply different) shortcomings. As such, the increasingly vast landscape of data available is an unavoidable source of information that must be accessed, and its issues addressed.

The late XX and early XXI centuries have led to some of the greatest technological breakthroughs ever achieved in a human being's lifespan, of which are particularly relevant, under the scope of this review, the advances in computational hardware and biomedical research. The last decade has seen computational power double roughly every year. Yet, when it comes to omics data integration, no computational power or memory seems to be enough. This tendency clearly highlights the importance of pursuing smart approaches with comprehensible outcomes, rather than simply using computational brute force on problems whose landscapes are significantly more complex than that to which humans are equipped to, unaided, understand.

Funding

This work was supported by the European Regional Development Fund through the COMPETE 2020—Operational Programme for Competitiveness and Internationalisation and Portuguese national funds via Fundação para a Ciência e a Tecnologia (FCT) POCI-01-0145-FEDER-031356, LA/P/0058/2020, PTDC/QUI-OUT/32243/2017, UIDB/04046/2020, DSAIPA/DS/0118/2020 and UIDP/04046/2020. The authors would also like to acknowledge STRATAGEM—New diagnostic and therapeutic tools against multidrug-resistant tumors, CA17104. A.J.P., and N.R.-F. were supported by FCT through PhD-scholarships SFRH/BD/144966/2019 and PD/BD/135179/2017.

References

Adam, G., Rampášek, L., Safikhani, Z., Smirnov, P., Haibe-Kains, B., & Goldenberg, A. (2020). Machine learning approaches to drug response prediction: Challenges and recent progress. *NPJ Precision Oncology*, *4*, 19. https://doi.org/10.1038/s41698-020-0122-1.

Aissa, A. F., Islam, A. B. M. M. K., Ariss, M. M., Go, C. C., Rader, A. E., Conrardy, R. D., et al. (2021). Single-cell transcriptional changes associated with drug tolerance and response to combination therapies in cancer. *Nature Communications*, *12*(1), 1–25. https://doi.org/10.1038/s41467-021-21884-z.

Akhoon, B. A., Tiwari, H., & Nargotra, A. (2019). In silico drug design methods for drug repurposing. In K. Roy (Ed.), *Silico drug design* (pp. 47–84). Academic Press. https://doi.org/10.1016/B978-0-12-816125-8.00003-1 (chapter 3).

Alarcon-Barrera, J. C., Kostidis, S., Ondo-Mendez, A., & Giera, M. (2022). Recent advances in metabolomics analysis for early drug development. *Drug Discovery Today*, *27*(6), 1763–1773. https://doi.org/10.1016/j.drudis.2022.02.018.

Andrews, T. S., Kiselev, V. Y., McCarthy, D., & Hemberg, M. (2021). Tutorial: Guidelines for the computational analysis of single-cell RNA sequencing data. *Nature Protocols*, *16*(1), 1–9. https://doi.org/10.1038/s41596-020-00409-w.

Argelaguet, R., Cuomo, A. S. E., Stegle, O., & Marioni, J. C. (2021). Computational principles and challenges in single-cell data integration. *Nature Biotechnology*, *39*(10), 1202–1215. https://doi.org/10.1038/s41587-021-00895-7.

Avey, D., Sankararaman, S., Yim, A. K. Y., Barve, R., Milbrandt, J., & Mitra, R. D. (2018). Single-cell RNA-Seq uncovers a robust transcriptional response to morphine by glia. *Cell Reports*, *24*(13), 3619–3629. https://doi.org/10.1016/j.celrep.2018.08.080.

Barabási, A.-L., Gulbahce, N., & Loscalzo, J. (2011). Network medicine: A network-based approach to human disease. *Nature Reviews. Genetics*, *12*(1), 56–68. https://doi.org/10.1038/nrg2918.

Barretina, J., Caponigro, G., Stransky, N., Venkatesan, K., Margolin, A. A., Kim, S., et al. (2012). The Cancer Cell Line Encyclopedia enables predictive modelling of anticancer drug sensitivity. *Nature*, *483*(7391), 603–607. https://doi.org/10.1038/nature11003.

Basu, A., Bodycombe, N. E., Cheah, J. H., Price, E. V., Liu, K., Schaefer, G. I., et al. (2013). An interactive resource to identify cancer genetic and lineage dependencies targeted by small molecules. *Cell*, *154*(5), 1151–1161. https://doi.org/10.1016/j.cell.2013.08.003.

Batool, M., Ahmad, B., & Choi, S. (2019). A structure-based drug discovery paradigm. *International Journal of Molecular Sciences*, *20*(11), 2783. https://doi.org/10.3390/ijms20112783.

Bell, C. C., Fennell, K. A., Chan, Y. C., Rambow, F., Yeung, M. M., Vassiliadis, D., et al. (2019). Targeting enhancer switching overcomes non-genetic drug resistance in acute myeloid leukaemia. *Nature Communications*, *10*(1), 1–15. https://doi.org/10.1038/s41467-019-10652-9.

Ben-David, U., Siranosian, B., Ha, G., Tang, H., Oren, Y., Hinohara, K., et al. (2018). Genetic and transcriptional evolution alters cancer cell line drug response. *Nature, 560*(7718), 325–330. https://doi.org/10.1038/s41586-018-0409-3.

Blondel, V. D., Guillaume, J. L., Lambiotte, R., & Lefebvre, E. (2008). Fast unfolding of communities in large networks. *Journal of Statistical Mechanics: Theory and Experiment, 2008*(10). https://doi.org/10.1088/1742-5468/2008/10/P10008.

Blundell, T. L., Sibanda, B. L., Montalvão, R. W., Brewerton, S., Chelliah, V., Worth, C. L., et al. (2006). Structural biology and bioinformatics in drug design: Opportunities and challenges for target identification and lead discovery. *Philosophical Transactions of the Royal Society. Series B, Biological Sciences, 361*(1467), 413–423. https://doi.org/10.1098/rstb.2005.1800.

Boodhun, N. (2018). Seeing is believing: Structures and functions of biological molecules. *BioTechniques, 64*(4), 143–146. https://doi.org/10.2144/btn-2017-0123.

Buenrostro, J. D., Wu, B., Litzenburger, U. M., Ruff, D., Gonzales, M. L., Snyder, M. P., et al. (2015). Single-cell chromatin accessibility reveals principles of regulatory variation. *Nature, 523*(7561), 486–490. https://doi.org/10.1038/nature14590.

Bueschbell, B., Caniceiro, A. B., Suzano, P. M. S., Machuqueiro, M., Rosário-Ferreira, N., & Moreira, I. S. (2022). Network biology and artificial intelligence drive the understanding of the multidrug resistance phenotype in cancer. *Drug Resistance Updates: Reviews and Commentaries in Antimicrobial and Anticancer Chemotherapy, 60*. https://doi.org/10.1016/j.drup.2022.100811, 100811.

Burkhardt, D. B., Stanley, J. S., Tong, A., Perdigoto, A. L., Gigante, S. A., Herold, K. C., et al. (2021). Quantifying the effect of experimental perturbations at single-cell resolution. *Nature Biotechnology, 39*(5), 619–629. https://doi.org/10.1038/s41587-020-00803-5.

Butler, A., Hoffman, P., Smibert, P., Papalexi, E., & Satija, R. (2018). Integrating single-cell transcriptomic data across different conditions, technologies, and species. *Nature Biotechnology, 36*(5), 411–420. https://doi.org/10.1038/nbt.4096.

Cha, J., & Lee, I. (2020). Single-cell network biology for resolving cellular heterogeneity in human diseases. *Experimental & Molecular Medicine, 52*(11), 1798–1808. https://doi.org/10.1038/s12276-020-00528-0.

Chen, H., Albergante, L., Hsu, J. Y., Lareau, C. A., Lo Bosco, G., Guan, J., et al. (2019). Single-cell trajectories reconstruction, exploration and mapping of omics data with STREAM. *Nature Communications, 10*(1), 1903. https://doi.org/10.1038/s41467-019-09670-4.

Chen, J., Wu, Z., Qi, R., Ma, A., Zhao, J., Xu, D., et al. (2021). Deep transfer learning of drug responses by integrating bulk and single-cell RNA-seq data. *BioRxiv*. https://doi.org/10.1101/2021.08.01.454654.

Chen, K. H., Boettiger, A. N., Moffitt, J. R., Wang, S., & Zhuang, X. (2015). Spatially resolved, highly multiplexed RNA profiling in single cells. Science, 348(6233), 1360–1363. https://doi.org/https://doi.org/10.1126/science.aaa6090.

Chen, X., Yan, C. C., Zhang, X., Zhang, X., Dai, F., Yin, J., et al. (2016). Drug-target interaction prediction: Databases, web servers and computational models. *Briefings in Bioinformatics, 17*(4), 696–712. https://doi.org/10.1093/bib/bbv066.

Cohen, I. R., & Harel, D. (2007). Explaining a complex living system: Dynamics, multi-scaling and emergence. *Journal of the Royal Society Interface, 4*(13), 175–182. https://doi.org/10.1098/rsif.2006.0173.

Dahabiyeh, L. A., Malkawi, A. K., Wang, X., Colak, D., Mujamammi, A. H., Sabi, E. M., et al. (2020). Dexamethasone-induced perturbations in tissue metabolomics revealed by chemical isotope labeling LC-MS analysis. *Metabolites, 10*(2), 42. https://doi.org/10.3390/metabo10020042.

Darmanis, S., Gallant, C. J., Marinescu, V. D., Niklasson, M., Segerman, A., Flamourakis, G., et al. (2016). Simultaneous multiplexed measurement of RNA and proteins in single cells. *Cell Reports, 14*(2), 380–389. https://doi.org/10.1016/j.celrep.2015.12.021.

Dele-Oni, D. Ô., Christianson, K. E., Egri, S. B., Vaca Jacome, A. S., DeRuff, K. C., Mullahoo, J., et al. (2021). Proteomic profiling dataset of chemical perturbations in multiple biological backgrounds. *Scientific Data*, *8*(1), 1–10. https://doi.org/10.1038/s41597-021-01008-4.

Di, L., Kerns, E. H., & Carter, G. T. (2009). Drug-like property concepts in pharmaceutical design. *Current Pharmaceutical Design*, *15*(19), 2184–2194. https://doi.org/10.2174/138161209788682479.

Dixit, A., Parnas, O., Li, B., Chen, J., Fulco, C. P., Jerby-Arnon, L., et al. (2016). Perturb-Seq: Dissecting molecular circuits with scalable single-cell RNA profiling of pooled genetic screens. *Cell*, *167*(7). https://doi.org/10.1016/j.cell.2016.11.038. 1853-1866.e17.

Dubuis, S., Ortmayr, K., & Zampieri, M. (2018). A framework for large-scale metabolome drug profiling links coenzyme A metabolism to the toxicity of anti-cancer drug dichloroacetate. *Communications Biology*, *1*(1), 1–11. https://doi.org/10.1038/s42003-018-0111-x.

Eng, C. H. L., Lawson, M., Zhu, Q., Dries, R., Koulena, N., Takei, Y., et al. (2019). Transcriptome-scale super-resolved imaging in tissues by RNA seqFISH+. *Nature*, *568*(7751), 235–239. https://doi.org/10.1038/s41586-019-1049-y.

Ferraro, M., Moroni, E., Ippoliti, E., Rinaldi, S., Sanchez-Martin, C., Rasola, A., et al. (2021). Machine learning of allosteric effects: The analysis of ligand-induced dynamics to predict functional effects in TRAP1. *The Journal of Physical Chemistry. B*, *125*(1), 101–114. https://doi.org/10.1021/acs.jpcb.0c09742.

Forcato, M., Romano, O., & Bicciato, S. (2021). Computational methods for the integrative analysis of single-cell data. *Briefings in Bioinformatics*, *22*(1), 20–29. https://doi.org/10.1093/bib/bbaa042.

Fustero-Torre, C., José Jiménez-Santos, M., García-Martín, S., Carretero-Puche, C., García-Jimeno, L., Domenico, T. D., et al. (2021). Beyondcell: Targeting cancer therapeutic heterogeneity in single-cell RNA-seq. *BioRxiv*. https://doi.org/10.1101/2021.04.08.438954.

Gangaev, A., Ketelaars, S. L. C., Isaeva, O. I., Patiwael, S., Dopler, A., Hoefakker, K., et al. (2021). Identification and characterization of a SARS-CoV-2 specific CD8 + T cell response with immunodominant features. *Nature Communications*, *12*(1), 2593. https://doi.org/10.1038/s41467-021-22811-y.

García-Cañaveras, J. C., Castell, J. V., Donato, M. T., & Lahoz, A. (2016). A metabolomics cell-based approach for anticipating and investigating drug-induced liver injury. *Scientific Reports*, *6*(1), 27239. https://doi.org/10.1038/srep27239.

Gearty, S. V., Dündar, F., Zumbo, P., Espinosa-Carrasco, G., Shakiba, M., Sanchez-Rivera, F. J., et al. (2022). An autoimmune stem-like CD8 T cell population drives type 1 diabetes. *Nature*, *602*(7895), 156–161. https://doi.org/10.1038/s41586-021-04248-x.

Gehring, J., Hwee Park, J., Chen, S., Thomsom, M., & Pachter, L. (2019). Highly multiplexed single-cell RNA-seq by DNA oligonucleotide tagging of cellular proteins. *Nature Biotechnology*, *38*(1), 35–38. https://doi.org/10.1038/s41587-019-0372-z.

Gentile, F., Yaacoub, J. C., Gleave, J., Fernandez, M., Ton, A.-T., Ban, F., et al. (2022). Artificial intelligence-enabled virtual screening of ultra-large chemical libraries with deep docking. *Nature Protocols*, *17*(3), 672–697. https://doi.org/10.1038/s41596-021-00659-2.

Güvenç Paltun, B., Mamitsuka, H., & Kaski, S. (2019). Improving drug response prediction by integrating multiple data sources: Matrix factorization, kernel and network-based approaches. *Briefings in Bioinformatics*, *22*(1), 346–359. https://doi.org/10.1093/bib/bbz153.

Hagai, T., Chen, X., Miragaia, R. J., Rostom, R., Gomes, T., Kunowska, N., et al. (2018). Gene expression variability across cells and species shapes innate immunity. *Nature*, *563*(1), 197–202. https://doi.org/10.1038/s41586-018-0657-2.

Hamamoto, R., Komatsu, M., Takasawa, K., Asada, K., & Kaneko, S. (2019). Epigenetics analysis and integrated analysis of multiomics data, including epigenetic data, using artificial intelligence in the era of precision medicine. *Biomolecules, 10*(1), 62. https://doi.org/10.3390/biom10010062.

Hart, T., Chandrashekhar, M., Aregger, M., Steinhart, Z., Brown, K. R., MacLeod, G., et al. (2015). High-resolution CRISPR screens reveal fitness genes and genotype-specific cancer liabilities. *Cell, 163*(6), 1515–1526. https://doi.org/10.1016/j.cell.2015.11.015.

Hartigan, J. A., & Wong, M. A. (2012). Algorithm AS 136 A K-Means Clustering Algorithm. *Journal of the Royal Statistical Society Series B Methodological, 28*(1), 100–108.

Hashmi, M. U., & Abbas, S. R. (2019). Single-cell omics for drug discovery and development. In D. Barh, & V. Azevedo (Eds.), *Single-cell omics* (pp. 197–220). Academic Press. https://doi.org/10.1016/B978-0-12-817532-3.00013-X (chapter 11).

Ho, Y. J., Anaparthy, N., Molik, D., Mathew, G., Aicher, T., Patel, A., et al. (2018). Single-cell RNA-seq analysis identifies markers of resistance to targeted BRAF inhibitors in melanoma cell populations. *Genome Research, 28*(9), 1353–1363. https://doi.org/10.1101/gr.234062.117.

Hughes, J. P., Rees, S., Kalindjian, S. B., & Philpott, K. L. (2011). Principles of early drug discovery. *British Journal of Pharmacology, 162*(6), 1239–1249. https://doi.org/10.1111/j.1476-5381.2010.01127.x.

Jagger, B. R., Kochanek, S. E., Haldar, S., Amaro, R. E., & Mulholland, A. J. (2020). Multiscale simulation approaches to modeling drug-protein binding. *Current Opinion in Structural Biology, 61*, 213–221. https://doi.org/10.1016/j.sbi.2020.01.014.

Jehan, Z. (2019). Single-cell omics: An overview. In D. Barh, & V. Azevedo (Eds.), *Single-cell omics* (pp. 3–19). Academic Press. https://doi.org/10.1016/B978-0-12-814919-5.00001-4 (chapter 1).

Jerby-Arnon, L., Shah, P., Cuoco, M. S., Rodman, C., Su, M. J., Melms, J. C., et al. (2018). A cancer cell program promotes T cell exclusion and resistance to checkpoint blockade. *Cell, 175*(4). https://doi.org/10.1016/j.cell.2018.09.006. 984-997.e24.

Ji, Y., Lotfollahi, M., Wolf, F. A., & Theis, F. J. (2021). Machine learning for perturbational single-cell omics. *Cell Systems, 12*(6), 522–537. https://doi.org/10.1016/j.cels.2021.05.016.

Kagohara, L. T., Zamuner, F., Davis-Marcisak, E. F., Sharma, G., Considine, M., Allen, J., et al. (2020). Integrated single-cell and bulk gene expression and ATAC-seq reveals heterogeneity and early changes in pathways associated with resistance to cetuximab in HNSCC-sensitive cell lines. *British Journal of Cancer, 123*(1), 101–113. https://doi.org/10.1038/s41416-020-0851-5.

Kang, H. M., Subramaniam, M., Targ, S., Nguyen, M., Maliskova, L., & McCarthy, E. (2018). Multiplexed droplet single-cell RNA-sequencing using natural genetic variation. *Nature Biotechnology, 36*(1), 89–94. https://doi.org/10.1038/nbt.4042.

Kato, H., Takahashi, S., & Saito, K. (2011). Omics and integrated omics for the promotion of food and nutrition science. *African Journal of Traditional, Complementary, and Alternative Medicines, 1*(1), 25–30. https://doi.org/10.1016/s2225-4110(16)30053-0.

Kinker, G. S., Greenwald, A. C., Tal, R., Orlova, Z., Cuoco, M. S., McFarland, J. M., et al. (2020). Pan-cancer single-cell RNA-seq identifies recurring programs of cellular heterogeneity. *Nature Genetics, 52*(11), 1208–1218. https://doi.org/10.1038/s41588-020-00726-6.

Kiselev, V. Y., Kirschner, K., Schaub, M. T., Andrews, T., Yiu, A., Chandra, T., et al. (2017). SC3: Consensus clustering of single-cell RNA-seq data. *Nature Methods, 14*(5), 483–486. https://doi.org/10.1038/nmeth.4236.

Kobak, D., & Linderman, G. C. (2021). Initialization is critical for preserving global data structure in both t-SNE and UMAP. *Nature Biotechnology, 39*(2), 156–157. https://doi.org/10.1038/s41587-020-00809-z.

Kong, S. L., Li, H., Tai, J. A., Courtois, E. T., Poh, H. M., Lau, D. P., et al. (2019). Concurrent single-cell RNA and targeted DNA sequencing on an automated platform for comeasurement of genomic and transcriptomic signatures. *Clinical Chemistry*, *65*(1), 272–281. https://doi.org/10.1373/clinchem.2018.295717.

Kumar, A., & Jernigan, R. L. (2021). Ligand binding introduces significant allosteric shifts in the locations of protein fluctuations. *Frontiers in Molecular Biosciences*, *8*. https://doi.org/10.3389/fmolb.2021.733148, 733148.

Lähnemann, D., Köster, J., Szczurek, E., McCarthy, D. J., Hicks, S. C., Robinson, M. D., et al. (2020). Eleven grand challenges in single-cell data science. *Genome Biology*, *21*, 1–35. https://doi.org/10.1186/s13059-020-1926-6.

Lake, B. B., Chen, S., Sos, B. C., Fan, J., Kaeser, G. E., Yung, Y. C., et al. (2018). Integrative single-cell analysis of transcriptional and epigenetic states in the human adult brain. *Nature Biotechnology*, *36*(1), 70–80. https://doi.org/10.1038/nbt.4038.

Lamb, J., Crawford, E. D., Peck, D., Modell, J. W., Blat, I. C., Wrobel, M. J., et al. (2006). The connectivity map: Using gene-expression signatures to connect small molecules, genes, and disease. *Science*, *313*(5795), 1929–1935. https://doi.org/10.1126/science.1132939.

Lee, J., Hyeon, D. Y., & Hwang, D. (2020). Single-cell multiomics: Technologies and data analysis methods. *Experimental & Molecular Medicine*, *52*(9), 1428–1442. https://doi.org/10.1038/s12276-020-0420-2.

Lee, S. Y., Song, M. Y., Kim, D., Park, C., Park, D. K., Kim, D. G., et al. (2020). A proteotranscriptomic-based computational drug-repositioning method for Alzheimer's disease. *Frontiers in Pharmacology*, *10*(January), 1–11. https://doi.org/10.3389/fphar.2019.01653.

Levine, J. H., Simonds, E. F., Bendall, S. C., Davis, K. L., Amir, E.-A. D., Tadmor, M. D., et al. (2015). Data-driven phenotypic dissection of AML reveals progenitor-like cells that correlate with prognosis. *Cell*, *162*(1), 184–197. https://doi.org/10.1016/j.cell.2015.05.047.

Li, B., Cao, Y., Westhof, E., & Miao, Z. (2020). Advances in RNA 3D structure modeling using experimental data. *Frontiers in Genetics*, *11*. https://doi.org/10.3389/fgene.2020.574485, 574485.

Little, D. R., Gerner-Mauro, K. N., Flodby, P., Crandall, E. D., Borok, Z., Akiyama, H., et al. (2019). Transcriptional control of lung alveolar type 1 cell development and maintenance by NK homeobox 2-1. *Proceedings of the National Academy of Sciences of the United States of America*, *116*(41), 20545–20555. https://doi.org/10.1073/pnas.1906663116.

Longo, S. K., Guo, M. G., Ji, A. L., & Khavari, P. A. (2021). Integrating single-cell and spatial transcriptomics to elucidate intercellular tissue dynamics. *Nature Reviews. Genetics*, *22*(10), 627–644. https://doi.org/10.1038/s41576-021-00370-8.

Lotfollahi, M., Naghipourfar, M., Theis, F. J., & Alexander Wolf, F. (2020). Conditional out-of-distribution generation for unpaired data using transfer VAE. *Bioinformatics*, *36*(26), I610–I617. https://doi.org/10.1093/bioinformatics/btaa800.

Lotfollahi, M., Susmelj, A. K., De Donno, C., Ji, Y., Ibarra, I. L., Alexander Wolf, F., et al. (2021). Learning interpretable cellular responses to complex perturbations in high-throughput screens. *BioRxiv*. https://doi.org/10.1101/2021.04.14.439903.

Lotfollahi, M., Susmelj, A. K., Donno, C. D., Ji, Y., Ibarra, I. L., Wolf, F. A., et al. (2021). Compositional perturbation autoencoder for single-cell response modeling. *BioRxiv*. https://doi.org/10.1101/2021.04.14.439903.

Lotfollahi, M., Wolf, F. A., & Theis, F. J. (2019). scGen predicts single-cell perturbation responses. *Nature Methods*, *16*(8), 715–721. https://doi.org/10.1038/s41592-019-0494-8.

Luecken, M. D., Büttner, M., Chaichoompu, K., Danese, A., Interlandi, M., Mueller, M. F., et al. (2022). Benchmarking atlas-level data integration in single-cell genomics. *Nature Methods*, *19*(1), 41–50. https://doi.org/10.1038/s41592-021-01336-8.

Luecken, M. D., & Theis, F. J. (2019). Current best practices in single-cell RNA-seq analysis: A tutorial. *Molecular Systems Biology, 15*(6), e8746. https://doi.org/10.15252/msb.20188746.

Luo, C., Keown, C. L., Kurihara, L., Zhou, J., He, Y., Li, J., et al. (2017). Single-cell methylomes identify neuronal subtypes and regulatory elements in mammalian cortex. *Science, 357*(6351), 600–604. https://doi.org/10.1126/science.aan3351.

Marquina-Sanchez, B., Fortelny, N., Farlik, M., Vieira, A., Collombat, P., Bock, C., et al. (2020). Single-cell RNA-seq with spike-in cells enables accurate quantification of cell-specific drug effects in pancreatic islets. *Genome Biology, 21*(1), 106. https://doi.org/10.1186/s13059-020-02006-2.

Marx, V. (2022). How single-cell multi-omics builds relationships. *Nature Methods, 19*(2), 142–146. https://doi.org/10.1038/s41592-022-01392-8.

Matthews, H., Hanison, J., & Nirmalan, N. (2016). "Omics"—Informed drug and biomarker discovery: Opportunities. *Challenges and Future Perspectives. Proteomes, 4*(3), 28. https://doi.org/10.3390/proteomes4030028.

McCallum, L., Lip, S., & Padmanabhan, S. (2014). Pharmacodynamic pharmacogenomics. In S. Padmanabhan (Ed.), *Handbook of pharmacogenomics and stratified medicine* (pp. 365–383). Academic Press. https://doi.org/10.1016/B978-0-12-386882-4.00018-9 (chapter 18).

McCarthy, D. J., Campbell, K. R., Lun, A. T. L., & Wills, Q. F. (2017). Scater: Pre-processing, quality control, normalization and visualization of single-cell RNA-seq data in R. *Bioinformatics, 33*(8), 1179–1186. https://doi.org/10.1093/bioinformatics/btw777.

McFarland, J. M., Paolella, B. R., Warren, A., Geiger-Schuller, K., Shibue, T., Rothberg, M., et al. (2020). Multiplexed single-cell transcriptional response profiling to define cancer vulnerabilities and therapeutic mechanism of action. *Nature Communications, 11*(1). https://doi.org/10.1038/s41467-020-17440-w.

McInnes, L., Healy, J., Saul, N., & Großberger, L. (2018). UMAP: Uniform manifold approximation and projection. *Journal of Open Source Software, 3*(29), 861. 10.21105/joss.00861.

Miao, Z., Humphreys, B. D., McMahon, A. P., & Kim, J. (2021). Multi-omics integration in the age of million single-cell data. *Nature Reviews. Nephrology, 17*(11), 710–724. https://doi.org/10.1038/s41581-021-00463-x.

Mitra, A. K., Mukherjee, U. K., Harding, T., Jang, J. S., Stessman, H., Li, Y., et al. (2016). Single-cell analysis of targeted transcriptome predicts drug sensitivity of single cells within human myeloma tumors. *Leukemia, 30*(5), 1094–1102. https://doi.org/10.1038/leu.2015.361.

Motlagh, H. N., Wrabl, J. O., Li, J., & Hilser, V. J. (2014). The ensemble nature of allostery. *Nature, 508*(7496), 331–339. https://doi.org/10.1038/nature13001.

Mulqueen, R. M., Pokholok, D., Norberg, S. J., Torkenczy, K. A., Fields, A. J., Sun, D., et al. (2018). Highly scalable generation of DNA methylation profiles in single cells. *Nature Biotechnology, 36*(5), 428–431. https://doi.org/10.1038/nbt.4112.

Nassar, S. F., Raddassi, K., & Wu, T. (2021). Single-cell multiomics analysis for drug discovery. *Metabolites, 11*(11), 729. https://doi.org/10.3390/metabo11110729.

Nguyen, T., Le, H., Quinn, T. P., Nguyen, T., Le, T. D., & Venkatesh, S. (2021). GraphDTA: Predicting drug-target binding affinity with graph neural networks. *Bioinformatics, 37*(8), 1140–1147. https://doi.org/10.1093/bioinformatics/btaa921.

Norman, T. M., Horlbeck, M. A., Replogle, J. M., Ge, A. Y., Xu, A., Jost, M., et al. (2019). Exploring genetic interaction manifolds constructed from rich single-cell phenotypes. *Science, 365*(6455), 786–793. https://doi.org/10.1126/science.aax4438.

Nosengo, N. (2016). Can you teach old drugs new tricks? *Nature, 534*(7607), 314–316. https://doi.org/10.1038/534314a.

Nyman, E., Stein, R. R., Jing, X., Wang, W., Marks, B., Zervantonakis, I. K., et al. (2020). Perturbation biology links temporal protein changes to drug responses in a melanoma cell line. *PLoS Computational Biology*, *16*(7), 1–26. https://doi.org/10.1371/journal.pcbi.1007909.

Olalekan, S., Xie, B., Back, R., Eckart, H., & Basu, A. (2021). Characterizing the tumor microenvironment of metastatic ovarian cancer by single-cell transcriptomics. *Cell Reports*, *35*(8). https://doi.org/10.1016/j.celrep.2021.109165, 109165.

Osorio, D., Zhong, Y., Li, G., Huang, J. Z., & Cai, J. J. (2020). scTenifoldNet: A machine learning workflow for constructing and comparing transcriptome-wide gene regulatory networks from single-cell data. *Patterns (New York, N.Y.)*, *1*(9). https://doi.org/10.1016/j.patter.2020.100139, 100139.

Ozturk, K., & Carter, H. (2021). Predicting functional consequences of mutations using molecular interaction network features. *Human Genetics*.. https://doi.org/10.1007/s00439-021-02329-5 (Advance online publication).

Paananen, J., & Fortino, V. (2020). An omics perspective on drug target discovery platforms. *Briefings in Bioinformatics*, *21*(6), 1937–1953. https://doi.org/10.1093/bib/bbz122.

Pal, S. (2020). Biomacromolecules. In *Fundamentals of molecular structural biology* (pp. 83–117). Academic Press.

Peterson, V. M., Zhang, K. X., Kumar, N., Wong, J., Li, L., Wilson, D. C., et al. (2017). Multiplexed quantification of proteins and transcripts in single cells. *Nature Biotechnology*, *35*(10), 936–939. https://doi.org/10.1038/nbt.3973.

Preto, A. J., Matos-Filipe, P., Mourão, J., & Moreira, A. I. S. (2021). *SynPred: Prediction of drug combination effects in cancer using full-agreement synergy metrics and deep learning*. Preprints.

Prieto-Martínez, F. D., López-López, E., Eurídice Juárez-Mercado, K., & Medina-Franco, J. L. (2019). Computational drug design methods—Current and future perspectives. In K. Roy (Ed.), *Silico drug design* (pp. 19–44). Academic Press. https://doi.org/10.1016/B978-0-12-816125-8.00002-X (chapter 2).

Raimundo, F., Meng-Papaxanthos, L., Vallot, C., & Vert, J. P. (2021). Machine learning for single-cell genomics data analysis. *Current Opinion in Systems Biology*, *26*, 64–71. https://doi.org/10.1016/j.coisb.2021.04.006.

Rampášek, L., Hidru, D., Smirnov, P., Haibe-Kains, B., & Goldenberg, A. (2019). Dr.VAE: Improving drug response prediction via modeling of drug perturbation effects. *Bioinformatics*, *35*(19), 3743–3751. https://doi.org/10.1093/bioinformatics/btz158.

Rappez, L., Stadler, M., Triana, S., Gathungu, R. M., Ovchinnikova, K., Phapale, P., et al. (2021). SpaceM reveals metabolic states of single cells. *Nature Methods*, *18*(7), 799–805. https://doi.org/10.1038/s41592-021-01198-0.

Rautenstrauch, P., Vlot, A. H. C., Saran, S., & Ohler, U. (2022). Intricacies of single-cell multi-omics data integration. *Trends in Genetics: TIG*, *38*(2), 128–139. https://doi.org/10.1016/j.tig.2021.08.012.

Regev, A., Teichmann, S., Lander, E., Amit, I., Benoist, C., Birney, E., et al. (2017). Science forum: The human cell atlas. *ELife*, *6:e27041*, 1–30. https://doi.org/10.7554/eLife.27041.

Rosário-Ferreira, N., Bonvin, A. M. J., & Moreira, I. S. (2022). Using machine-learning-driven approaches to boost hot-spot's knowledge. *WIREs Computational Molecular Science*, *e1602*. https://doi.org/10.1002/wcms.1602.

Rotem, A., Ram, O., Shoresh, N., Sperling, R. A., Goren, A., Weitz, D. A., et al. (2015). Single-cell ChIP-seq reveals cell subpopulations defined by chromatin state. *Nature Biotechnology*, *33*(11), 1165–1172. https://doi.org/10.1038/nbt.3383.

Rubin, A. J., Parker, K. R., Satpathy, A. T., Qi, Y., Wu, B., Ong, A. J., et al. (2019). Coupled single-cell CRISPR screening and epigenomic profiling reveals causal gene regulatory networks. *Cell*, *176*(1–2). https://doi.org/10.1016/j.cell.2018.11.022. 361-376.e17.

Santos, R., Ursu, O., Gaulton, A., Bento, A. P., Donadi, R. S., Bologa, C. G., et al. (2017). A comprehensive map of molecular drug targets. *Nature Reviews. Drug Discovery, 16*, 19–34. https://doi.org/10.1038/nrd.2016.230.

Schmitt, M., Sinnberg, T., Bratl, K., Zittlau, K., Garbe, C., Macek, B., et al. (2021). Proteogenomics reveals perturbed signaling networks in malignant melanoma cells resistant to braf inhibition. *Molecular & Cellular Proteomics: MCP, 20*. https://doi.org/10.1016/j.mcpro.2021.100163, 100163.

Schnepp, P. M., Shelley, G., Dai, J., Wakim, N., Jiang, H., Mizokami, A., et al. (2020). Single-cell transcriptomics analysis identifies nuclear protein 1 as a regulator of docetaxel resistance in prostate cancer cells. *Molecular Cancer Research: MCR, 18*(9), 1290–1301. https://doi.org/10.1158/1541-7786.MCR-20-0051.

Sharma, A., Cao, E. Y., Kumar, V., Zhang, X., Leong, H. S., Wong, A. M. L., et al. (2018). Longitudinal single-cell RNA sequencing of patient-derived primary cells reveals drug-induced infidelity in stem cell hierarchy. *Nature. Communications, 9*(1), 4931. https://doi.org/10.1038/s41467-018-07261-3.

Shim, J., Hong, Z.-Y., Sohn, I., & Hwang, C. (2021). Prediction of drug-target binding affinity using similarity-based convolutional neural network. *Scientific Reports, 11*(1), 4416. https://doi.org/10.1038/s41598-021-83679-y.

Squair, J. W., Skinnider, M. A., Gautier, M., Foster, L. J., & Courtine, G. (2021). Prioritization of cell types responsive to biological perturbations in single-cell data with Augur. *Nature Protocols, 16*(8), 3836–3873. https://doi.org/10.1038/s41596-021-00561-x.

Srivatsan, S. R., McFaline-Figueroa, J. L., Ramani, V., Saunders, L., Cao, J., Packer, J., et al. (2020). Massively multiplex chemical transcriptomics at single-cell resolution. *Science, 367*(6473), 45–51. https://doi.org/10.1126/science.aax6234.

Stewart, C. A., Gay, C. M., Xi, Y., Sivajothi, S., Sivakamasundari, V., Fujimoto, J., et al. (2020). Single-cell analyses reveal increased intratumoral heterogeneity after the onset of therapy resistance in small-cell lung cancer. *Nature Cancer, 1*(4), 423–436. https://doi.org/10.1038/s43018-019-0020-z.

Stoeckius, M., Hafemeister, C., Stephenson, W., Houck-Loomis, B., Chattopadhyay, P. K., Swerdlow, H., et al. (2017). Simultaneous epitope and transcriptome measurement in single cells. *Nature Methods, 14*(9), 865–868. https://doi.org/10.1038/nmeth.4380.

Stuart, T., Butler, A., Hoffman, P., Hafemeister, C., Papalexi, E., Mauck, W. M., 3rd, et al. (2019). Comprehensive integration of single-cell data. *Cell, 177*(7). https://doi.org/10.1016/j.cell.2019.05.031. 1888-1902.e21.

Stuart, T., & Satija, R. (2019). Integrative single-cell analysis. *Nature Reviews. Genetics, 20*(5), 257–272. https://doi.org/10.1038/s41576-019-0093-7.

Sturm, G., List, M., & Zhang, J. D. (2021). Tissue heterogeneity is prevalent in gene expression studies. *NAR Genomics and Bioinformatics, 3*(3), lqab077. https://doi.org/10.1093/nargab/lqab077.

Subramanian, A., Narayan, R., Corsello, S. M., Peck, D. D., Natoli, T. E., Lu, X., et al. (2017). A next generation connectivity map: L1000 platform and the first 1,000,000 profiles. *Cell, 171*(6). https://doi.org/10.1016/j.cell.2017.10.049. 1437-1452.e17.

Sun, D., Gao, W., Hu, H., & Zhou, S. (2022). Why 90% of clinical drug development fails and how to improve it? *Acta Pharmaceutica Sinica B, 12*(5). https://doi.org/10.1016/j.apsb.2022.02.002. In press.

Svensson, V., Vento-Tormo, R., & Teichmann, S. A. (2018). Exponential scaling of single-cell RNA-seq in the past decade. *Nature Protocols, 13*(4), 599–604. https://doi.org/10.1038/nprot.2017.149.

Tini, G., Marchetti, L., Priami, C., & Scott-Boyer, M. P. (2018). Multi-omics integration-A comparison of unsupervised clustering methodologies. *Briefings in Bioinformatics, 20*(4), 1269–1279. https://doi.org/10.1093/bib/bbx167.

Tirier, S. M., Mallm, J. P., Steiger, S., Poos, A. M., Awwad, M. H. S., Giesen, N., et al. (2021). Subclone-specific microenvironmental impact and drug response in refractory multiple myeloma revealed by single-cell transcriptomics. *Nature Communications*, *12*, 6960. https://doi.org/10.1038/s41467-021-26951-z.

Trapnell, C. (2015). Defining cell types and states with single-cell genomics. *Genome Research*, *25*(10), 1491–1498. https://doi.org/10.1101/gr.190595.115.

Vamathevan, J., Clark, D., Czodrowski, P., Dunham, I., Ferran, E., Lee, G., et al. (2019). Applications of machine learning in drug discovery and development. *Nature Reviews. Drug Discovery*, *18*(6), 463–477. https://doi.org/10.1038/s41573-019-0024-5.

Van den Berge, K., Roux de Bézieux, H., Street, K., Saelens, W., Cannoodt, R., Saeys, Y., et al. (2020). Trajectory-based differential expression analysis for single-cell sequencing data. *Nature Communications*, *11*(1), 1–13. https://doi.org/10.1038/s41467-020-14766-3.

van der Maaten, L., & Hinton, G. (2008). Visualizing Data using t-SNE. *Journal of Machine Learning Research: JMLR*, *9*(86), 2579–2605. http://jmlr.org/papers/v9/vandermaaten08a.html.

Van Emon, J. M. (2016). The omics revolution in agricultural research. *Journal of Agricultural and Food Chemistry*, *64*(1), 36–44. https://doi.org/10.1021/acs.jafc.5b04515.

Wang, X., Allen, W. E., Wright, M. A., Sylwestrak, E. L., Samusik, N., Vesuna, S., et al. (2018). Three-dimensional intact-tissue sequencing of single-cell transcriptional states. *Science*, *361*(6400), 380–381. https://doi.org/10.1126/science.aat5691.

Wang, X., Alnabati, E., Aderinwale, T. W., Subramaniya, M. V., & S. R., Terashi, G., & Kihara, D. (2021). Detecting protein and DNA/RNA structures in cryo-EM maps of intermediate resolution using deep learning. *Nature Communications*, *12*(1), 2302. https://doi.org/10.1038/s41467-021-22577-3.

Wang, X., Frederick, J., Wang, H., Hui, S., Backman, V., & Ji, Z. (2021). Spike-in normalization for single-cell RNA-seq reveals dynamic global transcriptional activity mediating anticancer drug response. *NAR Genomics and Bioinformatics*, *3*(2), 1–13. https://doi.org/10.1093/nargab/lqab054.

Williamson, M. P. (2013). Using chemical shift perturbation to characterise ligand binding. *Progress in Nuclear Magnetic Resonance Spectroscopy*, *73*, 1–16. https://doi.org/10.1016/j.pnmrs.2013.02.001.

Wolf, F. A., Angerer, P., & Theis, F. J. (2018). SCANPY: Large-scale single-cell gene expression data analysis. *Genome Biology*, *19*(1), 15. https://doi.org/10.1186/s13059-017-1382-0.

Wu, F., Fan, J., He, Y., Xiong, A., Yu, J., Li, Y., et al. (2021). Single-cell profiling of tumor heterogeneity and the microenvironment in advanced non-small cell lung cancer. *Nature Communications*, *12*(1), 1–11. https://doi.org/10.1038/s41467-021-22801-0.

Wu, Z., Lawrence, P. J., Ma, A., Zhu, J., Xu, D., & Ma, Q. (2020). Single-cell techniques and deep learning in predicting drug response. *Trends in Pharmacological Sciences*, *41*(12), 1050–1065. https://doi.org/10.1016/j.tips.2020.10.004.

Xi, N. M., & Li, J. J. (2021). Benchmarking computational doublet-detection methods for single-cell RNA sequencing data. *Cell Systems*, *12*(2). https://doi.org/10.1016/j.cels.2020.11.008. 176-194.e6.

Xia, F., Allen, J., Balaprakash, P., Brettin, T., Garcia-Cardona, C., Clyde, A., et al. (2022). A cross-study analysis of drug response prediction in cancer cell lines. *Briefings in Bioinformatics*, *23*(1). https://doi.org/10.1093/bib/bbab356.

Xu, J., Zhang, P., Huang, Y., Zhou, Y., Hou, Y., Bekris, L. M., et al. (2021). Multimodal single-cell/nucleus RNA sequencing data analysis uncovers molecular networks between disease-associated microglia and astrocytes with implications for drug repurposing in Alzheimer's disease. *Genome Research*, *31*(10), 1900–1912. https://doi.org/10.1101/gr.272484.120.

Yu, L., Zhou, D., Gao, L., & Zha, Y. (2021). Prediction of drug response in multilayer networks based on fusion of multiomics data. *Methods*, *192*, 85–92. https://doi.org/10.1016/j.ymeth.2020.08.006.
Yu, W., & MacKerell, A. D. (2017). Computer-aided drug design methods. *Antibiotics: Methods and Protocols*, 85–106. https://doi.org/10.1007/978-1-4939-6634-9_5.
Zappia, L., & Theis, F. J. (2021). Over 1000 tools reveal trends in the single-cell RNA-seq analysis landscape. *Genome Biology*, *22*(1), 1–18. https://doi.org/10.1186/s13059-021-02519-4.
Zhang, S. (2011). Computer-aided drug discovery and development. In S. D. Satyanarayanajois (Ed.), *Drug design and discovery: methods and protocols* (pp. 23–38). Humana Press. https://doi.org/10.1007/978-1-61779-012-6_2.
Zhang, X., Xia, P., Wang, P., Yang, J., & Baird, D. J. (2018). Omics advances in ecotoxicology. *Environmental Science & Technology*, *52*(7), 3842–3851. https://doi.org/10.1021/acs.est.7b06494.
Zhao, J., Cao, Y., & Zhang, L. (2020). Exploring the computational methods for protein-ligand binding site prediction. *Computational and Structural Biotechnology Journal*, *18*, 417–426. https://doi.org/10.1016/j.csbj.2020.02.008.
Zhao, W., Dovas, A., Spinazzi, E. F., Levitin, H. M., Banu, M. A., Upadhyayula, P., et al. (2021). Deconvolution of cell type-specific drug responses in human tumor tissue with single-cell RNA-seq. *Genome Medicine*, *13*(1), 82. https://doi.org/10.1186/s13073-021-00894-y.
Zhou, Y., Song, W. M., Andhey, P. S., Swain, A., Levy, T., Miller, K. R., et al. (2020). Human and mouse single-nucleus transcriptomics reveal TREM2-dependent and TREM2-independent cellular responses in Alzheimer's disease. *Nature Medicine*, *26*(1), 131–142. https://doi.org/10.1038/s41591-019-0695-9.
Zhou, Y., Zhao, J., Fang, J., Martin, W., Li, L., Nussinov, R., et al. (2021). My personal mutanome: A computational genomic medicine platform for searching network perturbing alleles linking genotype to phenotype. *Genome Biology*, *22*(1), 53. https://doi.org/10.1186/s13059-021-02269-3.

CHAPTER THREE

A review of bioinformatics tools and web servers in different microarray platforms used in cancer research

R. Hephzibah Cathryn[a], S. Udhaya Kumar[a], Salma Younes[b], Hatem Zayed[b], and C. George Priya Doss[a,*]

[a]Laboratory of Integrative Genomics, Department of Integrative Biology, School of Biosciences and Technology, Vellore Institute of Technology, Vellore, India
[b]Department of Biomedical Sciences, College of Health and Sciences, Qatar University, QU Health, Doha, Qatar
*Corresponding author: e-mail address: georgepriyadoss@vit.ac.in

Contents

Advances in Protein Chemistry and Structural Biology, Volume 131
ISSN 1876-1623
https://doi.org/10.1016/bs.apcsb.2022.05.002

Abstract

Over the past decade, conventional lab work strategies have gradually shifted from being limited to a laboratory setting towards a bioinformatics era to help manage and process the vast amounts of data generated by omics technologies. The present work outlines the latest contributions of bioinformatics in analyzing microarray data and their application to cancer. We dissect different microarray platforms and their use in gene expression in cancer models. We highlight how computational advances empowered the microarray technology in gene expression analysis. The study on protein-protein interaction databases classified into primary, derived, meta-database, and prediction databases describes the strategies to curate and predict novel interaction networks *in silico*. In addition, we summarize the areas of bioinformatics where neural graph networks are currently being used, such as protein functions, protein interaction prediction, and *in silico* drug discovery and development. We also discuss the role of deep learning as a potential tool in the prognosis, diagnosis, and treatment of cancer. Integrating these resources efficiently, practically, and ethically is likely to be the most challenging task for the healthcare industry over the next decade; however, we believe that it is achievable in the long term.

Abbreviations

BPCA Bayesian principal component analysis
DEGs differential expressed genes
DLBCL delineated subtypes of diffuse large B-cell lymphoma
GMC Gaussian mixture clustering
GNN graphical neural networks
KNN k nearest-neighbours
metaMISS Meta analysis for missing binary data
MIAME minimum information about microarray experiments
SKNN sequential k nearest-neighbours
SVD singular value decomposition
WeNNI weighed nearest neighbour imputation

1. Microarray and bioinformatics

The application of computational techniques for investigating the data concerned with biomolecules on a large scale is known as Bioinformatics. It allows researchers to utilize the available information to submit new entries since they are established by various tools and resources to examine the data and elucidate the results in a biologically practical approach (Gharajeh, 2018).

In modern biology and medicine, bioinformatics is essential for managing and processing the enormous amounts of data generated by omics technologies. Bioinformatics clinical applications include immediate clinical applications, which involve basic computational approaches to aid in diagnosis and treatment regimens. In addition, there are short-term clinical applications of bioinformatics, which involve discovering genes and shedding light on the pathogenic processes associated with a particular disease, leading the way for the development of specific therapies. Other clinical applications of bioinformatics can be viewed in the long term, such as using basic genetic tests and integrative bioinformatics analysis of genomic, pathologic, and clinical data to indicate possible adverse drug reactions in patients (Bayat, 2002).

Historical development and review of the microarray field of study were done by Edwin M. Southern (Southern, 2001). Ekins and Chu studied the origin and applications of microarray (Ekins & Chu, 1999). Mark Schena team published two review papers in 1996 and 1998 that described the use of gene expression microarrays as functional genomic discovery platforms (Schena, 1996, 1998). The study of principles of microarray technology applied to the cancer research field by utilizing the cDNA microarray technique and other different techniques developed in analyzing gene expression profiles (Granjeaud, Bertucci, & Jordan, 1999; Khan et al., 1999). The detection of the known mutation using allele-specific oligonucleotide assay (ASO) such as DNA microarrays, which is a development of automated microarray systems for mutational analysis (Ravine, 1999). A review on gene expression techniques and application of transcriptional profiling changes in gene expression in specific pathways using DNA microarray technology known to be a powerful tool in developing molecular biological research field was carried out (Harkin, 2000; Schulze & Downward, 2000; van Hal et al., 2000). Bednar has published broad reviews on microarray technologies and applications (Bednár, 2000).

In molecular genetics technology, DNA microarrays allow us to get a global perspective of the cell and evaluate the expression of tens of thousands of genes simultaneously (Alonso-Betanzos, Bolón-Canedo, Morán-Fernández, & Sánchez-Maroño, 2019; Piatetsky-Shapiro & Tamayo, 2003). Microarray databases include a significant amount of genetic data, which can help understand biology and medicine (Hira & Gillies, 2015).

The process of using bioinformatics techniques to determine groups and molecular mechanisms that define a phenotype, such as a human disease, using microarray data analysis is referred to as "microarray data mining" (Butte, 2002; Desany & Zhang, 2004; Yang et al., 2012). Microarray data mining is rapidly being considered to identify diagnostic or prognostic marker genes in response to today's increased demand for biomarker discoveries (He, 2006). The compilation of transcriptomic data, primarily from microarray studies, has been particularly substantial in the last decade, resulting in several major public data repositories such as Gene Expression Omnibus and Array Express (Tseng, Ghosh, & Feingold, 2012). Since 2000, the Gene Expression Omnibus (GEO) has served as a public repository for high-throughput molecular abundance experimental data, providing free distribution and shared access to comprehensive datasets. Many public and commercial software and online tools are available, from probe design to gene expression analysis.

In general, gene expression analysis and SNP analysis are the most common uses for DNA microarray hybridization. Other uses of microarray systems include pharmacogenomics research, infectious, genetic diseases, and forensic and genetic identifiers (Heller, 2002). Most importantly, microarray technology is used for cancer detection for gene expression analysis (Graveel, Jatkoe, Madore, Holt, & Farnham, 2001). DNA microarrays generate massive amounts of data, posing a significant bottleneck (Ermolaeva et al., 1998; van Hal et al., 2000). The efficiency of computational tools helps determine the novel subtypes of disease and their mechanism or response to drugs. However, some statistical challenges regarding the tools were also detected when interpreting the computational study (Slonim & Yanai, 2009).

Using microarray in cancer research is for two approaches. First, to understand the type or subtype of cancer, genetic makeup, and abnormal biological pathways. These can be achieved using mapping approaches, including pathways analysis and gene ontology (GO). Second, to classify the different subtypes of cancer, providing a powerful tool for cancer prognosis and potential diagnosis and treatment. This study summarizes the overall review (Fig. 1) of various microarray databases and programs used for microarray gene expression data, identifying differentially expressed genes (DEGs), protein interaction network, functional enrichment analysis, and expression of critical genes validated.

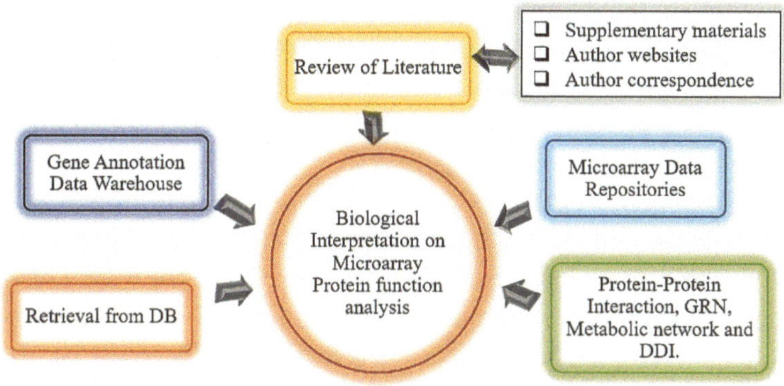

Fig. 1 Overall workflow represents the review on biological interpretation of microarray dataset and functional analysis using various computational tools.

2. Application of microarray data

Microarrays are becoming more reliable and affordable tools for genetic analysis of cancer (Conway & Gary, 2003; Nakanishi, Oka, & Akagi, 2001). Custom genotyping arrays intended to explore a small group of genes or particular pathways expanded the potential uses of microarray technology in analyzing different cancer subtypes (Bansal, 2005; Brown & Botstein, 1999; Nakanishi et al., 2001). This section highlights some of the technological developments and innovative uses of microarray and explores the different fields that have used microarray technologies. Together, these elements highlight the promise of microarray applications and their technical capabilities and provide real opportunities for further advances to be used in clinical settings.

Classifying the data by discriminant functional analysis for sample subtypes is one of the most promising areas of microarray applications, especially in diseases or infections (Call, Borucki, & Besser, 2003; Chandler et al., 2006; Dubois et al., 2004; Hanson, Niemeyer, Folio, Agan, & Rowley, 2004). For example, research workers at the NCI used microarray in classifying tissue samples from more than 200 patients with diffuse large-B-cell lymphoma in 2002. They discovered three subgroups with the varied expression of 17 unique genes, resulting in the development of a framework efficient for predicting overall survival rates after treatment

(Rosenwald et al., 2002). Another study employed microarrays to confirm prior classifying into non-pathogenic, low-pathogenic, and high-pathogenic *Yersinia enterocolitica* strains. Researchers found gene clusters indicative of each kind (i.e., present in one group but not another) and had functional implications (Howard et al., 2006).

Microarrays have proven effective in the discovery of novel genes and the identification of new microbes. Recently, researchers looked at an *E. coli* strain A49 that had a mutation alteration in the rnpA gene that rendered it thermally sensitive and unable to flourish at or above 43°C. According to the researchers, several genes were found to be variably expressed under varied growth conditions. According to a detailed investigation of the identified genes, RNase P, the mutated gene product, may have new functions in addition to those currently described, notably in the processing of precursor RNAs (Li, Cole, & Altman, 2003). Researchers constructed a customized array in 2003 that contained highly conserved patterns from fully sequenced viral genome in GenBank. The researchers then hybridized a SARS patient's viral sample on to array and found a previously novel coronavirus. Subsequent research using viral sequencing validated these findings and revealed the ability of tailored arrays to accelerate pathogen identification, which is critical in combating future epidemics (Wang et al., 2003).

Microarrays aid in the development of biological products, including vaccines and therapeutic proteins. Minor variations in a biological process, as outlined by multiple government rules, might result in a variety of end products, which will demand testing and validation (Bansal, 2005). Microarrays may avoid these techniques by developing criteria (for example, expression levels for a specific collection of genes) that can be used to assess consistency and replication. However, a substantial study would be required to create the essential needs first. Furthermore, in this application, microarray outputs must be viewed in terms of trends for a collection of genes instead of individual expression data (Bansal, 2005; Crowther, 2002). This is because the variability associated with a single gene might exceed the levels necessary to verify or validate biological processes, but the variation linked with groups of genes where an entire program is decoded is still far less (Crowther, 2002).

Another implementation of microarray technology that has been and continues to be helpful is pathway probing or the activation of biological pathways. Although microarray data alone cannot frequently explain the sequential steps necessary for a specific mechanism to occur, it does provide insight into which genes or proteins should be investigated further (Bodrossy & Sessitsch, 2004; Crowther, 2002). For example, a 2003 study used microarrays in association with other experimental approaches to investigate a system responsible for controlling the transcription of cyclooxygenase-2 (COX-2), a pro-inflammatory protein linked to arthritis and pain (Abulencia et al., 2003). A continuation of this study detailing the pathway and possible monitoring mechanisms with major therapeutic implications, as described in 2005 (Healy et al., 2005).

Combining genomics and proteomics with bioinformatics will increase the possibility of microarrays in clinical molecular diagnosis. This will help to create complete models of specific biological processes. Recent research, for example, used two-dimensional gel electrophoresis and microarrays to identify genes involved in cyanobacterial adaptation to changing visible light. Researchers have focused on *Fremyella diplosiphon* and identified around 80 proteins with differing numbers of cells growing in green light vs. red light, and 17 genes that were previously unheard of to be light-regulated. Following further examination, it was shown that several of these genes had homologs in other species. Their role, however, was not fully understood (Stowe-Evans, Ford, & Kehoe, 2004). Another study used proteomics and microarrays to look at an *Escherichia coli* mutant that produced higher hemolysin (HlyA) than parental strain. The mutant had lower quantities of tRNA-synthetase in comparison to the parent strain. The parent strain secreted more HlyA than the mutant strain when transformed with the altered *hlyA* gene (Lee & Lee, 2005). This led to an elegant design of a modified hlyA gene that incorporates codons that result in the same amino acid sequence to limit translation rate.

Several significant reviews on microarrays in cancer research and diagnostic tools have recently been published. These include Kallioniemi's more broad reviews on application of high-throughput biochip technologies in cancer research (Kallioniemi, 2001), review on diverse applications of microarray technology in breast cancer (Cooper, 2001), review study on pediatric cancer by gene expression profiling (GEP) utilizing DNA microarray methods (Triche, Schofield, & Buckley, 2001), for identification of prostate cancer biomarkers using microarray tools (Grouse, Munson, & Nelson, 2001), use of DNA microarray analysis to optimize brain cancer therapy (Mischel, Cloughesy, & Nelson, 2004), importance of microarray analysis and its clinical application to identify genetic characters in gastric cancer (D'Angelo, Di Rienzo, & Ojetti, 2014), gene expression in ovarian cancer (Lisowska et al., 2014), colorectal cancer (Nannini et al., 2009) and analysis of transcriptional profile in Muscle-invasive bladder cancer (MIBC) and identifying candidate genes and pathways (Isali et al., 2022) and gene expression profiling associated with epithelial growth factor receptor (EGFR) in breast cancer and NSCLC was studied in this review for analysis of cancer treatment and drug combination strategies (Subramaniyan et al., 2022). DNA microarray technology studies are available in Monograph titled DNA Microarrays: A Practical Approach has been edited by Mark Schena (Schena, 1999), Basic Methods in Molecular Biology: DNA arrays methods and protocols by Rampal JB (Davis, 2012). The broad areas of microarray applications include microarrays for gene expression studies (Bulyk, Huang, Choo, & Church, 2001; Slonim & Yanai, 2009; Ziauddin & Sabatini, 2001), analysis of gene expression for drug development, metabolism, and toxicity (Gerhold et al., 2001; Madden, Wang, & Landes, 2000), neuroscience application (Luo & Geschwind, 2001; Parrish et al., 2004; Sanfilippo & Di Rosa, 2021), for genotyping (Hacia et al., 2000; Hacia & Collins, 1999; Mahalingam & Fedoroff, 2001), pharmacogenomics (Chicurel & Dalma-Weiszhausz, 2002).

The present review enlightens the growth of microarray technology and its applicability to cancer research. Workflows were represented to understand the scope of standard web interface tools and academic tools and compiled user-friendly databases for researchers. Further, the bioinformatic methods and application of microarray analysis for differential gene expression, functional enrichment, protein-protein interaction (PPI), and hub gene expression level were analyzed for colorectal cancer samples from the microarray repository. This review will explore the present status of gene expression profiling in colon cancer utilizing microarray technology, various protein network analyses using microarray data, and potential future approaches in this field of study.

3. Classification of microarray data in cancer

DNA microarray technology is used in detecting gene expression, genotyping, and transcription factor binding analysis. Microarray-developed technology works under the hybridization principle (complementary binding sequences of DNA) between two sample groups, mainly in cancer research. DNA microarray is used to investigate the DNA of the samples or tissues of two groups parallelly. The microarray platform analysis (Fig. 2) and the microarray experiments are classified based on the mode of preparation, type, and nature of the probes used, to detect the target

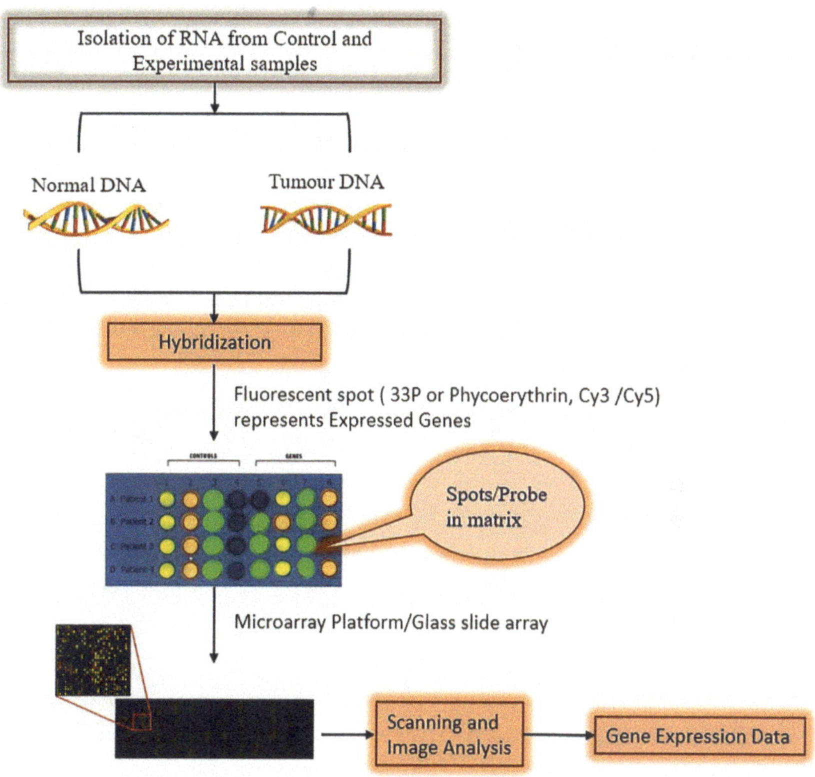

Fig. 2 The general analysis of microarray platform for gene expression analysis.

samples used for investigation (Agapito, 2019). The modern DNA array is classified into three basic arrays: spotted array, self-assembled, and *in situ* synthesized arrays. (i) Spotted arrays are used for high-density arrays with DNA of interest spotted using slotted pins in glass slides coated with poly-lysine, a silane, or chemically reactive with oligos or PCR products. The DNA samples are spotted on the glass slide and labeled using a fluorescent detector that gives two-colored samples on the same array expressing differential genes (DeRisi et al., 1996). (ii) Self-assembled arrays are fiber optic arrays with DNA synthesized on a small polystyrene bead mixture applied to fiber optic cable that results in a randomly assembled array. David Walt and colleagues at Tufts University pioneered this strategy (Ferguson, Steemers, & Walt, 2000; Michael, Taylor, Schultz, & Walt, 1998; Steemers, Ferguson, & Walt, 2000; Walt, 2000). (iii) *In situ* synthesized arrays are chemically synthesized methods on solid substrate combined photolabile group with photolithography (Fodor et al., 1991) or produced by inkjet oligo synthesis method (Blanchard, Kaiser, & Hood, 1996). These arrays are useful for analyzing expression, sequencing, and genotyping of genes (Chee et al., 1996; Lockhart et al., 1996; Wallraff et al., 1997).

3.1 Types of microarray

Microarray studies are categorized into printed microarray, *in situ* synthesis oligonucleotide microarray, high-density bead microarray (HDBM), electronic microarray (EM), and suspension bead microarray (SBA) based on the nature of the probes. They are described more below:

I. Printed microarray, known as the first microarrays developed and so-called is because printed or spotting of a probe on the microarray glass slide. Based on probe nature, it is known as double-stranded DNA (dsDNA) or Oligonucleotide microarray. The probe spot can be done by contact or non-contact printing. Contact printing allows the probe solution printed on a microarray glass slide. Non-contact printing involves placing the probe drop solution on the microarray glass slide using a bubble jet or inkjet.

II. In situ synthesized Oligonucleotide microarray is used for high-density oligonucleotide microarrays. It is familiar with iSSOM, an Affymetrix GeneChip (Gautier, Cope, Bolstad, & Irizarry, 2004).

III. Illumina develops High-density bead microarray (HDBM) technology for whole-genome genotyping and copy number variation (Fan et al., 2006; Shen et al., 2005).

IV. Electronic microarray (EM) involves the hybridization process of DNA is run by electric fields on a semiconductor chip developed by Nanogen (Blin, Cissé, & Bockelmann, 2015; Sosnowski, Tu, Butler, O'Connell, & Heller, 1997). They are flexible technology for diagnostic applications.

V. Suspension bead microarray (SBA) is a 3D microarray developed first in 1977 (Horan & Wheeless, 1977). SBA utilizes microsphere beads about 56 μm in diameter for simultaneous testing of various gene variants (Schwenk, Gry, Rimini, Uhlén, & Nilsson, 2008). SBA is regarded as the most alluring platform in clinical molecular diagnostics. MM chips, protein microarray, peptide microarray, tissue microarray, cellular microarray, chemical compound microarray, antibody microarray, carbohydrate array, phenotype array, reverse phase protein microarray, interferometric reflectance imaging sensor (IRIS), and DNA microarray are the other types of microarrays.

3.2 Platforms

The Affymetrix GeneChip and the Illumina Bead Array are the most advanced platforms common in microarray technology.

Affymetrix technology develops the high-density oligonucleotide arrays (Amaratunga & Cabrera, 2004) represented by a probe set of 11 different pairs of 25 bp oligo covering a transcribed region of the gene, each pair of the base is of perfect match (PM) and mismatch oligonucleotide (MM) (Jiang et al., 2008). MM is similar to PM, but the 13th base is complementary to the central base of PM. Affymetrix used the photolithography method to assemble GeneChip to measure the genes from samples by microarray experiment (Gautier et al., 2004; Irizarry et al., 2003). The sequences of each probe are undergone through light-sensitive masking agents across the whole array repeatedly, with no more than 100 repeats, thus completely arranged and becoming a GeneChip. The price of Affymetrix microarray is high. Some packages used for Affymetrix acquired datasets are *affy* (Gautier et al., 2004).

Illumina technology is a cutting-edge method for analyzing gene expression that provides high-performance, high-throughput programs that enable researchers to extend their experiments while decreasing the expense of large-scale research. It is composed of two formats: The Sentrix Array Matrix (SAM) and Sentrix Bead Chip (Fan et al., 2006; Steemers & Gunderson, 2005). Arrays are arranged on an optical fiber bundle that contains 50,000 fiber bundles (silica beads 3 μm in diameter) coated with ten to a hundred thousand oligonucleotides of a specific sequence into the hexagonal matrix. Some of the microarrays are arranged on silicon slides that have been processed by micro-electromechanical systems (MEMS) technology to include nano wells that support bead self-assembly in the BeadChip format. Illumina is cheap among other platforms (Fajriyah, 2021). Packages used for Illumina acquired datasets are *Lumi* (Du, Kibbe, & Lin, 2008), *bead array* (Dunning, Smith, Ritchie, & Tavaré, 2007), and *illuminaio* (Smith, Baggerly, Bengtsson, Ritchie, & Hansen, 2013). The limma package Ritchie et al. (2015) is used for other and agilent acquired datasets. The other platforms are PLIER (Therneau & Ballman, 2008), SigPathway (Tian et al., 2005), and mm74av1 mmentregcdf package (Entrez gene-based chip) for the MG-74AV1 Affymetrix platform.

4. Microarray data expression repositories

Some websites give more detailed information on bioinformatics software blogs for microarray gene expression analysis, such as https://mybiosoftware.com/category/microarray-analysis, and other expression repositories are provided in Table S1 in Supplementary Information in the online version at https://doi.org/10.1016/bs.apcsb.2022.05.002. The commonly used microarray repositories for gene expression are discussed below:

GEO is one of the major publicly accessible repositories managed by NCBI. It is a member of the Microarray Gene Expression Data Society (MGED) (Whetzel et al., 2006) as a part of the publication process to be one of the acceptable public repositories. It is divided into three components platform accession (GPL), sample accession (GSM), and series accession (GSE) (Barrett, 2004). GEO data repository is for functional genomics that allows the Minimum Information About Microarray Experiments (MIAME) guidelines compliant data contributions. Array and sequence-based datasets are accepted. GDS (dataset accession) and GEO2R may be used to detect (DEGs), execute R-based analysis of GEO data with existing R packages, and display findings as a table of genes sorted by relevance that can be viewed with GEO Profile graphics. GEO database is helpful to download high-throughput microarray or RNA-seq experiments and well-curated gene expression profiles, and RNA methylation profiling (Edgar, Domrachev, & Lash, 2002) ArrayExpress is one of the significant public repositories available online that stores functional genomic high-throughput experiments and is available to the research field for reuse and maintenance by the Bioinformatics Institute (EBI). Data annotation, standard data transfer formats, and public repository for such data, such as DDBJ/EMBL/GenBank for molecular sequence data, are the three crucial components (Brazma, 2003). It consists of an ArrayExpress microarray data repository supported by MIAME and an Array Express warehouse of gene expression profiles from the repository and annotated (Parkinson et al., 2007). Array express is one of the members of the MGED (Whetzel et al., 2006) as a part of the publication process to be one of the acceptable public repositories as GEO (Edgar, Domrachev, & Lash, 2002) and CIBEX (Ikeo, Ishi-i, Tamura, Gojobori, & Tateno, 2003).

Center for Information Biology Gene Expression (CIBEX) https://integbio.jp/dbcatalog/en/record/nbdc00029 is a public gene expression repository supported by MIAME and member of the MGED. This database serves high-throughput experiments on microarray measuring mRNA, Serial analysis of Gene Expressions (SAGE), mass spectrometry proteomics data, provides an intensity of gene expression through histogram, and obtains the result on scatter plot analysis. It also has the tool for cluster analysis of the genes. It works under Linux and windows with JAVA to download the CIBEX application (Ikeo et al., 2003).

The Cancer Genome Atlas (TCGA) https://portal.gdc.cancer.gov/ is a multi-agency innovation project financed by the National Institutes of Health (NIH), a division of the National Cancer Institute (NCI), which collects data from different microarray systems. Genomic Data Commons (GDC) is responsible for about 33 projects. Data Portal is a platform that helps researchers and bioinformaticians to download data on cancer analysis https://wiki.nci.nih.gov/display/TCGA/TCGA+Wiki+Home. It consists of the collection of whole-exome sequencing, mRNA and miRNA expression data, somatic mutations, copy number variations, etc. (Lee, 2016). To retrieve gene expression data of particular cancer samples, the RNAseq dataset with mRNA level and corresponding clinical information from patients in TCGA and biological information from tumor and adjacent normal tissue can be downloaded to calculate the expression mRNA. To process the DEG data from TCGA, an edgeR package is used to screen the DEGs using Bioconductor—TCGA Workflow or TCGA biolinks/R package (Colaprico et al., 2016; Silva et al., 2016).

Oncomine is a cancer microarray database and web-based platform that consists of 65 gene expression datasets and over 4700 microarray experiments. Coexpression analysis in the oncomine cancer profiling database can identify gene sets with synchronous expressional patterns. Hub gene validation and survival analysis of the DEGs can be analyzed utilizing oncomine. It draws datasets from Stanford Microarray Database (SMD) (Ball, 2004) and GEO (Barrett, 2004; Rhodes et al., 2004, 2007).

Gene Expression Profiling Interactive Analysis (GEPIA) is a web tool for gene expression analysis based on tumor and normal samples from TCGA and GTEx projects. GEPIA provides differential expression profiling, correlation analysis, survival analysis, similar gene detection, and dimension reduction analysis (Tang et al., 2017). For DEG's analysis, the results are presented in dot or box plots and also represent pathological stages based on TCGA annotations. The parameters and protocols for the GEPIA database can be studied using weblink http://gepia.cancer-pku.cn/help.html#top. Recently GEPIA has developed a standalone extension tool, GEPIA2021 allowing the users to perform multiple interactive analyses based on deconvolution results, and it is available online (Li, Tang, Zhang, Ye, & Liu, 2021).

GTEx (The Genotype-Tissue Expression) is a project available publicly to study tissue-specific gene expression and regulation. GTEx is designed by the NIH, expression quantitative trait locus (eQTL) approach in GTEx to interpret the genetic component altering the gene expressing

https://www.gtexportal.org/home/eqtlDashboardPage, eQTL are made accessible through NCBI GTEx database and GTEx data portal (Lonsdale et al., 2013).

WebArrayDB is an open-source integrated microarray database used to execute cross-platform microarray analysis supported by MIAME and data to be mined based on R-based tools. The website is freely available http://www.webarraydb.org/webarray/index.html, and it runs on a Linux server with Apache and MySQL (Xia, McClelland, & Wang, 2005).

RMAExpress is a stand-alone application for Windows, OS X, and Linux that may be downloaded to evaluate gene expression utilizing Robust Multichip Average expression summary and quality evaluation. For RMA analysis CEL and CDL files are needed, and it will read the data in datafiles. RMA analysis computes the results. The RMAExpress output files can also be loaded into R and use the features of Bioconductor for gene expression analysis (Bolstad, Irizarry, Astrand, & Speed, 2003).

BRB Array Tools is a fully accessible software program that visualizes and analyses DNA microarray gene expression data. It is available for download and usage as an Excel add-in as well as R setup for installing BRB array utilities (Simon et al., 2007).

5. Gene expression profiling (GEP)

GEP is an approach for rapidly investigating the expression of thousands of genes and is used precisely to categorize tumors from the perspective of cancer. Microarray technology was initially used in GEP in 1995 (Schena, Shalon, Davis, & Brown, 1995). This method allows for the simultaneous assessment of tens and thousands of mRNA sequences and could be used to analyze gene expression within a single sample or across many samples. The relative quantity of mRNA expressed in two or more experimental conditions is measured in GEP studies.

GEP has been utilized to assist in appropriately categorizing tumors in the aspect of cancer. The data collected by GEP is frequently utilized to estimate clinical outcomes. Historically, histology and histopathology dominated cancer categorization, intending to precisely identify tumor types using morphological markers. However, for some malignancies, these approaches are unable to distinguish between subtypes. GEP is an effective tool for cancer care because of its capacity to classify tumors into subtypes with different biological properties.

Delineated subtypes of diffuse large B-cell lymphoma (DLBCL) have mainly been unsuccessful because of inter-and intra-observer repeatability differences. Furthermore, the clinical results of DLBCL tumors are highly heterogeneous (The Non-Lymphoma Hodgkin's Classification Project, 1997), demonstrating that DLBCLs, notwithstanding their clinical diversity, cannot be distinguished using these histological markers. It is estimated that 60% of DLBCL patients do not respond to treatment regimens and have a low survival probability. Stanford scientists Ash Alizadeh and colleagues performed a study utilizing GEP to categorize DLBCL into two subtypes over two decades ago: germinal center B-like DLBCL and activating B-like DLBCL. The difference between the two groups is noteworthy in terms of patient overall survival. Over ten years, the chance of survival for patients with germinal center B-like DLBCLs was around 80%, but for those with active B-like DLBCLs, it was around 40% (Alizadeh et al., 2000). Even before the term "precision medicine" became widespread, this sparked a search for ways to utilize the different biological traits that underlie each subtype.

Another example in breast tumor classifications, in which establishing the subgroups based on histological predictions has generally failed (McGuire, 1991). Furthermore, a misinterpretation may cause patients to suffer as a result of unwanted side effects from non-targeted medications and higher healthcare costs. According to estimations, 70 to 80% of breast cancer patients who underwent chemotherapy based on typical histology projections would have survived if they had not received it (Polychemotherapy for Early Breast Cancer, 1998). Stanford scientists directed by C.M. Perou conducted a study examining that gene expression patterns across 8102 genes in 65 breast cancer samples. The tumors were classified into two groups based on clinical diagnoses of ER-positive and ER-negative malignancies. The ER-positive malignancies displayed a high expression level of genes seen in breast luminal cells. According to the experts, this greater differentiation may include at least two physiologically distinct types of cancer, each of which may require a different course of treatment (Perou et al., 2000). Microarray studies have been used to create gene expression patterns associated with prognosis and uncover genes associated with survival. Adjuvant treatment is commonly used to help patients with tumors with poor prognoses since it dramatically improves overall survival in breast cancer patients. Oncotype DX, Prosigna (PAM 50), EndoPredict, and MammaPrint are commercially available GEP tests. The most often used test is Oncotype DX, designed for ER+ and lymph node-negative tumors (Smith & Farrah, 2019).

In drug development, GEP may serve as an essential diagnostic tool to help in identifying mechanisms of toxicity in drug metabolism. GEP helps

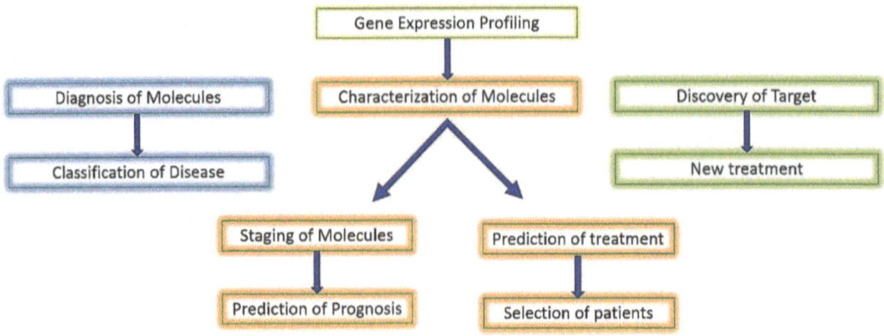

Fig. 3 The clinical application of gene expression profiling in oncology.

anticipate a patient's prognosis and uses DNA methylation to regulate expression among human populations (Zhong, Kim, Zhi, & Cui, 2019). GEP analysis using microarray technology has demonstrated a better prospect in cancer research and medical oncology (Fig. 3), modeling the expression of thousands of genes in a solitary tumor sample simultaneously and enabling the evaluation of complex gene expression profiles (Nannini et al., 2009). According to GEP studies, alteration in miRNA expression has been seen in various human diseases. Functional studies have linked miRNA dysregulation to disease development in several cases. It is critical to research the principles and mechanisms that determine complicated gene expression to comprehend better complex diseases, such as malignancies (Imam, 2016). RNA-Seq rapidly replaces microarrays as the preferred method for estimating gene expression levels in gene expression investigations. Researchers can also use RNA-Seq to investigate previously unavailable phenomena for microarrays, such as allele-specific expression and the discovery of previously unknown transcribed areas.

6. Tools available to analyze microarray data

AMDA (Automated Microarray Data Analysis) (Pelizzola, Pavelka, Foti, & Ricciardi-Castagnoli, 2006) is free software as an R Package https://cran.r-project.org/under General public license that can be downloaded, as it is based on the R/Bioconductor project. It provides effective bioinformatic microarray monitoring tools that provide scientists with trouble-free and integrated analysis of Affymetrix microarray experiments, and also simple to install on standard operating systems like Linux, and Mac OS X. It is written in the R programming language (Ihaka & Gentleman, 1996). AMDA performs all procedures, including data analysis, image analysis, quality controls, normalization, identification of DEGs, clustering, correspondence analysis, and functional assessment. The final report issued by the LaTEX document on the accomplished analysis comprises analysis outcomes and comments, as well as pointers to files for extensive education.

Birdsuite (Korn et al., 2008) is a freely accessible toolkit for detecting SNP alleles, common copy-number polymorphism (CNPs), and new genetic CNVs in Affymetrix Platform data. Birdsuite is designed explicitly for integrated analysis of Single Nucleotide Polymorphisms and Copy Number Variations. The latest version is only accessible for Linux-based computers. The Matlab version must be downloaded and used. The other two Birdsuite versions may be found at https://www.broadinstitute.org/birdsuite/birdsuite-downloads. Users must install the following elements using the bird suite: Java 1.5 or above (https://www.oracle.com/java/technologies/downloads/), Python (https://www.python.org/), and the NumPy python package (http://numpy.scipy.org/), R.2.4 https://cran.r-project.org/, mclust R package http://lib.stat.cmu.edu/R/CRAN/web/packages/mclust/index.html, Affymetrix power tools APT 1.8.6 or 1.10.2 binaries https://www.thermofisher.com/in/en/home/life-science/microarray-analysis/affymetrix as well as the Affymetrix library bundle for the Genome-wide Human SNP array 6.0.

CisGenome is a bioinformatic tool used to create ChIP-seq analysis pipelines (Ji et al., 2008). It is compatible with all Linux/Unix, Windows, and Mac OS platforms. It is available for free download for academic and non-commercial use at https://www.biostat.jhsph.edu/hji/cisgenome/index files/download.htm. CisGenome was created using the computer languages C and C++. A graphical user interface is also available for the desktop version, making the command-line version easier to use. CisGenome users may do data normalization, visualization, false discovery rate (FDR) computation, peak recognition, gene-peak correlation, sequence analysis, and motif assessment.

Cloud4SNP is a cloud-based application for simultaneous pre-processing and statistical analysis of pharmacogenomics SNP DMET (Drug Metabolizing Enzyme and Transporters) microarray data (Agapito et al., 2013). It is a cloud version of DMET Analyzer (Guzzi et al., 2012), a software environment for creating and implementing knowledge workflows on the cloud with the help of a data mining cloud system (Marozzo, Talia, & Trunfio, 2013). Cloud4SNP uses data parallelism and an improved technique to prevent running unnecessary fisher tests on probes with comparable SNP distribution.

DMET-Analyzer is a free utility distributed under the General Public License version 2.0 and can be found at https://sourceforge.net/projects/dmetanalyzer/ (Guzzi et al., 2012). This tool examines the relationship between genomic variation and clinical features in individuals for various drug reactions. It enables the automation of the DMET dataset analysis procedure. DMET is a Java-based application that allows users to understand DMET files generated by Affymetrix DMET-Console and is provided for Linux/Unix, Mac OS, and Windows platform systems. The input files must be in either excel or tab-delimited format. The output may be simply saved in text, HTML, and other formats.

DMET-Miner is a bioinformatic tool that combines allelic variant groups that respond to drugs. It is developed in Java and is studied by users using a graphical user interface (Agapito, Guzzi, & Cannataro, 2015). This program can be found at https://sites.google.com/site/dmetminer/get-software.

GSEA (Gene Set Enrichment Analysis) software is written in Java and R. It analyses sequentially and runs on compute clusters for massive datasets. It has extensions for gene expression data ".gct," phenotypes ".cls," gene set ".gmt," and chip annotation ".chip."

MIDAS (Microarray Data Analysis System) can be downloaded for free at https://sourceforge.net/projects/midas-tm4/files/latest/download?source=files. It is written in Java and suits Linux, Windows, and Mac OS. The GUI creates the workflow and filtering for users. Scatter and box plots are used to depict the process analysis. The output may be exported as "tav" or pdf/plain text files (Agapito, 2019).

TM4 (http://www.tm4.org/) is freely downloadable software; designed for cDNA chips and also adapts Affymetrix arrays. It is a toolkit comprised of four programs that are Microarray Data Manager (MADAM), TIGR spot finder, MIDAS, and Multi Experiment Viewer (MeV) are some of the tools available.

SNOMAD (Standardization and Normalization of Microarray Data) http://pevsnerlab.kennedykrieger.org/snomad.htm is a simple and easy-to-use web-based application for analyzing the standardization and normalizing of gene expression data from biological and technology sources. It is written in the R programming language (Colantuoni, Henry, Zeger, & Pevsner, 2002).

SAM (Significance Analysis of Microarrays) software is accessed through the web link https://tibshirani.su.domains/SAM/. SAM analyzes based on the change in expression level in the gene it measures repeatedly assigning a score to each gene. If genes reach the configurable threshold, SAM uses permutations to calculate the proportion of genes found and the false discovery rate (FDR) (Tusher, Tibshirani, & Chu, 2001). SAM can also be used as an add-on in excel.

Genesis software can be downloaded and retrieved by the user. This software is helpful to identify gene expression by high throughput analysis using the cDNA microarray technology approach. Genesis is also developed to cluster genes into groups with related expression patterns from multiple experiments and analyze the microarray data by implementing hierarchical and non-hierarchical clustering algorithms (Sturn, Quackenbush, & Trajanoski, 2002).

GEMS (Gene Expression Model Selector) evaluate high-quality cancer diagnostic models and discovers biomarkers from microarray data. It is available as free software and can be downloaded. GEMS is classified based on investigated outcome data and develops a model from cancer patient gene expression data to predict the outcome of unknown data (Statnikov, Tsamardinos, Dosbayev, & Aliferis, 2005).

GenMAPP (Gene Microarray Pathway Profiler) is a free online tool designed to visualize the expression data on a biological pathway, including color code that coincides with the user's parameters (Dahlquist, Salomonis, Vranizan, Lawlor, & Conklin, 2002). Another accessory program, MAPPFinder, is used to identify the global expression of gene profiler for annotations of GO and pathways afflicted by gene list (Doniger et al., 2003).

Onto-Express (OE) and Pathway Express (PE) are free and can be downloaded. OE software creates functional profiles based on GO concepts for the following groups: biochemical activity, BP, CC, MF, cellular activity, and chromosomal location, and computes unique values for each. Using DEGs as input, PE identifies notably perturbed pathways and evaluates perturbation factors for each gene that reflect the genes' correlation (Khatri et al., 2005, 2007).

DAVID (Database for Annotation, Visualization, and Integrated Discovery) is an online tool that combines functional genomic annotations by providing the list of genes DAVID annotates and summarizes for GO, protein domain, and biochemical pathways with graphical summaries (Dennis et al., 2003) and EASE (Expression Analysis Systematic Explorer) http://home.cc.umanitoba.ca/~psgendb/birchhomedir/doc/MeV/manual/ease.html application can be downloaded another version of DAVID with added features implemented with MeV framework. EASE is a programmable application that allows for rapid elucidation of genes derived from microarray and high throughput genomic data analysis and converts functional genomics from genes to "themes" (Ford, Xu, Gates, Jiang, & Ford, 2006; Hosack, Dennis, Sherman, Lane, & Lempicki, 2003).

7. Biological networks

DNA, RNA, proteins, and metabolites all play important roles in the molecular systems that control the cellular activities of life. Understanding the structure and interactions is critical for a variety of applications, including drug development and pathway detection. The structure and interactions of molecules are represented in a graph, consisting of nodes such as atoms and a set of edges that are bonds between atoms, expressing the connections between nodes. Biological networks provide a theoretical and accessible framework for studying, modeling, characterizing, and comprehending complex interactions between different components in a biological system. Network biology investigates the biological system's interactome, a collection of direct and indirect molecular interactions. A biological network, therefore, represents the molecular networking layout of a cell's information processing model. Biological networks are effective representations for visualizing and interpreting biomolecule functions and interactions (Zhang, Tian, & Zhang, 2014).

7.1 Analysis of PPI databases

The protein interaction databases collect PPI interactions directly from authors' contributions to the literature or from other databases. To access online the PPI data, many criteria are utilized to filter, annotate, and categorize the gathered interactions in databases, as well as provide a diversified user experience. In some cases, methods for data analysis are also provided. Some databases only include data on a single or a few species. Databases perform a significant function by evaluating PPI to identify gene association. PPIs discovered *in vitro* using current approaches have a large percentage of false positives or interactions that show inappropriate binding or are not detected *in vivo* (von Mering et al., 2002).

There are two types of PPI databases: primary databases and derived or integrated databases and meta-database (Patil, 2019), and web servers are enumerated (Tables 1 and 2).

Table 1 Collection of various databases that are used to interpret PPI networks, categorized based on primary, derived, and meta-PPI interaction databases.

Tool name	Description	URL	References
BindML+/ BindML	Predicting permanent and transient type PPI interface residues and binding site prediction by maximum likelihood	https://kiharalab.org/bindml/plus/info.html	Wei, La, and Kihara (2017)
PIC	Protein Interactions Calculator	http://pic.mbu.iisc.ernet.in/	Tina, Bhadra, and Srinivasan (2007)
MAPPIS	Multiple 3D Alignment of Protein-Protein Interfaces	http://bioinfo3d.cs.tau.ac.il/MAPPIS/	Shulman-Peleg, Shatsky, Nussinov, and Wolfson (2008)
PRISM	Hot spot prediction server for protein interfaces	http://prism.ccbb.ku.edu.tr/hotpoint	Tuncbag, Keskin, and Gursoy (2010)
PSOPIA	Prediction server for Protein-Protein interactions	https://mizuguchilab.org/PSOPIA/	Murakami and Mizuguchi (2014)
APID	Aigle Protein Interaction Data Analyzer	http://cicblade.dep.usal.es:8080/APID/init.action	Alonso-López et al. (2019)
ProtorP	Protein-protein interaction analysis server. It provides proteins associations in 3D structures	http://www.bioinformatics.sussex.ac.uk/protorp	Jones (2009)
iFrag	A protein-protein interface prediction server based on sequence fragments	http://sbi.imim.es/web/index.php/research/servers/iFrag?	Garcia-Garcia et al. (2017)
Interactome3D	Structural annotations and details of PPI networks	https://interactome3d.irbbarcelona.org/	Mosca, Céol, and Aloy (2013)
PRED_PPI	Predicting PPI	http://cic.scu.edu.cn/bioinformatics/predict_ppi	Guo et al. (2010)

Continued

Table 1 Collection of various databases that are used to interpret PPI networks, categorized based on primary, derived, and meta-PPI interaction databases.—cont'd

Tool name	Description	URL	References
PCRPi	Predicting hotspot residues in protein interfaces	http://www.bioinsilico.org/PCRPi/	Assi, Tanaka, Rabbitts, and Fernandez-Fuentes (2010)
PPCheck	Estimates the strength of the protein-protein interface	http://caps.ncbs.res.in/ppcheck/	Sukhwal and Sowdhamini (2015)
GraphPPIS	Structure-based PPI site prediction using deep graph convolutional network	https://github.com/biomed-AI/GraphPPIS, https://biomed.nscc-gz.cn/apps/GraphPPIS	Yuan, Chen, Zhao, Zhou, and Yang (2021, 2022)
cons-PPISP	consensus PPI site predictor	https://pipe.rcc.fsu.edu/ppisp.html	Chen and Zhou (2005)
meta-PPISP	metaserver for PPI site prediction	https://pipe.rcc.fsu.edu/meta-ppisp.html	Qin and Zhou (2007)
SPIDDER	Solvent accessibility Protein-Protein Interface identification and Recognition	http://sppider.cchmc.org/	Porollo and Meller (2007)
InterProSurf	Predict interacting residues on the monomeric protein surface	http://curie.utmb.edu/prosurf.html	Negi, Schein, Oezguen, Power, and Braun (2007)
TRI_tool	Transcriptional Regulation Interaction	https://www.vin.bg.ac.rs/180/tools/tfpred_info.php	Perovic et al. (2017)
HIVsemi	Predicting Interactions between HIV-1 and Human Proteins by semi-supervised multi-task learning	http://www.cs.cmu.edu/~qyj/HIVsemi/	Qi, Tastan, Carbonell, Weston, and Klein-Seetharaman (2010)

Table 1 Collection of various databases that are used to interpret PPI networks, categorized based on primary, derived, and meta-PPI interaction databases.—cont'd

Tool name	Description	URL	References
ChiPPI	Chimeric Protein-Protein Interaction Server	http://chippi.md.biu.ac.il/	Frenkel-Morgenstern et al. (2017)
HIPPIE	Human Integrated Protein-Protein Interaction rEference	http://cbdm-01.zdv.uni-mainz.de/~mschaefer/hippie/	Alonso-Betanzos et al. (2019); Schaefer et al. (2012)
InteroPorc	Automatic prediction tool to infer molecular PPI networks	http://biodev.extra.cea.fr/interoporc/	Michaut et al. (2008)
PrISE	Local structural similarity-based computational methods for predicting protein-protein interface residues	http://prise.cs.iastate.edu/	Jordan, EL-Manzalawy, Dobbs, and Honavar (2012)
InterPreTS	Interaction Prediction through Tertiary Structure of protein	http://www.russell.embl.de/interprets	Aloy and Russell (2003)
iLoops Server	Protein structural features (protein loops) to discover any pair of proteins interact or not	http://aleph.upf.edu/iLoopsServer/	Planas-Iglesias, Marin-Lopez, Bonet, Garcia-Garcia, and Oliva (2013)
Struct2Net	Structure bases computational predictions of PPIs	http://cb.csail.mit.edu/cb/struct2net/webserver/	Singh, Park, Xu, Hosur, and Berger (2010)
PRISM2.0	Protein Interaction by Structural Matching and models 3D complexes	http://cosbi.ku.edu.tr/prism/	Baspinar, Cukuroglu, Nussinov, Keskin, and Gursoy (2014)
SAPIN	Structural Analysis of Protein Interaction Networks	http://sapin.crg.es/	Yang, Campagna, et al. (2012)
GWIDD	Genome-Wide Docking Database is an integrated resource for structural PPIs on a genome-wide scale	http://gwidd.bioinformatics.ku.edu/	Kundrotas, Zhu, and Vakser (2012)

Continued

Table 1 Collection of various databases that are used to interpret PPI networks, categorized based on primary, derived, and meta-PPI interaction databases.—cont'd

Tool name	Description	URL	References
PPI Spider	interpretation of proteomics data in the context of PPI networks	http://www.bioprofiling.de/PPI_spider.html	Antonov, Dietmann, Rodchenkov, and Mewes (2009)
Path2PPI	Predicts PPI networks based on proteins belonging to a certain pathway or biological process from model organisms.	http://bioconductor.org/packages/release/bioc/html/Path2PPI.html	Philipp, Osiewacz, and Koch (2016)
POINet	protein interactome with sub-network analysis and hub prioritization that combines PPI and tissue-specific expression data from multiple resources.	http://poinet.bioinformatics.tw/	Lee et al. (2009)
RedNemo	Repeated Diffusion with Neighbourhood Modifications	http://webprs.khas.edu.tr/~cesim/RedNemo.tar.gz	Alkan and Erten (2017)
TissueNet V.2	Human Tissue associated PPIs of proteins	https://netbio.bgu.ac.il/tissuenet/	Basha et al. (2017)
HuRI	Human Reference Interactome	http://www.interactome-atlas.org/	Luck et al. (2020)
GeneMANIA	Gene interactions	https://genemania.org/	Warde-Farley et al. (2010)
DIP	database of interacting proteins	https://dip.doe-mbi.ucla.edu/dip/Main.cgi	Xenarios et al. (2000)
HPRD	The Human Protein Reference Database	http://www.hprd.org/	Goel, Harsha, Pandey, and Prasad (2012)
IntAct	molecular interactions from selected literature or collected directly from authors	https://www.ebi.ac.uk/intact/home	Orchard et al. (2014)

Table 1 Collection of various databases that are used to interpret PPI networks, categorized based on primary, derived, and meta-PPI interaction databases.—cont'd

Tool name	Description	URL	References
MINT	Molecular Interaction database	https://mint.bio.uniroma2.it/mint/	Ceol et al. (2010); Chatr-aryamontri et al. (2007)
MIPS	Mammalian Protein-Protein Interaction database	https://mips.helmholtz-muenchen.de/proj/ppi/	Pagel et al. (2005)
BIND	Biomolecular Interaction Network Database	http://bind.ca/	Bader (2001)
BioGRID	The biological general repository for interaction datasets	https://thebiogrid.org	Chatr-aryamontri et al. (2017)
MPIDB	Microbial Protein Interaction Database	https://www.re3data.org/repository/	Goll et al. (2008)
PINA	Protein Interaction Network Analysis	https://omics.bjcancer.org/pina/	Cowley et al. (2012), Du et al. (2021), p. 3; Wu et al. (2009)
MiMI	Michigan Molecular Interactions	https://dbgroup.eecs.umich.edu/project/mimi/	Jayapandian et al. (2007)
PIPS	Human protein-protein interactions predictor using naïve Bayesian classifier.	http://www.compbio.dundee.ac.uk/www-pips/.	McDowall, Scott, and Barton (2009)
OPHID	Online Predicted Human Interaction Database	http://ophid.utoronto.ca/ophidv2.204/	Brown and Jurisica (2005)
STRING	Known and predicted PPI interactions involve physical and functional associations	https://string-db.org/	Szklarczyk et al. (2015, 2017)
UniHI	The Unified Human Interactome database builts tissue-specific networks and inquire pathways	http://www.unihi.org/	Chaurasia et al. (2009); Kalathur et al. (2014)

Continued

Table 1 Collection of various databases that are used to interpret PPI networks, categorized based on primary, derived, and meta-PPI interaction databases.—cont'd

Tool name	Description	URL	References
HitPredict	It provides experimental information on sequence, structure, and functional annotations of interactions	http://www.hitpredict.org/	López, Nakai, and Patil (2015) and Patil, Nakai, and Nakamura (2011)
mentha	The interactome browser for integrated PPI networks	http://mentha.uniroma2.it/	Calderone, Castagnoli, and Cesareni (2013)
InnateDB	Experimentally curated molecular interaction by Gene/Protein	http://www.innatedb.com/	Breuer et al. (2013)
IID	Integrated Interaction Database	http://iid.ophid.utoronto.ca/	Kotlyar, Pastrello, Sheahan, and Jurisica (2016), Kotlyar, Pastrello, Malik, and Jurisica (2019), and Kotlyar et al. (2022)
HPIDB	Host-Pathogen Interaction DataBase	https://hpidb.igbb.msstate.edu/	Ammari, Gresham, McCarthy, and Nanduri (2016) and Kumar and Nanduri (2010)
Consensus PathDB	binary and complex PPI and biochemical pathways.	http://cpdb.molgen.mpg.de/	Kamburov, Stelzl, Lehrach, and Herwig (2013)
HAPPI	Human Annotated Protein-Protein Interaction	http://discovery.informatics.uab.edu/HAPPI/	Chen, Bardes, Aronow, and Jegga (2009) and Chen, Pandey, and Nguyen (2017)
MPIDB	Microbial Protein Interaction Database	https://www.re3data.org/repository/	Goll et al. (2008)
CORUM	mammalian protein complexes that have been carefully annotated	https://mips.helmholtz-muenchen.de/corum/	Ruepp et al. (2010)

Table 1 Collection of various databases that are used to interpret PPI networks, categorized based on primary, derived, and meta-PPI interaction databases.—cont'd

Tool name	Description	URL	References
MatrixDB	focuses on extracellular protein and polysaccharide interactions obtained from literature	http://matrixdb.univ-lyon1.fr/	Chautard, Fatoux-Ardore, Ballut, Thierry-Mieg, and Ricard-Blum (2011) and Launay, Salza, Multedo, Thierry-Mieg, and Ricard-Blum (2015)
Wiki-Pi	It is a web server of annotated human PPI that provides information about protein sequence, structure, and biological functions.	https://hagrid.dbmi.pitt.edu/wiki-pi/index.php/	Orii and Ganapathiraju (2012)
Network analyst	gene expression profiling & network visual analytics	https://www.networkanalyst.ca/	Xia, Gill, and Hancock (2015), Xia, Benner, and Hancock (2014) and Zhou et al. (2019)
Osprey	visualization of PPI and complex interaction networks	http://biodata.mshri.on.ca/osprey/	Breitkreutz, Stark, and Tyers (2003)
GeneNetwork	Gene network database that integrates details on genes, between genes from KEGG, BIND. Reactome, HPRD,	https://www.genenetwork.nl/	Franke et al. (2006)

7.1.1 Primary PPI databases

PPI primary databases include the collection of literature directly or through submissions from authors that contain small-scale studies or large-scale high throughput tests. The following are some of the primary databases for PPI:

CORUM is a database of mammalian protein complexes that have been carefully annotated (Ruepp et al., 2010), and it is freely available for users https://mips.helmholtz-muenchen.de/corum/. This database contains

Table 2 Webservers for PPI that are commonly used and available online are gathered which are user-friendly to researchers.

web servers	Description	URL	References
Coev2Net	Predicts, assesses, and boosts confidence in individual interactions inferred from high throughput experiment	http://cb.csail.mit.edu/cb/coev2net/	Hosur et al. (2012)
PrePPI	PPI prediction for human proteome, both structural and non-structural	https://bhapp.c2b2.columbia.edu/PrePPI/	Zhang, Petrey, Garzón, Deng, and Honig (2013)
iWRAP	Threading based PPI prediction and identifying assumed interacting residues	http://cb.csail.mit.edu/cb/iwrap/	Hosur et al. (2011)
SPRINT	Scoring Protein INTeraction is an ultrafast PPI prediction of the entire human interactome	https://github.com/lucian-ilie/SPRINT	Li and Ilie (2017)
SPPS	Sequence-based Protein Partner search	http://mdl.shsmu.edu.cn/SPPS/	Liu et al. (2012)
OrthoMCL DB	Genome-scale algorithm for grouping orthologous protein sequences	https://orthomcl.org/orthomcl/app	Chen and Rajewsky (2006)
P-POD	Princeton Protein Orthology Database provides a way to find and visualize orthologs to query sequences in eukaryotes	http://ppod.princeton.edu/	Heinicke et al. (2007)
MirrorTree Server	A graphical and interactive study of the co-evolution of two protein families and assess their interactions in a taxonomic context.	http://csbg.cnb.csic.es/mtserver/	Ochoa and Pazos (2010)

Table 2 Webservers for PPI that are commonly used and available online are gathered which are user-friendly to researchers.—cont'd

web servers	Description	URL	References
G-NEST	Gene Neighborhood scoring tool	https://github.com/dglemay/G-NEST	Lemay et al. (2012)
COGs	Phylogenetic classification of proteins encoded in complete genomes	https://www.ncbi.nlm.nih.gov/research/cog-project/	Tatusov, Koonin, and Lipman (1997)
TSEMA	Interactive prediction of mapping two families of interacting proteins.	https://tsema.bioinfo.cnio.es/	Izarzugaza et al. (2006)
OncoPPI	Translates genomics data into PPI networks for the therapeutic process	http://oncoppi.emory.edu/	Li and Ilie (2017)
MEGADOCK-WEB	It is an integrated database of Comprehensive PPI predictions that evaluates based on docking results with pathways and it is freely accessible.	http://www.bi.cs.titech.ac.jp/megadock-web/	Hayashi, Matsuzaki, Yanagisawa, Ohue, and Akiyama (2018)
interPro	A database for classifying protein family and protein domain	https://www.ebi.ac.uk/interpro/	Blum et al. (2021)

limited interactions collected from the literature and hand-curated by the author. Because these interactions are generated entirely from narrowly focused research, they are more reliable than those found in large databases, including many interactions from high-throughput experiments. CORUM data is often used in PPI quality reviews as a reference standard, or proven true, dataset.

DIP—The database of interacting proteins, compared to Intact and BioGRID, is the much smaller database that keeps track of curated PPI found in works of literature (Salwinski, 2004). It is easily accessible online.

IntAct is a freely accessible repository of molecular interactions derived from chosen literature or directly by scientists (Orchard et al., 2014). It is one of the few PPI databases containing a lot of annotations and proteins mapped to a lot of different database identifiers. The vast majority of interactions in IntAct are physiological associations established from high-throughput research, with humans being most prevalent.

BioGRID is a general biological repository for interaction datasets as a primary database that can be open accessed that collects and stores PPI from literature high-throughput research (Chatr-aryamontri et al., 2017). There are two types of interactions in BioGRID, physical and genetic interactions. It also keeps track of protein post-translational modifications and has the most curated interaction of yeast species. BioGRID data is publicly available through collaborator model organism databases and meta-databases, and it is immediately accessible in a variety of standard formats.

MINT Molecular Interaction Database is a public repository that collects and curates PPI data on a smaller scale than IntAct and BioGRID (Ceol et al., 2010). The complete dataset is freely accessible online in both interactive and batch modes via web-based interfaces and an FTP server. The IntAct database currently hosts this information. However, some interactions in MINT have not yet been moved to IntAct. HomoMINT, a database of human protein interactions derived from studies with ortholog proteins in model animals, has been merged into MINT (Chatr-aryamontri et al., 2007).

HPRD (The Human Protein Reference Database) is one of the most comprehensive databases on human physical PPI interaction (Prasad, Kandasamy, & Pandey, 2009). HPRD is a comprehensive database of experimentally validated human protein characteristics. The HPRD database includes PPI interactions, post-translational modifications, enzyme/substrate interconnections, disease associations, tissue expression, and human protein subcellular localization (Goel et al., 2012).

MatrixDB is a publicly available database that emphasizes extracellular protein and polysaccharide interactions found in the literature (Launay et al., 2015).

MIPS (Mammalian Protein-Protein Interaction) Database is a comprehensive online interface that contains high-quality PPI that has been carefully picked from the scientific literature by skilled curators. MIPS is also used as a reference dataset for binary PPIs (Pagel et al., 2005).

Wiki-Pi is a web database that allows users to search for interactions based on biomedical parameters like disease, pathway, drug, and biological function associations. It gives information on protein-centric, not PPI-centric (Orii & Ganapathiraju, 2012).

BIND (The Biomolecular Interaction Network Database) is a publicly available database that describes PPI, molecular complex, and pathways with high-throughput proteomics, including post-translational modifications (Bader, 2001). It has various curated experimental data and protein interaction types, protein-molecule interaction, and protein-nucleic interactions.

7.1.2 Derived PPI databases

PPI data from primary interaction databases are combined, filtered, and annotated in derived databases. Several of the derived databases also include some amount of curation. Some databases additionally include hypothesized interactions and interactions with 3D structural annotations. As a result, users can frequently obtain their interaction data from derived databases for additional study.

STRING is a functional relationship database (Szklarczyk et al., 2015, 2017). It is accessible to users via the internet. It is regarded as one of the most comprehensive derived PPI datasets. STRING data is currently accessible for network analysis via a plugin in the program Cytoscape (Cline et al., 2007). For reliability of all interactions in STRING, multiple pieces of evidence are used to score them.

HitPredict is available online for users. It is a centralized database of physical PPIs that have been experimentally identified and have confidence scores to reflect their reliability. The use of network and pathway analysis to examine genes and their interrelationships necessitates high-quality PPI data with a higher confidence level (López et al., 2015; Patil et al., 2011).

Mentha can be accessed and analyzed by the users. It is a web service that gathers and merges data from multiple databases. Its data comes from IMEx (International Molecular Exchange) consortium established PPI databases that have been manually curated. The scope and aim behind mentha differ from databases like STRING, which combine data acquired using text mining and prediction methods (Calderone et al., 2013). mentha includes eight interactomes (*Homo sapiens, Arabidopsis thaliana, Caenorhabditis elegans, Drosophila melanogaster, Escherichia coli K12, Mus musculus, Rattus norvegicus, and Saccharomyces cerevisiae*) as well as a global network that includes all organisms.

InnateDB is a publicly available database that contains the genes, proteins, experimentally proven interactions, and signaling pathways engaged in the innate immune system to microbial infection in humans, rats, and livestock. The database increases the innate immunity interactome coverage by merging known interactions and pathways from massive public databases with manually-curated data in a single repository. Physical and indirect interactions between proteins, DNA, and RNA are known. InnateDB does not assign a score to interactions (Breuer et al., 2013).

IID (Integrated Interaction Database) is a public database of PPI discovered and predicted in 18 species, including humans, 5 model organisms, and 12 domesticated species. It gathers experimentally discovered interactions from primary databases (Kotlyar et al., 2016). It also comprises PPIs that are particular to specific tissues. IID does not offer a reliability score for interactions.

HPIDB (Host-Pathogen Interaction DataBase) is a database that provides protein interactions between pathogens and their hosts (Ammari et al., 2016). PPIs are gathered without concern for validity.

Consensus PathDB is an online tool that includes binary and complex PPI and biochemical pathways. It also provides gene enrichment analysis. The PPIs are curated from literature with confidence scores (Kamburov et al., 2013). The protocol for Consensus PathDB and interpretations utilizing the database for the users are also given (Herwig, Hardt, Lienhard, & Kamburov, 2016).

HAPPI (Human Annotated Protein-Protein Interaction) is an extensive experimental and predicted PPI database derived from primary sources (Chen et al., 2017). The PPI is all handpicked and scored.

HIPPIE (Human Integrated Protein-Protein Interaction rEference) is a web-based application that provides confidence scores and functionally annotated PPI from several databases. Allows researchers to conduct network analysis on probable true PPI sets by building subnetworks surrounding the protein of interest at a certain confidence level (Alanis-Lobato, Andrade-Navarro, & Schaefer, 2017; Schaefer et al., 2012).

Interactome3D is a publicly available resource that aggregates 12,000 PPI from main databases used for three-dimensional annotation (Mosca et al., 2013) in eight model species ranging from *E.coli* to humans. It demonstrates the use of Interactome3D by structurally annotating the complement cascade pathway, which rationalizes a probable common mode of action proposed for numerous disease-causing mutations.

7.1.3 PPI meta-databases

An experimentally validated PPIs produced through consistent integration of numerous primary databases are included in PPI meta-databases listed in Table 1.

APID (Aigle Protein Interactomes Dataserver) is a public database that contains a comprehensive and curated collection of "protein interactomes" for over 1100 organisms, including 30 species with over 500 interactions. The PPIs are experimentally detected physical interactions from primary databases with experimental evidence (Alonso-López et al., 2019).

MPIDB (Microbial Protein Interaction Database) provides physical microbial PPIs. It provides well-curated interactions from literature and determines interactions for 191 bacterial strains that are detected experimentally and can be downloaded (Goll et al., 2008).

PINA (Protein Interaction Network Analysis) for PPI construction is a public database with curated interactions from literature (Wu et al., 2009). It provides filtering and analysis of biological and functional insights. Protein annotations were obtained on the go in the original version of PINA using the UniProt web application, which proved clunky for building an extensive PPI network with hundreds of interacting proteins. We stored the UniProt annotations further into the PINA annotation database in PINA v2.0, dramatically improving query speed for extensive networks. The PINA web tool has also been upgraded to make them easier to use and responsive.

7.1.4 PPI prediction database

Prediction databases integrate experimentally proven PPIs with anticipated PPIs produced using various computational methodologies. Different methods in predicting PPI partners are discussed, as well as recent advances in the prediction of specific domains that mediate PPI were studied (Shoemaker & Panchenko, 2007), illustrated in Table 1.

MIMI (Michigan Molecular Interactions) is a publicly assessed database that allows the users to retrieve PPI over 3.7 million, 1288 pathways, 3.5 million genes, and 19,2 million molecules from different databases at once, giving complementary and conflicting information (Jayapandian et al., 2007; Sharma, Virk, Khurana, & Kaur, 2018).

PIPs (https://www.compbio.dundee.ac.uk/www-pips/) is an online database of predicted human PPI prediction. The predictions are made using the Bayesian method to calculate the score of PPI. It collects information from expressions, domains, orthologs, post-translation modifications, and subcellular location (McDowall et al., 2009).

OPHID (Online Predicted Human Interaction Database) is an online database of predicted human protein interactions. It integrates predictions from *S. cerevisiae, Caenorhabditis elegans, Drosophila melanogaster*, and *M. musculus* with literature-derived human PPI from BIND, HPRD, and

MINT. It was organized to enhance the human interactome data and give a repository for known experimentally derived human PPIs (Brown & Jurisica, 2005).

UniHI (Unified Human Interactome) http://www.unihi.org/ is an online database that allows users to search for, analyze, and access human molecular interaction networks. UniHI aims to reduce data duplication while incorporating the advantages of single databases in terms of meticulous curation and annotation of PPIs. UniHI 7 (Kalathur et al., 2014), the latest database update, has about ~350,000 molecular interactions between genes, proteins, and drugs and various additional data categories like gene expression and functional annotations.

8. Pathway databases and analysis

Analyzing pathways for DEGs is critical for identifying disease mechanisms and gene-to-gene interactions. Biological pathway analysis, also known as "functional enrichment analysis," considers both upregulated and downregulated gene expressions. Pathway analyses are categorized based on Omics data, systems biology, functional biological units, informatics units, and pathway databases (Amaral & Ottino, 2004; García-Campos, Espinal-Enríquez, & Hernández-Lemus, 2015; Huang, 2000; Mitrea et al., 2013).

The pathway databases retrieve information and structure data from various sources. The experimental proofs are curated from literature by specialized scientists. A website known as Pathguide (http://pathguide.org/) (Bader, Cary, & Sander, 2006) consists of pathway databases listing around 702 biological pathway databases, segregated into nine categories based on the interaction types, almost more than 2.5 million pathways in total. The pathway data can be retrieved utilizing the web tool Pathway Common (Cerami et al., 2011) version 12 (http://www.pathwaycommons.org/), consisting of 5772 pathways 2,424,055 interactions, and 22 databases. Some standard formats used in pathway databases are SBML (Systems Biology Markup Language) (Hucka et al., 2003) and SBGN (Systems Biology Graphical Notation) (Novère et al., 2009), and BioPAX (Demir et al., 2010).

The pathway analysis consists of three main phases: input, analysis, and output. The input phase consists of two datasets, the high throughput biological datasets and pathway data from the pathway database, which are further used to analyze. The analysis phase consists of statistical and computational approaches based on the datasets above. The output phase consists of visualization by heatmaps and results, represented in the ranked list of pathways by confidence level (*P*-value or q-value) and direct acyclic graph (GO category). For example, stand-alone software, web-based applications, and programming packages for pathway analysis are listed in Table S3 in Supplementary Information in the online version at https://doi.org/10.1016/bs.apcsb.2022.05.002.

The pathway analysis is classified into first-generation over-representation analysis, such as GSEA (Subramanian et al., 2005), second-generation functional class scoring method, and third-generation pathway topology-based method. The pathway-topology-based method got increased attention for its interaction analysis integrated within the pathway-topology analysis. Mitrea et al. studied pathway-topology methods and interpreted almost 22 pathway-topology analyses or third-generation analyses for signaling or metabolic pathways (Mitrea et al., 2013). Some topology-based pathway commercial tools are available, such as Pathway-Guide (and MetaCore). A survey done by Huang et al. listed 68 bioinformatic enrichment tools for large gene sets and divided them into three classes based on algorithms: Singular enrichment analysis, gene set enrichment analysis and modular enrichment analysis (Huang, Sherman, & Lempicki, 2009). The visualization for the pathway and enrichment analysis are done using certain websites such as ImageGP (ImageGP | ImageGP (ehbio.com)), Interactivenn (http://www.interactivenn.net/), Venny 2.1 (https://bioinfogp.cnb.csic.es/tools/venny/), UCSC genome browser (https://genome.ucsc.edu/), SangerBox tool (http://sangerbox.com/Tool), MExpress (https://mexpress.be/), shinyGEO (http://gdancik.github.io/shinyGEO/), ggplot2 (https://github.com/tidyverse/ggplot2.git), JMP Genomics from SAS (https://www.jmp.com/en_in/software/genomics-data-analysis-software/), Partek Genomic suite (https://www.partek.com/) and Systems biological Graphical Notation (SBGN) (https://sbgn.github.io/).

9. Gene and transcriptional regulatory networks

Gene regulation refers to the processes that control gene expression and its significance. The interactions between genes and their regulatory molecules, mainly transcription factors, are represented as maps of potential transcriptional regulation, referred to as gene regulatory networks (GRNs). The microarray data provide sufficient qualitative and quantitative information on the interaction between transcription factors and DNA. Since microarrays do not directly demonstrate functional interactions between genes, the binding of transcription factors (TFs) and their binding sites (TFBSs) to specific gene promoters regulate transcription and modify RNA synthesis, demonstrating the first degree of functional interaction. GRN should be evaluated to comprehend the cell's working mechanism and the complexities of gene interactions. Numerous ways for deriving GRN have been developed. Modeling and implying gene regulatory networks are critical challenges in systems biology.

The discovery of functional aspects of TFBS on a whole-genome level is challenging for studying genome science and gene regulation. Various bioinformatics approaches include partial differential equations, ordinary differential equations, Bayesian networks (Zou & Conzen, 2005), and Boolean networks (Martin, Zhang, Martino, & Faulon, 2007). In GEP, transcription factors serve as key molecular switches. Identifying and characterizing transcription factor binding sites is critical for annotating genomic regulatory areas and understanding gene-regulatory networks (Pritsker, Liu, Beer, & Tavazoie, 2004). Several groups have taken advantage of this problem and identified probable binding sites in the promoter regions of co-expressed genes (Chowdhary, Bajic, Dong, Wong, & Liu, 2010). The commonly used tools for TFBS are listed in Table 3.

9.1 Prediction of miRNA targets using bioinformatic approach

The discovery of miRNAs as gene expression regulators has been described as a paradigm shift in biology and medicine. Validated targets are the best biomarker(s) for determining the efficacy of a miRNA replica or inhibitor for scientists developing miRNA therapies. miRNA targets have been discovered using three different methods: bioinformatic target prediction, biochemical isolation of miRNA/mRNA complexes, and transcriptomic/proteomic analysis in the presence and absence of miRNA (Hammond, 2015). Various methods have been designed, the most prominent of which being TargetScan (https://www.targetscan.org/vert_80/) (McGeary et al., 2019), miRanda (https://bioweb.pasteur.fr/packages/pack@miRanda@3.3a) (Enright et al., 2003), and PicTar (https://pictar.mdc-berlin.de/) (Chen & Rajewsky, 2006) and the other standard miRNA network prediction tools are given in Table 3. Quantitative proteomic analysis directly analyzes the impact of a miRNA on protein synthesis. It may indicate the actual target list, but it is also more technically challenging; thus, simplified transcriptome analysis can be performed since most miRNA targets have diminished mRNA steady-state levels (Eichhorn et al., 2014). Various mathematical algorithms

Table 3 Various databases used in gene regulatory network analysis, transcriptomic analysis, metabolic network, and drug-drug interaction.

GRN database names and category	Weblinks
(GRAND) Gene Regulatory Network Database	https://grand.networkmedicine.org/
RegNetwork: Regulatory Network Repository	http://www.regnetworkweb.org/
GRNdb (Gene Regulatory networks)	http://www.grndb.com/
PAINT (A Promoter Analysis and Interaction Network Generation Tool)	http://www.dbi.tju.edu/dbi/tools/paint/
iRegulon	http://iregulon.aertslab.org/
rSNPBase 3.1	http://rsnp3.psych.ac.cn/
BRANE Clust	https://www-syscom.univ-mlv.fr/~pirayre/Codes-GRN-BRANE-clust.html
GeneReg	https://cran-archive.r-project.org/web/checks/2019/2019-01-26_check_results_GeneReg.html
Star Net 2	http://vanburenlab.medicine.tamhsc.edu/starnet2.html
STITCH v 5.0 (Search Tool for Interactions of chemicals)	http://stitch.embl.de/
BNArray	http://www.cls.zju.edu.cn/binfo/BNArray/
FastNCA (Fast Network component Analysis)	http://www.eee.hku.hk/~cqchang/FastNCA.htm
RegulonDB	https://regulondb.ccg.unam.mx/
Onco-Regulon	http://www.scfbio-iitd.res.in/software/onco/NavSite/index.htm
Gene Network	http://genenetwork.sbl.bc.sinica/
DREAM 5	http://gnw.sourceforge.net/dreamchallenge.html
Human MicroRNA Disease Database (HMDD)	http://www.cuilab.cn/hmdd
MiREDiBase	https://ncrnaome.osumc.edu/miredibase

Continued

Table 3 Various databases used in gene regulatory network analysis, transcriptomic analysis, metabolic network, and drug-drug interaction.—cont'd

GRN database names and category	Weblinks
Prediction of miRNA binding sites	
Circular RNA Interactome	https://circinteractome.nia.nih.gov/
Cancer specific Circular RNA	http://geneyun.net/CSCD2/
Prediction of miRNA Target sites	
Target scan	https://www.targetscan.org/vert_80/
miRTarBase	https://mirtarbase.cuhk.edu.cn/~miRTarBase/miRTarBase_2022/php/index.php
miRDB	http://mirdb.org/
DIANA - microT-CDS	http://diana.imis.athena-innovation.gr/DianaTools/index.php?r=MicroT_CDS/index
TransmiR	http://cmbi.bjmu.edu.cn/transmir
miRGTF-net	https://github.com/s-a-nersisyan/miRGTF-net
Transcriptional factor binding	
JASPAR	https://jaspar.genereg.net/
TRANSFAC	https://genexplain.com/transfac/
oPOSSUM	http://www.cisreg.ca/oPOSSUM/
MATCH	http://gene-regulation.com/pub/programs.html#match
Whole Exome rVISTA	https://genome.lbl.gov/vista/index.shtml
TFSCAN	http://mobyle.pasteur.fr/cgi-bin/portal.py?form=tfscan
TFSEARCH	https://www.cbrc.jp/research/db/TFSEARCH.html/
Con Tra	http://bioit.dmbr.ugent.be/ConTra/index.php
TransFind	http://transfind.sys-bio.net/
CircuitsDB	http://biocluster.di.unito.it/circuits/

and techniques give quantitative insights into the role of miRNA in gene expression regulation (Lai, Wolkenhauer, & Vera, 2016). The regulatory network of miRNAs comprises interactions with mRNA, protein, and other non-coding RNA. Because of the network's complexity, miRNAs can perform various biological functions while also having high therapeutic potential. The inhibition or replacement therapeutic strategies, for example, can be carried out based on the overexpression or downregulation of miRNAs in cancerous cells (Si, Shen, Zheng, & Fan, 2019).

10. Metabolic networks

Even though metabolic networks have been a widely used analytic approach in bioinformatics for some years, the topic of how to effectively design them from experimental evidence remains unresolved. A technique for constructing metabolic networks from microarray data that use an upgraded version of the KEGG database (Manyam, Birerdinc, & Baranova, 2015), with the resulting networks being directed graphs generated from the KEGG database's link of array probes and enzymes. The flow of catalytic processes throughout the connected pathways determines the network's directionality. This approach is new in that the resultant-directed metabolic networks are generated without the bias created by mapping metabolite data onto a pathway's pre-existing structure (Easton, 2013). The collection of all chemical events that occur within a live creature to preserve life is depicted graphically by metabolic networks (Muzio, O'Bray, & Borgwardt, 2021). Metabolic networks are often divided into pathways, which comprise a sequence of chemical events that fulfill a specific metabolic function. Wang and his colleagues created a lipid metabolism network and discovered lipid sub-networks and clusters that are thought to play a role in cardiovascular disease pathogenesis (Wang et al., 2018). A comprehensive multi-omics integrative strategy based on metabolic networks and expression data has been developed to identify cardiomyopathy-related core genes (Rong et al., 2022). The list of databases used in metabolic network analysis is summarized in Table 3.

11. Drug-drug interaction networks

The interaction between different drugs is modeled by DDI (Hu & Hayton, 2011). The drugs are displayed as nodes and provide interactions as edges in DDI. DDI studies help gain knowledge on drug interactions and are widely used in polypharmacy research (Zhang, Zhang, (Derek), Zhao, & Huang, 2009). Due to the challenges posed by laboratory screening of DDI, computational approaches to predicting DDI have captured the interest of researchers and bioinformaticists from multiple disciplines. There are various Graph Convolutional Network (Defferrard, Bresson, & Vandergheynst, 2017) (GCN—a subset of Graphical neural networks (GNN) (Wu et al., 2021) that performs graph-structured data into image convolution by deep learning methods of network analysis in the graph domain). There are several innovative methods, such as Decagon (Zitnik, Agrawal, & Leskovec, 2018) (http://snap.stanford.edu/decagon/) for connecting protein and drugs via different types of edges, according to the type of interactions, GAMENet (Graph Augmented Memory Networks for Recommending Medication Combination) (https://github.com/sjy1203/GAMENet) (Shang, Xiao, Ma, Li, & Sun, 2019), Compnet (Wang et al., 2019), DCDB (Drug Combination Database toward Network medicine and combination therapy) (http://drugcombdb.denglab.org/) (Deng, Zou, Zhang, & Liu, 2018), Hypergraph Link Prediction (HLP) (Vaida & Purcell, 2019) is an approach that encodes the multilinked structure of graphs. The other commonly used databases are given in Table 3.

12. Clustering, classification, and comparison of microarray data

Microarray datasets are usually considerable, and diagnostical precision is affected by the extent of variables. Therefore, narrowing down the datasets for top genes distinguished between normal and tumor groups is beneficial. Expression level in the gene is measured by the intensity ratio between two sample experiments that consider genes with above fold change to be significant. Nevertheless, this requires occasional replicates of arrays to be performed. Thus, several statistical techniques have been developed (Bhandari, Walambe, Kotecha, & Khare, 2022). The techniques for analyzing expression microarray pipeline, interpreting the analysis of missing values, normalization, feature selection, and learning/microarray data analysis (Fig. 4). The imputations of missing value have various approaches such as Global (SVDimpute (Troyanskaya et al., 2001), BPCA (Oba et al., 2003)), Local approach [(KNN impute) (Troyanskaya et al., 2001), SKNNimpute (Kim, Kim, & Yi, 2004), IKNNimpute (Brás & Menezes, 2007), GMCimpute (Ouyang, Welsh, & Georgopoulos, 2004), etc.)], Hybrid (LinCmb (Jörnsten, Wang, Welsh, & Ouyang, 2005)) and

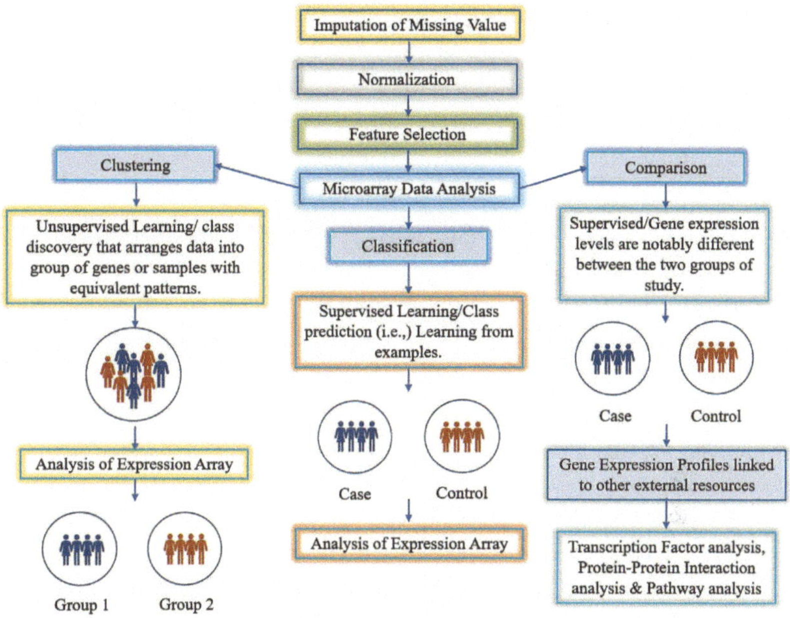

Fig. 4 The general pipeline for various approaches and classification of microarray gene expression data in case and control sample groups.

Knowledge-assisted approach [(GOimpute (Tuikkala, Elo, Nevalainen, & Aittokallio, 2006), HAIimpute (Xiang et al., 2008), WeNNI (Johansson & Häkkinen, 2006), iMISS (Hu, Li, Waterman, & Zhou, 2006) and metaMISS (Jörnsten, Ouyang, & Wang, 2007)]. Feature selection involves filter, wrapper, embedded, and Hybrid approaches. An open-source tool available such as geneCBR (http://www.genecbr.org/) can perform clustering, comparison, and classification in cancer research based on microarray data.

The learning and analysis of the microarray method are divided into clustering, comparison, and classifications which are discussed below:

The expression levels in genes are remarkably different between normal and tumor groups, which is the first chore of in-depth microarray analysis (Wei, Li, & Bumgarner, 2004). To identify the changes in the gene expression or relative expression from microarray data using univariate statistical methods including nonparametric t-test (Troyanskaya, Garber, Brown, Botstein, & Altman, 2002), significance analysis of microarray (SAM) (Tusher et al., 2001), F-statistic method (Fs) test (Cui, Hwang, Qiu, Blades, & Churchill, 2005), Bayesian method for microarray analysis

(Baldi & Long, 2001), Analysis of Variance (ANOVA) used for more complex microarray datasets (Kerr, Martin, & Churchill, 2000). Different software utilizes the methods mentioned above and is free to analyze the gene expressions.

Unsupervised learning or pattern analysis that arranges data into a group of genes or samples with matching patterns is called clustering analysis in gene expression data. It is of three types (i) Hierarchical clustering involves a similar gene expression profile, (ii) k-means/non-Hierarchical clustering that partitions expression data into k (number of clusters) groups, and (iii) self-organizing map (SOMS) are also a non-hierarchical neural network-based clustering method that is also related to k-mean cluster approach (Brazma & Vilo, 2000; Svrakic, 2003). To find similar genes with correlated expression patterns, the Principle Component Analysis (PCA) technique is implemented with k-means or SOM clustering by the user to specify the multiple clusters (Quackenbush, 2001). The PCA technique given its mathematical equation https://users.ics.aalto.fi/jhollmen/dippa/node30.html and Multidimensional scaling approaches with its mathematical description given in http://www.statsoft.com/textbook/stmulsca.html are used in differential expression of genes to give more robust data. The singular Value Decomposition approach applies in the arrangement of the microarray data in a rectangle matrix that consists of rows of genes and columns of experiments constitute based on the equation Singular Value Decomposition (SVD) tutorial (mit.edu) (Weeraratna & Taub, 2007). Various software is available for cluster analysis of microarray data (Table 4).

Supervised learning or class prediction or discriminant analysis, i.e., Learning from examples known to classify microarray data analysis using several classifiers such as support vector machine (SVM), probabilistic neural network (PNN), k-nearest neighbor (KNN), artificial neural network (ANN), radial basis function network (RBFN), naïve Bayes (NB) and logistic regression (LR) and decision tree that splits samples into nodes repeatedly until termination (Kumar & Rath, 2016; Selvaraj & Natarajan, 2011; Zhang, Yu, Singer, & Xiong, 2001). The classification of microarray data analysis steps are provided in Table 5.

13. Methodology and results

13.1 Microarray data processing

The public repository GEO, managed by the NCBI, was utilized to access the GSE74602 data, which contains gene expression profiles by the array.

Table 4 Microarray analysis based on cluster method using various tools that are enumerated with website links.

Softwares for cluster analysis	Weblinks
Cluster and Treeview	http://jtreeview.sourceforge.net/
dChip	http://biosun1.harvard.edu/~cli/dchip.exe
MeV	https://mev.tm4.org/
MAGIC TOOLS	http://www.bio.davidson.edu/projects/magic/magic.html
CAGED (Cluster Analysis of Gene Expression Analysis	https://open.bu.edu/handle/2144/1290?show=full
WebGimm	http://eh3.uc.edu/gimm/webgimm/index.html
VISDA (Visual and Statistical Analyser)	https://www.cbil.ece.vt.edu/caBIG-VISDA.htm
Genesis web	http://genome.tugraz.at/genesisserver/genesisserver_description.shtml
JustClust	https://paccanarolab.org/justclust
OC-Cluster analysis program	http://www.compbio.dundee.ac.uk/downloads/oc/
GIMM (Gaussian Infinite Mixture Models)	http://eh3.uc.edu/gimm/gimm/index.html

Table 5 Microarray analysis based on classification method using various tools that are enumerated with website links.

Software's for classification	Web links
Weka (MACHINE LEARNING SOFTWARE)	https://www.cs.waikato.ac.nz/ml/weka/
SAS	https://support.sas.com/documentation/onlinedoc/faw/5.2/dclug.pdf
SVMlight	https://www.cs.cornell.edu/people/tj/svm_light/
LIBSVM	https://www.csie.ntu.edu.tw/~cjlin/libsvmtools/

Datasets from diverse experiments are deposited in this database, allowing users to download gene expression profiles collected in GEO. GSE74602 contains sixty samples obtained from the platform GPL6104 Illumina human Ref-8 v2.0 expression bead chip. Thirty tumor and normal colorectal samples. To find the GEO datasets for gene expression profiles, the keywords "colorectal cancer," "Microarray," and "Homo sapiens." The colorectal adenocarcinoma dataset was analyzed using several computational tools and the validation of critical genes (Fig. 5). Differentially regulated genes were analyzed in patients with normal non-cancerous tissue samples and primary tumor tissues from the Singapore tissue bank.

13.2 Data pre-processing and screening of DEGs in colorectal adenocarcinoma

GEO2R web tools were used to identify DEGs between two or more sample groups in the GEO series (Barrett et al., 2013). Analysis of DEGs was done using edgeR and DESeq2, and interactions were studied using STRING, with enrichment analysis by DAVID, for GO and KEGG, and PPI was performed, providing a significant molecular signature for idiopathetic pulmonary fibrosis (Mishra et al., 2021).

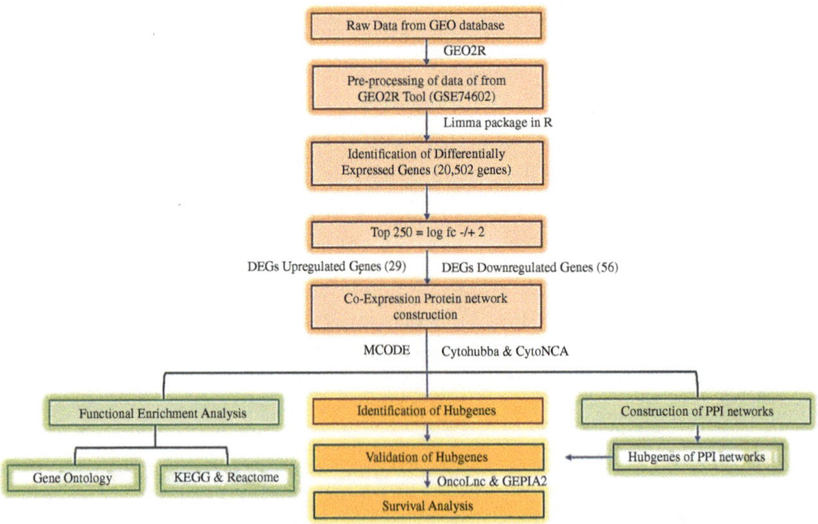

Fig. 5 The workflow for the analysis of colorectal adenocarcinoma cancer microarray dataset (GSE74602) from GEO database the DEG's are identified and analyzed using various tools and computational methods to validate the key gene expressions.

It displays a table of genes arranged by significance, which may be seen using the GEO Profile data (Table S4 in Supplementary Information in the online version at https://doi.org/10.1016/bs.apcsb.2022.05.002). GEO data analysis with established GEO query and limma R packages was performed between colorectal tumor and control samples. The Benjamin-Hochberg false discovery rate method (Hardcastle, 2016) was utilized for adjusting the raw *P*-values into FDR on an excellent balance of microarray data. The adjusted *P*-value <0.05 and log2 fold change $|\log2FC| \geq 2$ were kept as thresholds for DEG identification.

The gene expression profile of GSE74602 with 30 CRC samples and 30 control samples was downloaded from the GEO database. The platform GPL6104 Illumina human Ref-8 v2.0 expression bead chip was used in this study of COAD samples. The GEO2R tool was used to identify the top 250 DEG is between healthy and COAD tumor samples with log 2 fold change ≥ 2 and *P*-value 0.05 that was evaluated based on inbuilt R/Bioconductor version 3.2.3 and limma package version 3.26.8. The top 250 DEGs were selected with 63 upregulated and 181 downregulated genes, indicated in the volcano plot and heat map (Fig. 6A and B).

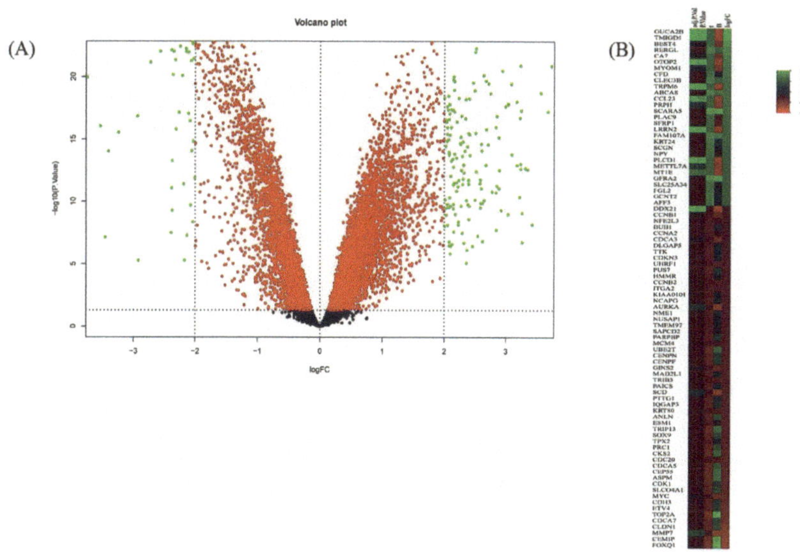

Fig. 6 (A) volcano plot for top 250 DEG's from colorectal adenocarcinoma sample dataset (GSE74602) with red dots representing x-axis, log FC; y-axis $-10\log$ *P*-value with *P* values <0.05 in red dots with logFC ≥ 2 and logFC ≤ 2, the significant genes are represented in green dots (B) Microarray heat map for top 250 DEG's with upregulated genes in red and downregulated genes in green.

13.3 GO and functional enrichment analysis of DEGs

The biological databases, such as DAVID https://david.ncifcrf.gov/, for GO functional annotation such as biological process and molecular function (Dennis et al., 2003). KEGG pathway (KEGG PATHWAY Database (genome.jp)) and Reactome pathway analysis (https://reactome.org/PathwayBrowser/) (Croft et al., 2011; Manyam et al., 2015) to identify the role of DEGs in colorectal cancer at the functional stage, GO, KEGG, and Reactome pathway analysis was performed utilizing DAVID database for high throughput gene analysis (Dennis et al., 2003). The microarray data analysis for identifying DEGs, hub genes, and possible pathway analysis by using DAVID, GO, and KEGG in detecting molecular pathogenesis of ovarian cancer was studied (Kumar, Kumar, Siva, Doss, & Zayed, 2019). Gene expression profiling study on systemic lupus erythematosus performed PPI networks using STRING and Cytoscape by functional annotation using ClueGO or CluePedia, and GO using Metacore was done (Udhaya Kumar et al., 2020). The protein interaction was performed using STRING and Cytoscape, Enrichment by GO, and molecular pathway enrichment analysis using Cluego and KEGG, REACTOME for familial hypercholesterolemia was analyzed (Udhaya Kumar et al., 2020).

The background parameter is the human genome with P-value <0.05 and threshold >2 as statistically significant. The other databases that were used in this study were Enrichr (Fig. S1A in Supplementary Information in the online version at https://doi.org/10.1016/bs.apcsb.2022.05.002), Funrich (Fig. S1B in Supplementary Information in the online version at https://doi.org/10.1016/bs.apcsb.2022.05.002), and Metascape (Fig. S1C in Supplementary Information in the online version at https://doi.org/10.1016/bs.apcsb.2022.05.002) for functional enrichment analysis.

GO, KEGG, and Reactome pathway analysis results for 85 DEGs of COAD samples were processed using the DAVID online tool. The top 10 significant findings of the GO biological process from the DEGs were cell division (GO:0051301), mitotic spindle organization (GO:0007052), mitotic spindle assembly checkpoint (GO:0007094), mitotic cell cycle phase transition (GO:0044772), mitotic chromosome condensation (GO:0007076), anaphase-promoting complex-dependent catabolic process (GO:0031175), chromosome segregation (GO:0007059), regulation of G2/M transition of the mitotic cell cycle (GO:0010389), negative regulation of gene expression (GO:0010629) and positive regulation of cell proliferation (GO:0008284) (Fig. 7A). The top 10 significant GO molecular

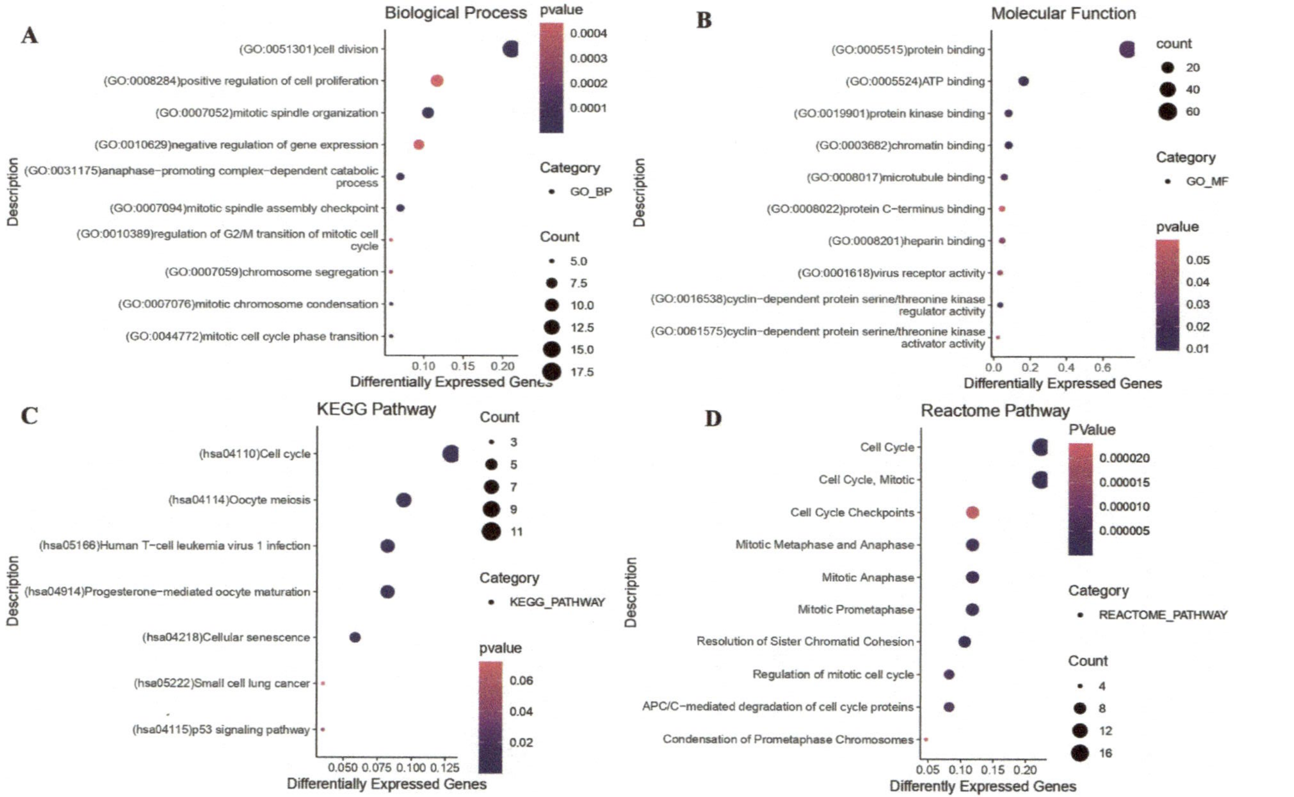

Fig. 7 DAVID functional annotations of GO, KEGG, and reactome pathway enrichment analysis of familiar DEG's (A) the top 10 significant GO terms in biological process with the color of dot representing *P*-value, size of the dot indicates count of the genes. (B) The top 10 significant GO terms in molecular function with the color of dot representing *P*-value, size of the dot indicate the count of the genes (C) top 7 pathways for KEGG (Kyoto encyclopedia of genes and genomes) enrichment analysis of DEG's with fold change >2 as cut-off value (D) the top 10 Reactome biological pathway analysis performed for DEGs.

function from DEGs of COAD samples carried out was protein binding (GO:0005515), ATP binding (GO:0005524), protein kinase binding (GO:0019901), chromatin binding (GO:0003682), microtubule-binding (GO:0008017), protein C-terminus binding (GO:0008022), heparin-binding (GO:0008201), virus receptor activity (GO:0001618), cyclin-dependent protein serine/threonine kinase regulator activity (GO:0016538), and cyclin-dependent protein serine/threonine kinase activator activity (GO:0061575) (Fig. 7B). KEGG pathway analysis showed pathways in cell cycle (hsa04110), oocyte meiosis (hsa04114), progesterone-mediated oocyte maturation (hsa04914), T-cell leukemia virus 1 infection in humans (hsa05166), cellular senescence (hsa04218), p53 signaling pathway (hsa04115) and in small cell lung cancer (hsa05222) (Fig. 7C) were top 7 significant KEGG pathways involved in DEG. The top 10 significant reactome biological pathway analyses were performed for DEGs showed mitotic cell cycle (R-HSA-69278), cell cycle (R-HSA-1640170), resolution of sister chromatid cohesion (R-HSA-2500257), prometaphase of mitosis (R-HSA-68877), anaphase of mitosis (R-HSA-68882), mitotic metaphase and anaphase (R-HSA-2555396), regulation of mitotic cell cycle (R-HSA-453276), cell cycle proteins are degraded by APC (R-HSA-174143), checkpoints in the cell cycle (R-HSA-69620) and prometaphase chromosome condensation (R-HSA-2514853) (Fig. 7D).

13.4 Protein-protein network interaction analysis and module analysis

The search tool for the Retrieval of Interacting Genes (STRING) version 11.5 tool evaluated the PPI interaction. Various studies based on network pharmacology were done for treating COVID-19 by gene-gene interaction in determining enrichment and pathway analysis method (Thirumal Kumar et al., 2022). A study on drug resistance using gene expression profiles for discovering the processes and pathways in tuberculosis (Udhaya Kumar et al., 2021). Analysis between emphysema and alpha-1 antitrypsin deficiency-related emphysema DEG were screened and interactions performed by STRING and Cytoscape, DEG's backbone network method were identified by network analyzer in Cytoscape, enrichment analysis incorporating ClueGO and GeneMANIA server with correlational analysis for quick gene-gene interaction and secondary gene relationships were studied (Udhaya Kumar et al., 2021).

The DEGs were incorporated in a string to map their interaction with a medium score >0.4. further to analyze and visualize the hub genes.

To evaluate the hub genes in the network cytoHubba (Chin et al., 2014) and CytoNCA (Tang, Li, Wang, Pan, & Wu, 2015), Cytoscape plugins were applied to detect hub genes in the PPI network. The Molecular Complex Detection (MCODE) (Bader & Hogue, 2003) plugin in Cytoscape was used with a maximum depth of 100, node score cut to identify the highly interconnected clusters in the network of PPI off as 0.2, and K-core is 2 as threshold parameters.

The interactive details amid DEGs and the protein network were obtained utilizing an online STRING database with 85 genes, including 29 upregulated and 56 downregulated genes. The combined medium score of >0.4 was considered significant, and interactions were visualized using Cytoscape software v.3.8.2. with 59 nodes and 624 edges of PPI (Fig. 8). The MCODE plugin in Cytoscape software was performed for the top three clusters from the PPI network for closely interlinked areas (Fig. 9). Module one of the PPI networks had 34 nodes and 543 edges with a score of 32.909,

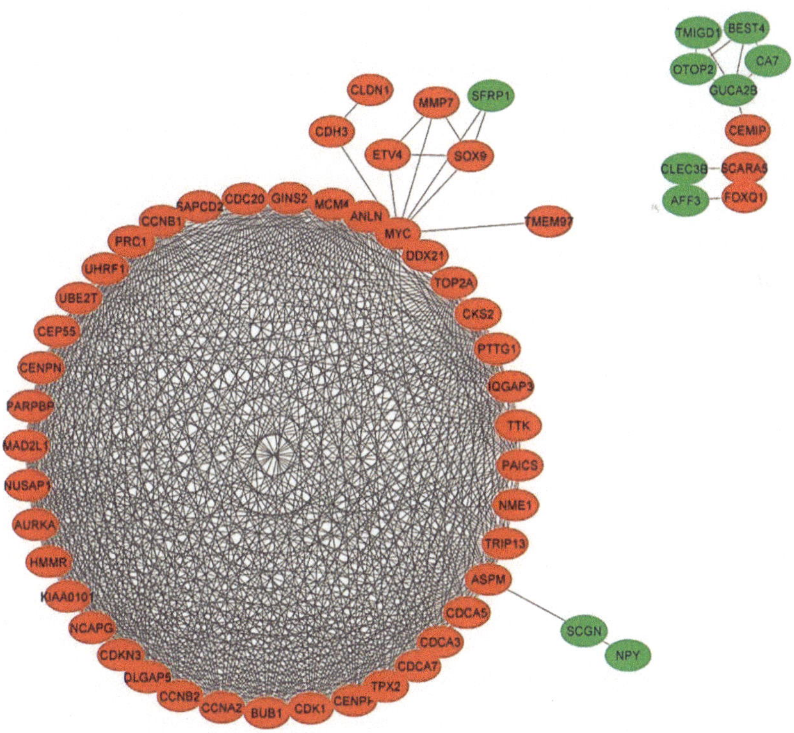

Fig. 8 A PPI network of top 250 DEGs of COAD, the color red represents downregulating genes and the color green as upregulating genes generated using Cytoscape.

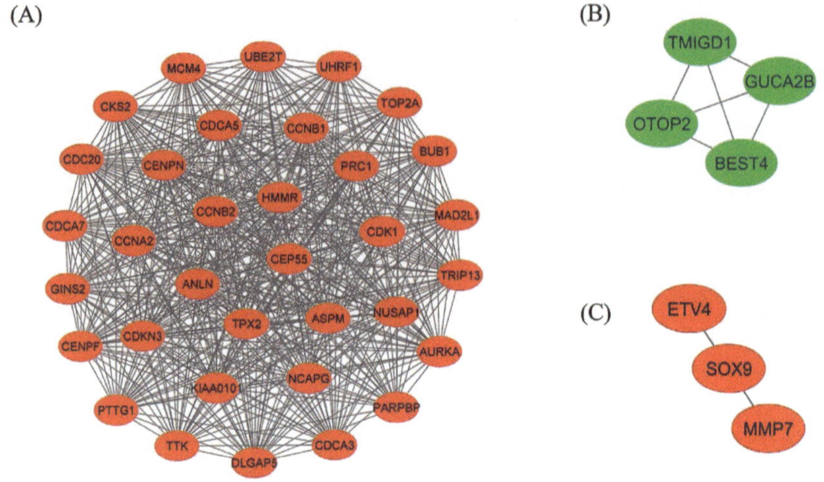

Fig. 9 Top three clusters identified from PPI network with Cytoscape using MCODE plugin with cluster score more than three (A) MCODE cluster 1, (B) MCODE cluster 2, and (C) MCODE cluster 3.

Table 6 The MCODE plugin in Cytoscape software was performed for the top three clusters from the PPI network.

Cluster	Score (density*#nodes)	Nodes	Edges	Node IDs
1	32.909	34	543	*CENPF, PRC1, TPX2, HMMR, CCNA2, CCNB2, MAD2L1, NUSAP1, CDC20, TTK, CDKN3, PTTG1, CDCA5, UHRF1, KIAA0101, CKS2, UBE2T, CENPN, PARPBP, CDCA7, CDCA3, MCM4, CCNB1, NCAPG, DLGAP5, GINS2, AURKA, TRIP13, ANLN, BUB1, CEP55, ASPM, CDK1, TOP2A*
2	4	4	6	*TMIGD1, BEST4, GUCA2B, OTOP2*
3	3	3	3	*MMP7, SOX9, ETV4*

module two consisted of 4 nodes and 6 edges with a score of 4, and module three with 3 nodes and 3 edges of score 3 (Table 6). The cytoHubba (Degree, Eigenvector, Betweenness, and Closeness) showed 57 genes (Fig. 10A), and CytoNCA showed 53 genes (MCC, EPC, MNC, and Stress) (Fig. 10B) plugins in the Cytoscape database were used for centrality

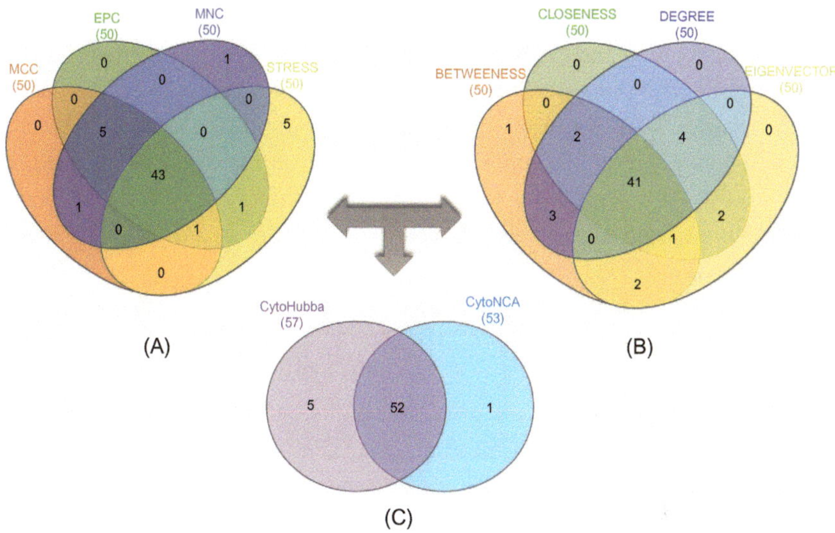

Fig. 10 Cytoscape database with CytoHubba and CytoNCA plugin was used for centrality analysis of four calculation methods for dominant nodes or edges (A) The cytoHubba (Degree, Eigenvector, Betweenness, and Closeness) showed 57 genes. (B) The CytoNCA showed 53 genes (MCC, EPC, MNC, and Stress). (C) A Venn diagram of 52 overlapping genes calculating centrality analysis between Cytohubba (EPC, MCC, MNC, STRESS) and CytoNCA (Betweenness, closeness, Degree, Eigenvector) of which 12 genes from our list belonged to MCODE module 1.

analysis of four calculation methods for dominant nodes or edges in PPI network based on top 50 genes (Table S5 in Supplementary Information in the online version at https://doi.org/10.1016/bs.apcsb.2022.05.002). A Venn diagram was drawn using InteractiVenn (Heberle, Meirelles, da Silva, Telles, & Minghim, 2015) (http://www.interactivenn.net/) to identify hub genes that are similar in all groups, of which 52 genes were overlapped (Fig. 10C), among them, 12 essential genes were identified based on genes involved in MCODE module 1 (Tables S5 and S6 in Supplementary Information in the online version at https://doi.org/10.1016/bs.apcsb.2022.05.002).

13.5 Expression level analysis of hub genes and survival analysis

To evaluate the hub gene data, the online tool OncoLnc (http://www.oncolnc.org/) (Anaya, 2016) was used to detect the survival data with mRNA expression levels linked to TCGA data and GEPIA 2 (http://gepia2.cancer-pku.cn/) for overall survival analysis and disease-free survival rate analysis. Pearson correlation coefficient was examined between the hub genes using TCGA and GTEx data available in GEPIA 2 (Tang et al., 2017) to compare mRNA expression levels between colorectal tumor and paired normal tissues. The Kaplan-Meier analysis was done for COAD patients' samples divided into 50 upper and 50 lower percentiles and submitted. As a result, Cox regression analysis for each hub gene with a log-rank *P*-value is evaluated, and the Kaplan plot appears in pdf format and is downloaded.

The survival analysis was performed in OncoLnc and GEPIA for the 12 hub genes (*CENPF, CDKN3, CCCNB2, AURKA, TOP2A, KIAA0101/PCLAF, CDK1, ASPM, BUB1, CCNBI, CDC20, and CCNA2*) of which the cox-regression results based on log-rank *P*-value <0.05 genes between both OncoLnc and GEPIA tools revealed *TOP2A, CCNBI,* and *CCNA2* as more expressed genes in survival analysis of COAD samples (Fig. 11A and B). The disease-free survival analysis and the stage plot for pathogenicity of the 12 hub genes were analyzed and interpreted in supplementary Figs. S2 and S3 in Supplementary Information in the online version at https://doi.org/10.1016/bs.apcsb.2022.05.002.

14. Limitations

The translation of microarray analysis into everyday practice is still aspirational. This is mainly due to the lack of comparison and interference of results drawn from each study and the technical variability in the collection of sample and preparation, platform types, and data analysis. In addition, small-scale study due to reduced patient samples poses a significant challenge, particularly when selecting appropriate data that will be useful for precise clinical application. Thus, overcoming these restrictions should be the primary goal in the future, to translate hypothetical data into usable clinical information. According to Weir et al., systematic knowledge of the molecular basis of every type of cancer involves at least three steps: comprehensive characterization of repeated genomic aberrations, elaboration of their biological role in cancer pathogenesis, and analysis of their utility for diagnostics, prognostics, and therapeutics. To uncover novel therapeutic targets and expand the number of treatment choices for CRC, an entire, complete proteome and genomic characterization, including modern technologies such as microarrays, comparative genomic hybridization, SNPs arrays, and miRNA, is required (Weir et al., 2007).

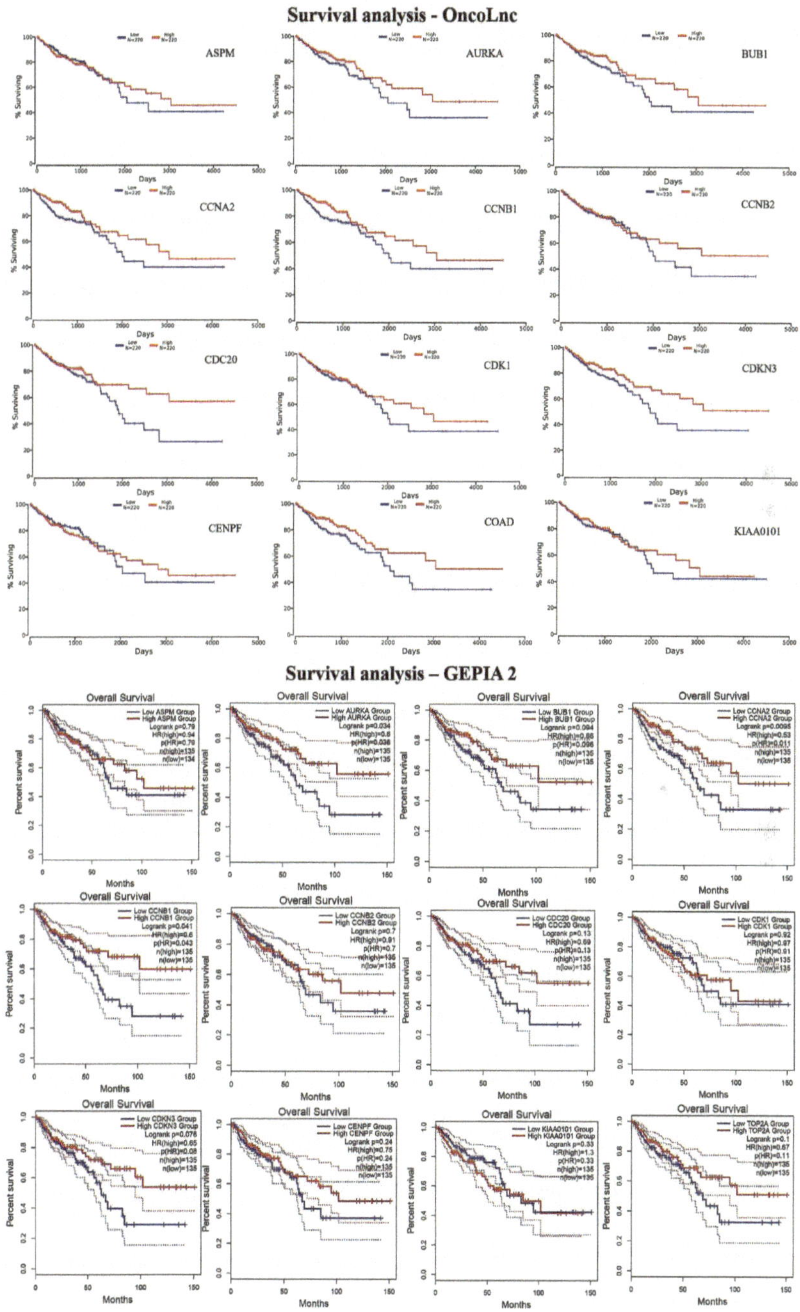

Fig. 11 Survival analysis of DEGs of 12 hub genes (*CENPF, CDKN3, CCCNB2, AURKA, TOP2A, KIAA0101/PCLAF, CDK1, ASPM, BUB1, CCNBI, CDC20, and CCNA2*). The red line represents samples highly expressed and the clue line indicates low expressed genes. (A) Oncolnc used for cox regression results by Kaplan plot for 12 hub genes in COAD based on *P*-value <0.05. (B) Survival analysis of 12 hub genes in colorectal adenocarcinoma based on TCGA data in GEPIA 2.

15. Conclusion and future perspectives

In conclusion, network analysis in bioinformatics using microarrays platforms has emerged to identify DEGs or pathways using biological knowledge and expression data relevant to cancer. The gene interaction network and its accompanying analyses are thought to play a role in searching for high-scoring subnetworks. Gene expression profiling employing microarray technology has offered a wide range of potential therapeutic applications in cancer, from examining tumor formation mechanisms to discovering gene signatures with predictive and prognostic value and finding effective tumor targets. Many studies have been conducted in recent years demonstrating that a whole-genome approach might radically transform the management strategies of cancer patients, but their translation into clinical research is still a long way off and requires more progress in cancer research.

We reviewed current work on the analysis of microarray gene expression data using traditional and statistical computational methods such as retrieval of cancer-related datasets from gene expression repositories, usage of various online webservers, software, and programs for analyzing microarray data, functional annotation, interaction network, and pathway analysis for further validation and visualization. A general pipeline for microarray gene expression data analysis was given with various online bioinformatic webservers and computational tools for colon adenocarcinoma patient sample datasets and interpreted.

In public health, the US Food and Drug Administration (FDA) leads to the growth of novel technologies such as microarrays, which may improve and speed assessments of pharmaceutical product safety and effectiveness. Within the next 10 years, it is believed that microarrays will be widely used in screening, evaluation, and the growth of the pharmaceutical sector, as well as diagnosis and monitoring of complex diseases, and guiding therapeutic approaches methods. Microarray-based diagnostics have the potential to be used in clinical practice by the FDA. According to the FDA, these new technologies have the potential to provide more effective methods of medical treatment, disease prevention, and management.

With the revolutionary introduction of high-throughput transcriptomics approaches, such as microarray-based on hybridization and RNA-sequencing (RNA-seq) technology [next-generation sequencing (NGS)-based], gene expression studies have become a routine practice in this modern scientific time. Since 2005, NGS technology has proliferated, eclipsing the microarray technique due to its speed, sensitivity, and throughput benefits.

For gene expression studies, co-expression network analysis, alternative splicing analysis, specific gene simple sequence repeats analysis, and pathway analysis, the RNA-seq technique is currently widely employed. Despite all new technologies and medical advances, cancer remains a mystery and is perceived as one of the most life-threatening malignant diseases. The introduction of RNA sequencing as a high-throughput method of transcriptome analysis has made it possible to analyze diseases and their hereditary origins at the molecular level. Single-cell RNA sequencing, or scRNA-seq, is being used to investigate the heterogeneity of single malignant cells to determine the source of malignancies. Studying miRNAs in the context of pharmacotranscriptomics is essential. However, studies should ideally include miRNA and mRNA assessments in the same individuals to better understand the molecular processes involved in drug response that include modified miRNA-dependent expression regulation. However, because most studies looked at either mRNAs or miRNAs, this approach has not yet been widely used.

Microarray studies allow for the analysis of genetic modifications that influence variations in drug pharmacokinetics or pharmacodynamics, resulting in a comprehensive understanding of gene activities, regulation, and interaction. Our *in silico* method identified hub genes involved in the pathogenesis of colon adenocarcinoma and the prognosis of patients that can employ therapeutic targets for this malignancy. Network analysis methods have also been utilized, including serum, regulatory T-cells, and lymphocyte data with a specific target for the immune system and cancer research using human microarray datasets.

Acknowledgments

The authors would like to thank the management of the Vellore Institute of Technology, India, and Qatar University, Qatar, for providing the necessary research facilities and encouragement to carry out this study.

Author contributions

H.C.R. contributed to manuscript writing, interpretation of datasets, and bioinformatic analysis. U.K.S., S.Y., G.P.D.C., and H.Z. supervised the entire study and were involved in study design, the acquisition, analysis, and manuscript drafting. The manuscript was reviewed and approved by all the authors.

Conflict of interest

The authors declare that the study done has no conflict of interest.

Funding

There was no funding provided for this study.

References

Abulencia, J. P., Gaspard, R., Healy, Z. R., Gaarde, W. A., Quackenbush, J., & Konstantopoulos, K. (2003). Shear-induced cyclooxygenase-2 via a JNK2/c-Jun-dependent pathway regulates prostaglandin receptor expression in chondrocytic cells. *The Journal of Biological Chemistry, 278*(31), 28388–28394. https://doi.org/10.1074/jbc.m301378200.

Agapito, G. (2019). Computer tools to analyze microarray data. In V. Bolón-Canedo, & A. Alonso-Betanzos (Eds.), *Vol. 1986. Microarray bioinformatics* (pp. 267–282). Springer New York. https://doi.org/10.1007/978-1-4939-9442-7_13.

Agapito, G., Cannataro, M., Guzzi, P. H., Marozzo, F., Talia, D., & Trunfio, P. (2013). Cloud4SNP: Distributed analysis of SNP microarray data on the cloud. In *Proceedings of the International Conference on Bioinformatics, Computational Biology and Biomedical Informatics, September* (pp. 468–475).

Agapito, G., Guzzi, P. H., & Cannataro, M. (2015). DMET-Miner: Efficient discovery of association rules from pharmacogenomic data. *Journal of Biomedical Informatics, 56*, 273–283. https://doi.org/10.1016/j.jbi.2015.06.005.

Alanis-Lobato, G., Andrade-Navarro, M. A., & Schaefer, M. H. (2017). HIPPIE v2.0: Enhancing meaningfulness and reliability of protein–protein interaction networks. *Nucleic Acids Research, 45*(D1), D408–D414. https://doi.org/10.1093/nar/gkw985.

Alizadeh, A. A., Eisen, M. B., Davis, R. E., Ma, C., Lossos, I. S., Rosenwald, A., et al. (2000). Distinct types of diffuse large B-cell lymphoma identified by gene expression profiling. *Nature, 403*(6769), 503–511. https://doi.org/10.1038/35000501.

Alkan, F., & Erten, C. (2017). RedNemo: Topology-based PPI network reconstruction via repeated diffusion with neighborhood modifications. *Bioinformatics (Oxford, England), 33*(4), 537–544. https://doi.org/10.1093/bioinformatics/btw655.

Alonso-Betanzos, A., Bolón-Canedo, V., Morán-Fernández, L., & Sánchez-Maroño, N. (2019). A review of microarray datasets: Where to find them and specific characteristics. In V. Bolón-Canedo, & A. Alonso-Betanzos (Eds.), *Vol. 1986. Microarray bioinformatics* (pp. 65–85). Springer New York. https://doi.org/10.1007/978-1-4939-9442-7_4.

Alonso-López, D., Campos-Laborie, F. J., Gutiérrez, M. A., Lambourne, L., Calderwood, M. A., Vidal, M., et al. (2019). APID database: Redefining protein–protein interaction experimental evidences and binary interactomes. *Database, 2019*. https://doi.org/10.1093/database/baz005.

Aloy, P., & Russell, R. B. (2003). InterPreTS: Protein interaction prediction through tertiary structure. *Bioinformatics (Oxford, England), 19*(1), 161–162. https://doi.org/10.1093/bioinformatics/19.1.161.

Amaral, L. A. N., & Ottino, J. M. (2004). Complex networks. *The European Physical Journal B, 38*(2), 147–162. https://doi.org/10.1140/epjb/e2004-00110-5.

Amaratunga, D., & Cabrera, J. (2004). *Exploration and analysis of DNA microarray and protein array data*. Wiley. https://doi.org/10.1002/9780470317129.

Ammari, M. G., Gresham, C. R., McCarthy, F. M., & Nanduri, B. (2016). HPIDB 2.0: A curated database for host–pathogen interactions. *Database, 2016*, baw103. https://doi.org/10.1093/database/baw103.

Anaya, J. (2016). OncoLnc: Linking TCGA survival data to mRNAs, miRNAs, and lncRNAs. *PeerJ Computer Science, 2*, e67. https://doi.org/10.7717/peerj-cs.67.

Antonov, A. V., Dietmann, S., Rodchenkov, I., & Mewes, H. W. (2009). PPI spider: A tool for the interpretation of proteomics data in the context of protein-protein interaction networks. *Proteomics*, *9*(10), 2740–2749. https://doi.org/10.1002/pmic.200800612.

Assi, S. A., Tanaka, T., Rabbitts, T. H., & Fernandez-Fuentes, N. (2010). PCRPi: Presaging critical residues in protein interfaces, a new computational tool to chart hot spots in protein interfaces. *Nucleic Acids Research*, *38*(6). https://doi.org/10.1093/nar/gkp1158, e86.

Bader, G. D. (2001). BIND—The biomolecular interaction network database. *Nucleic Acids Research*, *29*(1), 242–245. https://doi.org/10.1093/nar/29.1.242.

Bader, G. D., Cary, M. P., & Sander, C. (2006). Pathguide: A pathway resource list. *Nucleic Acids Research*, *34*(Suppl. 1), D504–D506. https://doi.org/10.1093/nar/gkj126.

Bader, G. D., & Hogue, C. W. (2003). An automated method for finding molecular complexes in large protein interaction networks. *BMC Bioinformatics*, *4*, 2. https://doi.org/10.1186/1471-2105-4-2.

Baldi, P., & Long, A. D. (2001). A Bayesian framework for the analysis of microarray expression data: Regularized t-test and statistical inferences of gene changes. *Bioinformatics*, *17*(6), 509–519. https://doi.org/10.1093/bioinformatics/17.6.509.

Ball, C. A. (2004). The Stanford Microarray Database accommodates additional microarray platforms and data formats. *Nucleic Acids Research*, *33*(Database issue), D580–D582. https://doi.org/10.1093/nar/gki006.

Bansal, A. K. (2005). Bioinformatics in microbial biotechnology—A mini review. *Microbial Cell Factories*, *4*(1), 19. https://doi.org/10.1186/1475-2859-4-19.

Barrett, T. (2004). NCBI GEO: Mining millions of expression profiles—Database and tools. *Nucleic Acids Research*, *33*(Database issue), D562–D566. https://doi.org/10.1093/nar/gki022.

Barrett, T., Wilhite, S. E., Ledoux, P., Evangelista, C., Kim, I. F., Tomashevsky, M., et al. (2013). NCBI GEO: Archive for functional genomics data sets—Update. *Nucleic Acids Research*, *41*(Database issue), D991–D995. https://doi.org/10.1093/nar/gks1193.

Basha, O., Barshir, R., Sharon, M., Lerman, E., Kirson, B. F., Hekselman, I., et al. (2017). The TissueNet v.2 database: A quantitative view of protein-protein interactions across human tissues. *Nucleic Acids Research*, *45*(D1), D427–D431. https://doi.org/10.1093/nar/gkw1088.

Baspinar, A., Cukuroglu, E., Nussinov, R., Keskin, O., & Gursoy, A. (2014). PRISM: A web server and repository for prediction of protein–protein interactions and modeling their 3D complexes. *Nucleic Acids Research*, *42*(W1), W285–W289. https://doi.org/10.1093/nar/gku397.

Bayat, A. (2002). Science, medicine, and the future: Bioinformatics. *BMJ*, *324*(7344), 1018–1022. https://doi.org/10.1136/bmj.324.7344.1018.

Bednár, M. (2000). DNA microarray technology and application. *Medical Science Monitor: International Medical Journal of Experimental and Clinical Research*, *6*(4), 796–800.

Bhandari, N., Walambe, R., Kotecha, K., & Khare, S. P. (2022). A comprehensive survey on computational learning methods for analysis of gene expression data in genomics. *ArXiv*, abs/2202.02958.

Blanchard, A. P., Kaiser, R. J., & Hood, L. E. (1996). High-density oligonucleotide arrays. *Biosensors and Bioelectronics*, *11*(6–7), 687–690. https://doi.org/10.1016/0956-5663(96)83302-1.

Blin, A., Cissé, I., & Bockelmann, U. (2015). Electronic hybridization detection in microarray format and DNA genotyping. *Scientific Reports*, *4*(1), 4194. https://doi.org/10.1038/srep04194.

Blum, M., Chang, H.-Y., Chuguransky, S., Grego, T., Kandasaamy, S., Mitchell, A., et al. (2021). The InterPro protein families and domains database: 20 years on. *Nucleic Acids Research*, *49*(D1), D344–D354. https://doi.org/10.1093/nar/gkaa977.

Bodrossy, L., & Sessitsch, A. (2004). Oligonucleotide microarrays in microbial diagnostics. *Current Opinion in Microbiology*, 7(3), 245–254. https://doi.org/10.1016/j.mib.2004.04.005.

Bolstad, B. M., Irizarry, R. A., Astrand, M., & Speed, T. P. (2003). A comparison of normalization methods for high density oligonucleotide array data based on variance and bias. *Bioinformatics (Oxford, England)*, *19*(2), 185–193. https://doi.org/10.1093/bioinformatics/19.2.185.

Brás, L. P., & Menezes, J. C. (2007). Improving cluster-based missing value estimation of DNA microarray data. *Biomolecular Engineering*, *24*(2), 273–282. https://doi.org/10.1016/j.bioeng.2007.04.003.

Brazma, A. (2003). ArrayExpress—A public repository for microarray gene expression data at the EBI. *Nucleic Acids Research*, *31*(1), 68–71. https://doi.org/10.1093/nar/gkg091.

Brazma, A., & Vilo, J. (2000). Gene expression data analysis. *FEBS Letters*, *480*(1), 17–24. https://doi.org/10.1016/S0014-5793(00)01772-5.

Breitkreutz, B.-J., Stark, C., & Tyers, M. (2003). Osprey: A network visualization system. *Genome Biology*, *4*(3), R22. https://doi.org/10.1186/gb-2003-4-3-r22.

Breuer, K., Foroushani, A. K., Laird, M. R., Chen, C., Sribnaia, A., Lo, R., et al. (2013). InnateDB: Systems biology of innate immunity and beyond—Recent updates and continuing curation. *Nucleic Acids Research*, *41*(D1), D1228–D1233. https://doi.org/10.1093/nar/gks1147.

Brown, P. O., & Botstein, D. (1999). Exploring the new world of the genome with DNA microarrays. *Nature Genetics*, *21*(1 Suppl), 33–37. https://doi.org/10.1038/4462.

Brown, K. R., & Jurisica, I. (2005). Online predicted human interaction database. *Bioinformatics*, *21*(9), 2076–2082. https://doi.org/10.1093/bioinformatics/bti273.

Bulyk, M. L., Huang, X., Choo, Y., & Church, G. M. (2001). Exploring the DNA-binding specificities of zinc fingers with DNA microarrays. *Proceedings of the National Academy of Sciences*, *98*(13), 7158–7163. https://doi.org/10.1073/pnas.111163698.

Butte, A. (2002). The use and analysis of microarray data. *Nature Reviews Drug Discovery*, *1*(12), 951–960. https://doi.org/10.1038/nrd961.

Calderone, A., Castagnoli, L., & Cesareni, G. (2013). mentha: A resource for browsing integrated protein-interaction networks. *Nature Methods*, *10*(8), 690–691. https://doi.org/10.1038/nmeth.2561.

Call, D. R., Borucki, M. K., & Besser, T. E. (2003). Mixed-genome microarrays reveal multiple serotype and lineage-specific differences among strains of Listeria monocytogenes. *Journal of Clinical Microbiology*, *41*(2), 632–639. https://doi.org/10.1128/JCM.41.2.632-639.2003.

Ceol, A., Chatr Aryamontri, A., Licata, L., Peluso, D., Briganti, L., Perfetto, L., et al. (2010). MINT, the molecular interaction database: 2009 update. *Nucleic Acids Research*, *38*(Suppl. 1), D532–D539. https://doi.org/10.1093/nar/gkp983.

Cerami, E. G., Gross, B. E., Demir, E., Rodchenkov, I., Babur, Ö., Anwar, N., et al. (2011). Pathway commons, a web resource for biological pathway data. *Nucleic Acids Research*, *39*(Suppl. 1), D685–D690. https://doi.org/10.1093/nar/gkq1039.

Chandler, D. P., Alferov, O., Chernov, B., Daly, D. S., Golova, J., Perov, A., et al. (2006). Diagnostic oligonucleotide microarray fingerprinting of *Bacillus* Isolates. *Journal of Clinical Microbiology*, *44*(1), 244–250. https://doi.org/10.1128/JCM.44.1.244-250.2006.

Chatr-aryamontri, A., Ceol, A., Palazzi, L. M., Nardelli, G., Schneider, M. V., Castagnoli, L., et al. (2007). MINT: The Molecular INTeraction database. *Nucleic Acids Research*, *35*(Database), D572–D574. https://doi.org/10.1093/nar/gkl950.

Chatr-aryamontri, A., Oughtred, R., Boucher, L., Rust, J., Chang, C., Kolas, N. K., et al. (2017). The BioGRID interaction database: 2017 update. *Nucleic Acids Research*, *45*(D1), D369–D379. https://doi.org/10.1093/nar/gkw1102.

Chaurasia, G., Malhotra, S., Russ, J., Schnoegl, S., Hänig, C., Wanker, E. E., et al. (2009). UniHI 4: New tools for query, analysis and visualization of the human protein–protein

interactome. *Nucleic Acids Research*, *37*(Suppl. 1), D657–D660. https://doi.org/10.1093/nar/gkn841.
Chautard, E., Fatoux-Ardore, M., Ballut, L., Thierry-Mieg, N., & Ricard-Blum, S. (2011). MatrixDB, the extracellular matrix interaction database. *Nucleic Acids Research*, *39*(Database issue), D235–D240. https://doi.org/10.1093/nar/gkq830.
Chee, M., Yang, R., Hubbell, E., Berno, A., Huang, X. C., Stern, D., et al. (1996). Accessing genetic information with high-density DNA arrays. *Science (New York, N.Y.)*, *274*(5287), 610–614. https://doi.org/10.1126/science.274.5287.610.
Chen, J., Bardes, E. E., Aronow, B. J., & Jegga, A. G. (2009). ToppGene Suite for gene list enrichment analysis and candidate gene prioritization. *Nucleic Acids Research*, *37*(Web Server issue), W305–W311. https://doi.org/10.1093/nar/gkp427.
Chen, J. Y., Pandey, R., & Nguyen, T. M. (2017). HAPPI-2: A comprehensive and high-quality map of human annotated and predicted protein interactions. *BMC Genomics*, *18*(1), 182. https://doi.org/10.1186/s12864-017-3512-1.
Chen, H., & Zhou, H.-X. (2005). Prediction of interface residues in protein–protein complexes by a consensus neural network method: Test against NMR data. *Proteins: Structure, Function, and Bioinformatics*, *61*(1), 21–35. https://doi.org/10.1002/prot.20514.
Chen, K., & Rajewsky, N. (2006). Natural selection on human microRNA binding sites inferred from SNP data. *Nature Genetics*, *38*(12), 1452–1456. https://doi.org/10.1038/ng1910.
Chicurel, M. E., & Dalma-Weiszhausz, D. D. (2002). Microarrays in pharmacogenomics—Advances and future promise. *Pharmacogenomics*, *3*(5), 589–601. https://doi.org/10.1517/14622416.3.5.589.
Chin, C. H., Chen, S. H., Wu, H. H., Ho, C. W., Ko, M. T., & Lin, C. Y. (2014). cytoHubba: Identifying hub objects and sub-networks from complex interactome. *BMC Systems Biology*, *8*(Suppl 4), S11. https://doi.org/10.1186/1752-0509-8-S4-S11.
Chowdhary, R., Bajic, V. B., Dong, D., Wong, L., & Liu, J. S. (2010). Genome-wide analysis of regions similar to promoters of histone genes. *BMC Systems Biology*, *4*(Suppl. 1), S4. https://doi.org/10.1186/1752-0509-4-S1-S4.
Cline, M. S., Smoot, M., Cerami, E., Kuchinsky, A., Landys, N., Workman, C., et al. (2007). Integration of biological networks and gene expression data using Cytoscape. *Nature Protocols*, *2*(10), 2366–2382. https://doi.org/10.1038/nprot.2007.324.
Colantuoni, C., Henry, G., Zeger, S., & Pevsner, J. (2002). SNOMAD (Standardization and NOrmalization of MicroArray Data): Web-accessible gene expression data analysis. *Bioinformatics*, *18*(11), 1540–1541. https://doi.org/10.1093/bioinformatics/18.11.1540.
Colaprico, A., Silva, T. C., Olsen, C., Garofano, L., Cava, C., Garolini, D., et al. (2016). TCGAbiolinks: An R/bioconductor package for integrative analysis of TCGA data. *Nucleic Acids Research*, *44*(8), 1–11. https://doi.org/10.1093/nar/gkv1507.
Conway, T., & Gary, K. (2003). Microarray expression profiling: Capturing a genome-wide portrait of the transcriptome. *Molecular Microbiology*, *47*(4), 879–889. https://doi.org/10.1046/j.1365-2958.2003.03338.x.
Cooper, C. S. (2001). Applications of microarray technology in breast cancer research. *Breast Cancer Research*, *3*(3), 158. https://doi.org/10.1186/bcr291.
Cowley, M. J., Pinese, M., Kassahn, K. S., Waddell, N., Pearson, J. V., Grimmond, S. M., et al. (2012). PINA v2.0: Mining interactome modules. *Nucleic Acids Research*, *40*(D1), D862–D865. https://doi.org/10.1093/nar/gkr967.
Croft, D., O'Kelly, G., Wu, G., Haw, R., Gillespie, M., Matthews, L., et al. (2011). Reactome: A database of reactions, pathways and biological processes. *Nucleic Acids Research*, *39*(Suppl. 1), D691–D697. https://doi.org/10.1093/nar/gkq1018.
Crowther, D. J. (2002). Applications of microarrays in the pharmaceutical industry. *Current Opinion in Pharmacology*, *2*(5), 551–554. https://doi.org/10.1016/s1471-4892(02)00207-2.

Cui, X., Hwang, J. T. G., Qiu, J., Blades, N. J., & Churchill, G. A. (2005). Improved statistical tests for differential gene expression by shrinking variance components estimates. *Biostatistics*, *6*(1), 59–75. https://doi.org/10.1093/biostatistics/kxh018.

D'Angelo, G., Di Rienzo, T., & Ojetti, V. (2014). Microarray analysis in gastric cancer: A review. *World Journal of Gastroenterology: WJG*, *20*(34), 11972–11976. https://doi.org/10.3748/wjg.v20.i34.11972.

Dahlquist, K. D., Salomonis, N., Vranizan, K., Lawlor, S. C., & Conklin, B. R. (2002). GenMAPP, a new tool for viewing and analyzing microarray data on biological pathways. *Nature Genetics*, *31*(1), 19–20. https://doi.org/10.1038/ng0502-19.

Davis, L. (2012). *Basic Methods in Molecular Biology*. Elsevier.

Defferrard, M., Bresson, X., & Vandergheynst, P. (2017). Convolutional neural networks on graphs with fast localized spectral filtering. *ArXiv*, *29*. ArXiv:1606.09375 [Cs, Stat] http://arxiv.org/abs/1606.09375.

Demir, E., Cary, M. P., Paley, S., Fukuda, K., Lemer, C., Vastrik, I., et al. (2010). The BioPAX community standard for pathway data sharing. *Nature Biotechnology*, *28*(9), 935–942. https://doi.org/10.1038/nbt.1666.

Deng, L., Zou, B., Zhang, W., & Liu, H. (2018). *DrugCombDB: A comprehensive database of drug combinations toward network medicine and combination therapy* (p. 9). bioRxiv.

Dennis, G., Sherman, B. T., Hosack, D. A., Yang, J., Gao, W., Lane, H. C., et al. (2003). DAVID: Database for annotation, visualization, and integrated discovery. *Genome Biology*, *4*(9), R60. https://doi.org/10.1186/gb-2003-4-9-r60.

DeRisi, J., Penland, L., Brown, P. O., Bittner, M. L., Meltzer, P. S., Ray, M., et al. (1996). Use of a cDNA microarray to analyse gene expression patterns in human cancer. *Nature Genetics*, *14*(4), 457–460. https://doi.org/10.1038/ng1296-457.

Desany, B., & Zhang, Z. (2004). Bioinformatics and cancer target discovery. *Drug Discovery Today*, *9*(18), 795–802. https://doi.org/10.1016/S1359-6446(04)03224-6.

Doniger, S. W., Salomonis, N., Dahlquist, K. D., Vranizan, K., Lawlor, S. C., & Conklin, B. R. (2003). MAPPFinder: Using gene ontology and GenMAPP to create a global gene-expression profile from microarray data. *Genome Biology*, *4*(1), R7. https://doi.org/10.1186/gb-2003-4-1-r7.

Du, P., Kibbe, W. A., & Lin, S. M. (2008). lumi: A pipeline for processing Illumina microarray. *Bioinformatics (Oxford, England)*, *24*(13), 1547–1548. https://doi.org/10.1093/bioinformatics/btn224.

Du, Y., Cai, M., Xing, X., Ji, J., Yang, E., & Wu, J. (2021). PINA 3.0: Mining cancer interactome. *Nucleic Acids Research*, *49*(D1), D1351–D1357. https://doi.org/10.1093/nar/gkaa1075.

Dubois, J. W., Hill, S., England, L. S., Edge, T., Masson, L., Trevors, J. T., et al. (2004). The development of a DNA microarray-based assay for the characterization of commercially formulated microbial products. *Journal of Microbiological Methods*, *58*(2), 251–262. https://doi.org/10.1016/j.mimet.2004.04.011.

Dunning, M. J., Smith, M. L., Ritchie, M. E., & Tavaré, S. (2007). beadarray: R classes and methods for Illumina bead-based data. *Bioinformatics*, *23*(16), 2183–2184. https://doi.org/10.1093/bioinformatics/btm311.

Easton, J. M. (2013). *Vol. 1. Constructing directed metabolic networks from microarray data.* EPSRC. https://www.technologynetworks.com/tn/go/lc/view-poster-232604.

Edgar, R., Domrachev, M., & Lash, A. E. (2002). Gene Expression Omnibus: NCBI gene expression and hybridization array data repository. *Nucleic Acids Research*, *30*(1), 207–210. https://doi.org/10.1093/nar/30.1.207.

Eichhorn, S. W., Guo, H., McGeary, S. E., Rodriguez-Mias, R. A., Shin, C., Baek, D., et al. (2014). MRNA destabilization is the dominant effect of mammalian microRNAs by

the time substantial repression ensues. *Molecular Cell, 56*(1). https://doi.org/10.1016/j.molcel.2014.08.028.
Ekins, R., & Chu, F. W. (1999). Microarrays: Their origins and applications. *Trends in Biotechnology, 17*(6), 217–218. https://doi.org/10.1016/S0167-7799(99)01329-3.
Enright, A. J., John, B., Gaul, U., Tuschl, T., Sander, C., & Marks, D. S. (2003). MicroRNA targets in Drosophila. *Genome Biology, 5*(1), R1. https://doi.org/10.1186/gb-2003-5-1-r1.
Ermolaeva, O., Rastogi, M., Pruitt, K. D., Schuler, G. D., Bittner, M. L., Chen, Y., et al. (1998). Data management and analysis for gene expression arrays. *Nature Genetics, 20*(1), 19–23. https://doi.org/10.1038/1670.
Fajriyah, R. (2021). Paper review: An overview on microarray technologies. *Bulletin of Applied Mathematics and Mathematics Education, 1*(1), 21. https://doi.org/10.12928/bamme.v1i1.3854.
Fan, J.-B., Gunderson, K. L., Bibikova, M., Yeakley, J. M., Chen, J., Wickham Garcia, E., et al. (2006). Illumina universal bead arrays. *Methods in Enzymology, 410*, 57–73. https://doi.org/10.1016/S0076-6879(06)10003-8.
Ferguson, J. A., Steemers, F. J., & Walt, D. R. (2000). High-density fiber-optic DNA random microsphere array. *Analytical Chemistry, 72*(22), 5618–5624. https://doi.org/10.1021/ac0008284.
Fodor, S. P., Read, J. L., Pirrung, M. C., Stryer, L., Lu, A. T., & Solas, D. (1991). Light-directed, spatially addressable parallel chemical synthesis. *Science (New York, N.Y.), 251*(4995), 767–773. https://doi.org/10.1126/science.1990438.
Ford, G., Xu, Z., Gates, A., Jiang, J., & Ford, B. D. (2006). Expression Analysis Systematic Explorer (EASE) analysis reveals differential gene expression in permanent and transient focal stroke rat models. *Brain Research, 1071*(1), 226–236. https://doi.org/10.1016/j.brainres.2005.11.090.
Franke, L., van Bakel, H., Fokkens, L., de Jong, E. D., Egmont-Petersen, M., & Wijmenga, C. (2006). Reconstruction of a functional human gene network, with an application for prioritizing positional candidate genes. *American Journal of Human Genetics, 78*(6), 1011–1025. https://doi.org/10.1086/504300.
Frenkel-Morgenstern, M., Gorohovski, A., Tagore, S., Sekar, V., Vazquez, M., & Valencia, A. (2017). ChiPPI: A novel method for mapping chimeric protein–protein interactions uncovers selection principles of protein fusion events in cancer. *Nucleic Acids Research, 45*(12), 7094–7105. https://doi.org/10.1093/nar/gkx423.
García-Campos, M. A., Espinal-Enríquez, J., & Hernández-Lemus, E. (2015). Pathway analysis: State of the art. *Frontiers in Physiology, 6*. https://www.frontiersin.org/article/10.3389/fphys.2015.00383.
Garcia-Garcia, J., Valls-Comamala, V., Guney, E., Andreu, D., Muñoz, F. J., Fernandez-Fuentes, N., et al. (2017). iFrag: A protein-protein interface prediction server based on sequence fragments. *Journal of Molecular Biology, 429*(3), 382–389. https://doi.org/10.1016/j.jmb.2016.11.034.
Gautier, L., Cope, L., Bolstad, B. M., & Irizarry, R. A. (2004). affy—Analysis of Affymetrix GeneChip data at the probe level. *Bioinformatics (Oxford, England), 20*(3), 307–315. https://doi.org/10.1093/bioinformatics/btg405.
Gerhold, D., Lu, M., Xu, J., Austin, C., Caskey, C. T., & Rushmore, T. (2001). Monitoring expression of genes involved in drug metabolism and toxicology using DNA microarrays. *Physiological Genomics, 5*(4), 161–170. https://doi.org/10.1152/physiolgenomics.2001.5.4.161.
Gharajeh, M. S. (2018). Biological big data analytics. In *Vol. 109. Advances in computers* (pp. 321–355). Elsevier. https://doi.org/10.1016/bs.adcom.2017.08.002.

Goel, R., Harsha, H. C., Pandey, A., & Prasad, T. S. K. (2012). Human Protein Reference Database and Human Proteinpedia as resources for phosphoproteome analysis. *Molecular BioSystems*, *8*(2), 453–463. https://doi.org/10.1039/c1mb05340j.

Goll, J., Rajagopala, S. V., Shiau, S. C., Wu, H., Lamb, B. T., & Uetz, P. (2008). MPIDB: The microbial protein interaction database. *Bioinformatics*, *24*(15), 1743–1744. https://doi.org/10.1093/bioinformatics/btn285.

Granjeaud, S., Bertucci, F., & Jordan, B. R. (1999). Expression profiling: DNA arrays in many guises. *BioEssays*, *21*(9), 781–790. https://doi.org/10.1002/(SICI)1521-1878(199909)21:9<781::AID-BIES10>3.0.CO;2-2.

Graveel, C. R., Jatkoe, T., Madore, S. J., Holt, A. L., & Farnham, P. J. (2001). Expression profiling and identification of novel genes in hepatocellular carcinomas. *Oncogene*, *20*(21), 2704–2712. https://doi.org/10.1038/sj.onc.1204391.

Grouse, L. H., Munson, P. J., & Nelson, P. S. (2001). Sequence databases and microarrays as tools for identifying prostate cancer biomarkers. *Urology*, *57*(4 Suppl. 1), 154–159. https://doi.org/10.1016/S0090-4295(00)00963-8.

Guo, Y., Li, M., Pu, X., Li, G., Guang, X., Xiong, W., et al. (2010). PRED_PPI: A server for predicting protein-protein interactions based on sequence data with probability assignment. *BMC Research Notes*, *3*(1), 145. https://doi.org/10.1186/1756-0500-3-145.

Guzzi, P. H., Agapito, G., Di Martino, M. T., Arbitrio, M., Tassone, P., Tagliaferri, P., et al. (2012). DMET-Analyzer: Automatic analysis of Affymetrix DMET Data. *BMC Bioinformatics*, *13*(1), 258. https://doi.org/10.1186/1471-2105-13-258.

Hacia, J. G., & Collins, F. S. (1999). Mutational analysis using oligonucleotide microarrays. *Journal of Medical Genetics*, *36*(10), 730–736. https://doi.org/10.1136/jmg.36.10.730.

Hacia, J. G., Edgemon, K., Fang, N., Mayer, R. A., Sudano, D., Hunt, N., et al. (2000). Oligonucleotide microarray based detection of repetitive sequence changes. *Human Mutation*, *16*(4), 354–363. https://doi.org/10.1002/1098-1004(200010)16:4<354::AID-HUMU8>3.0.CO;2-V.

Hammond, S. M. (2015). An overview of microRNAs. *Advanced Drug Delivery Reviews*, *87*, 3–14. https://doi.org/10.1016/j.addr.2015.05.001.

Hanson, E. H., Niemeyer, D. M., Folio, L., Agan, B. K., & Rowley, R. K. (2004). Potential use of microarray technology for rapid identification of central nervous system pathogens. *Military Medicine*, *169*(8), 594–599. https://doi.org/10.7205/MILMED.169.8.594.

Hardcastle, T. J. (2016). Generalized empirical Bayesian methods for discovery of differential data in high-throughput biology. *Bioinformatics (Oxford, England)*, *32*(2), 195–202. https://doi.org/10.1093/bioinformatics/btv569.

Harkin, D. P. (2000). Uncovering functionally relevant signaling pathways using microarray-based expression profiling. *The Oncologist*, *5*(6), 501–507. https://doi.org/10.1634/theoncologist.5-6-501.

Hayashi, T., Matsuzaki, Y., Yanagisawa, K., Ohue, M., & Akiyama, Y. (2018). MEGADOCK-Web: An integrated database of high-throughput structure-based protein-protein interaction predictions. *BMC Bioinformatics*, *19*(Suppl 4), 62. https://doi.org/10.1186/s12859-018-2073-x.

He, Y. D. (2006). Genomic approach to biomarker identification and its recent applications. *Cancer Biomarkers*, *2*(3–4), 103–133. https://doi.org/10.3233/CBM-2006-23-404.

Healy, Z. R., Lee, N. H., Gao, X., Goldring, M. B., Talalay, P., Kensler, T. W., et al. (2005). Divergent responses of chondrocytes and endothelial cells to shear stress: Cross-talk among COX-2, the phase 2 response, and apoptosis. *Proceedings of the National Academy of Sciences*, *102*(39), 14010–14015. https://doi.org/10.1073/pnas.0506620102.

Heberle, H., Meirelles, G. V., da Silva, F. R., Telles, G. P., & Minghim, R. (2015). InteractiVenn: A web-based tool for the analysis of sets through Venn diagrams. *BMC Bioinformatics*, *16*(1), 169. https://doi.org/10.1186/s12859-015-0611-3.

Heinicke, S., Livstone, M. S., Lu, C., Oughtred, R., Kang, F., Angiuoli, S. V., et al. (2007). The Princeton Protein Orthology Database (P-POD): A comparative genomics analysis tool for biologists. *PLoS ONE, 2*(8), e766. https://doi.org/10.1371/journal.pone.0000766.

Heller, M. J. (2002). DNA microarray technology: Devices, systems, and applications. *Annual Review of Biomedical Engineering, 4*(1), 129–153. https://doi.org/10.1146/annurev.bioeng.4.020702.153438.

Herwig, R., Hardt, C., Lienhard, M., & Kamburov, A. (2016). Analyzing and interpreting genome data at the network level with ConsensusPathDB. *Nature Protocols, 11*(10), 1889–1907. https://doi.org/10.1038/nprot.2016.117.

Hira, Z. M., & Gillies, D. F. (2015). A review of feature selection and feature extraction methods applied on microarray data. *Advances in Bioinformatics, 2015*, 1–13. https://doi.org/10.1155/2015/198363.

Horan, P. K., & Wheeless, L. L. (1977). Quantitative single cell analysis and sorting. *Science (New York, N.Y.), 198*(4313), 149–157. https://doi.org/10.1126/science.905822.

Hosack, D. A., Dennis, G., Sherman, B. T., Lane, H. C., & Lempicki, R. A. (2003). Identifying biological themes within lists of genes with EASE. *Genome Biology, 4*(10), R70. https://doi.org/10.1186/gb-2003-4-10-r70.

Hosur, R., Xu, J., Bienkowska, J., & Berger, B. (2011). iWRAP: An interface threading approach with application to prediction of cancer-related protein-protein interactions. Journal of molecular biology, 405(5), 1295–1310. https://doi.org/10.1016/j.jmb.2010.11.025.

Hosur, R., Peng, J., Vinayagam, A., Stelzl, U., Xu, J., Perrimon, N., et al. (2012). A computational framework for boosting confidence in high-throughput protein-protein interaction datasets. *Genome Biology, 13*(8), R76. https://doi.org/10.1186/gb-2012-13-8-r76.

Howard, S. L., Gaunt, M. W., Hinds, J., Witney, A. A., Stabler, R., & Wren, B. W. (2006). Application of comparative phylogenomics to study the evolution of yersinia enterocolitica and to identify genetic differences relating to pathogenicity. *Journal of Bacteriology, 188*(10), 3645–3653. https://doi.org/10.1128/JB.188.10.3645-3653.2006.

Hu, T.-M., & Hayton, W. L. (2011). Architecture of the drug-drug interaction network. *Journal of Clinical Pharmacy and Therapeutics, 36*(2), 135–143. https://doi.org/10.1111/j.1365-2710.2009.01103.x.

Hu, J., Li, H., Waterman, M. S., & Zhou, X. J. (2006). Integrative missing value estimation for microarray data. *BMC Bioinformatics*, 7(1), 449. https://doi.org/10.1186/1471-2105-7-449.

Huang, S. (2000). The practical problems of post-genomic biology. *Nature Biotechnology, 18*(5), 471–472. https://doi.org/10.1038/75235.

Huang, D. W., Sherman, B. T., & Lempicki, R. A. (2009). Bioinformatics enrichment tools: Paths toward the comprehensive functional analysis of large gene lists. *Nucleic Acids Research, 37*(1), 1–13. https://doi.org/10.1093/nar/gkn923.

Hucka, M., Finney, A., Sauro, H. M., Bolouri, H., Doyle, J. C., Kitano, H., et al., & Wang, J. (2003). The systems biology markup language (SBML): A medium for representation and exchange of biochemical network models. *Bioinformatics, 19*(4), 524–531. https://doi.org/10.1093/bioinformatics/btg015.

Ihaka, R., & Gentleman, R. (1996). R: A language for data analysis and graphics. *Journal of Computational and Graphical Statistics, 5*(3), 299. https://doi.org/10.2307/1390807.

Ikeo, K., Ishi-i, J., Tamura, T., Gojobori, T., & Tateno, Y. (2003). CIBEX: Center for Information Biology gene EXpression database. *Comptes Rendus Biologies, 326*(10–11), 1079–1082. https://doi.org/10.1016/j.crvi.2003.09.034.

Imam, N. (2016). Computational analysis of human cancer related RNA-Seq data: A review. *Journal of Applied Computing, 1*, 30–37.

Irizarry, R. A., Bolstad, B. M., Collin, F., Cope, L. M., Hobbs, B., & Speed, T. P. (2003). Summaries of Affymetrix GeneChip probe level data. *Nucleic Acids Research*, *31*(4), e15. https://doi.org/10.1093/nar/gng015.

Isali, I., McClellan, P., Calaway, A., Prunty, M., Abbosh, P., Mishra, K., et al. (2022). Gene network profiling in muscle-invasive bladder cancer: A systematic review and meta-analysis. *Urologic Oncology: Seminars and Original Investigations*. https://doi.org/10.1016/j.urolonc.2021.11.003.

Izarzugaza, J. M., Juan, D., Pons, C., Ranea, J. A., Valencia, A., & Pazos, F. (2006). TSEMA: Interactive prediction of protein pairings between interacting families. *Nucleic Acids Research*, *34*(Web Server issue), W315–W319. https://doi.org/10.1093/nar/gkl112.

Jayapandian, M., Chapman, A., Tarcea, V. G., Yu, C., Elkiss, A., Ianni, A., et al. (2007). Michigan Molecular Interactions (MiMI): Putting the jigsaw puzzle together. *Nucleic Acids Research*, *35*(Database), D566–D571. https://doi.org/10.1093/nar/gkl859.

Ji, H., Jiang, H., Ma, W., Johnson, D. S., Myers, R. M., & Wong, W. H. (2008). An integrated software system for analyzing ChIP-chip and ChIP-seq data. *Nature Biotechnology*, *26*(11), 1293–1300. https://doi.org/10.1038/nbt.1505.

Jiang, N., Leach, L. J., Hu, X., Potokina, E., Jia, T., Druka, A., et al. (2008). Methods for evaluating gene expression from Affymetrix microarray datasets. *BMC Bioinformatics*, *9*(1), 284. https://doi.org/10.1186/1471-2105-9-284.

Johansson, P., & Häkkinen, J. (2006). Improving missing value imputation of microarray data by using spot quality weights. *BMC Bioinformatics*, 7(1), 306. https://doi.org/10.1186/1471-2105-7-306.

Jones, S. (2009). ProtorP: A protein-protein interaction analysis server. *Bioinformatics*, *25*(3), 413–414. https://doi.org/10.1093/bioinformatics/btn584.

Jordan, R. A., EL-Manzalawy, Y., Dobbs, D., & Honavar, V. (2012). Predicting protein-protein interface residues using local surface structural similarity. *BMC Bioinformatics*, *13*(1), 41. https://doi.org/10.1186/1471-2105-13-41.

Jörnsten, R., Ouyang, M., & Wang, H.-Y. (2007). A meta-data based method for DNA microarray imputation. *BMC Bioinformatics*, *8*(1), 109. https://doi.org/10.1186/1471-2105-8-109.

Jörnsten, R., Wang, H.-Y., Welsh, W. J., & Ouyang, M. (2005). DNA microarray data imputation and significance analysis of differential expression. *Bioinformatics*, *21*(22), 4155–4161. https://doi.org/10.1093/bioinformatics/bti638.

Kalathur, R. K. R., Pinto, J. P., Hernández-Prieto, M. A., Machado, R. S. R., Almeida, D., Chaurasia, G., et al. (2014). UniHI 7: An enhanced database for retrieval and interactive analysis of human molecular interaction networks. *Nucleic Acids Research*, *42*(D1), D408–D414. https://doi.org/10.1093/nar/gkt1100.

Kallioniemi, O.-P. (2001). Biochip technologies in cancer research. *Annals of Medicine*, *33*(2), 142–147. https://doi.org/10.3109/07853890109002069.

Kamburov, A., Stelzl, U., Lehrach, H., & Herwig, R. (2013). The ConsensusPathDB interaction database: 2013 update. *Nucleic Acids Research*, *41*(D1), D793–D800. https://doi.org/10.1093/nar/gks1055.

Kerr, M. K., Martin, M., & Churchill, G. A. (2000). Analysis of variance for gene expression microarray data. *Journal of Computational Biology*, 7(6), 819–837. https://doi.org/10.1089/10665270050514954.

Khan, J., Saal, L. H., Bittner, M. L., Chen, Y., Trent, J. M., & Meltzer, P. S. (1999). Expression profiling in cancer using cDNA microarrays. *Electrophoresis*, *20*(2), 223–229. https://doi.org/10.1002/(SICI)1522-2683(19990201)20:2<223::AID-ELPS223>3.0.CO;2-A.

Khatri, P., Sellamuthu, S., Malhotra, P., Amin, K., Done, A., & Draghici, S. (2005). Recent additions and improvements to the Onto-Tools. *Nucleic Acids Research*, *33*(Web Server issue), W762–W765. https://doi.org/10.1093/nar/gki472.

Khatri, P., Voichita, C., Kattan, K., Ansari, N., Khatri, A., Georgescu, C., et al. (2007). Onto-Tools: New additions and improvements in 2006. *Nucleic Acids Research*, *35*(Web Server issue), W206–W211. https://doi.org/10.1093/nar/gkm327.

Kim, K.-Y., Kim, B.-J., & Yi, G.-S. (2004). Reuse of imputed data in microarray analysis increases imputation efficiency. *BMC Bioinformatics*, *5*(1), 160. https://doi.org/10.1186/1471-2105-5-160.

Korn, J. M., Kuruvilla, F. G., McCarroll, S. A., Wysoker, A., Nemesh, J., Cawley, S., et al. (2008). Integrated genotype calling and association analysis of SNPs, common copy number polymorphisms and rare CNVs. *Nature Genetics*, *40*(10), 1253–1260. https://doi.org/10.1038/ng.237.

Kotlyar, M., Pastrello, C., Ahmed, Z., Chee, J., Varyova, Z., & Jurisica, I. (2022). IID 2021: Towards context-specific protein interaction analyses by increased coverage, enhanced annotation and enrichment analysis. *Nucleic Acids Research*, *50*(D1), D640–D647. https://doi.org/10.1093/nar/gkab1034.

Kotlyar, M., Pastrello, C., Malik, Z., & Jurisica, I. (2019). IID 2018 update: Context-specific physical protein-protein interactions in human, model organisms and domesticated species. *Nucleic Acids Research*, *47*(D1), D581–D589. https://doi.org/10.1093/nar/gky1037.

Kotlyar, M., Pastrello, C., Sheahan, N., & Jurisica, I. (2016). Integrated interactions database: Tissue-specific view of the human and model organism interactomes. *Nucleic Acids Research*, *44*(D1), D536–D541. https://doi.org/10.1093/nar/gkv1115.

Kumar, S. U., Kumar, D. T., Siva, R., Doss, C. G. P., & Zayed, H. (2019). Integrative bioinformatics approaches to map potential novel genes and pathways involved in ovarian cancer. *Frontiers in Bioengineering and Biotechnology*, 7, 391. https://doi.org/10.3389/fbioe.2019.00391.

Kumar, R., & Nanduri, B. (2010). HPIDB - a unified resource for host-pathogen interactions. *BMC Bioinformatics*, *11*(6), S16. https://doi.org/10.1186/1471-2105-11-S6-S16.

Kumar, M., & Rath, S. K. (2016). Feature selection and classification of microarray data using machine learning techniques. In *Emerging trends in applications and infrastructures for computational biology, bioinformatics, and systems biology* (pp. 213–242). Elsevier. https://doi.org/10.1016/B978-0-12-804203-8.00015-8.

Kundrotas, P., Zhu, Z., & Vakser, I. (2012). GWIDD: A comprehensive resource for genome-wide structural modeling of protein-protein interactions. *Human Genomics*, *6*, 7. https://doi.org/10.1186/1479-7364-6-7.

Lai, X., Wolkenhauer, O., & Vera, J. (2016). Understanding microRNA-mediated gene regulatory networks through mathematical modelling. *Nucleic Acids Research*, *44*(13), 6019–6035. https://doi.org/10.1093/nar/gkw550.

Launay, G., Salza, R., Multedo, D., Thierry-Mieg, N., & Ricard-Blum, S. (2015). MatrixDB, the extracellular matrix interaction database: Updated content, a new navigator and expanded functionalities. *Nucleic Acids Research*, *43*(D1), D321–D327. https://doi.org/10.1093/nar/gku1091.

Lee, J.-S. (2016). Exploring cancer genomic data from the cancer genome atlas project. *BMB Reports*, *49*(11), 607–611. https://doi.org/10.5483/BMBRep.2016.49.11.145.

Lee, P. S., & Lee, K. H. (2005). Engineering HlyA hypersecretion in Escherichia coli based on proteomic and microarray analyses. *Biotechnology and Bioengineering*, *89*(2), 195–205. https://doi.org/10.1002/bit.20342.

Lee, S.-A., Chan, C.-H., Chen, T.-C., Yang, C.-Y., Huang, K.-C., Tsai, C.-H., et al. (2009). POINeT: Protein interactome with sub-network analysis and hub prioritization. *BMC Bioinformatics*, *10*(1), 114. https://doi.org/10.1186/1471-2105-10-114.

Li, Y., Cole, K., & Altman, S. (2003). The effect of a single, temperature-sensitive mutation on global gene expression in Escherichia coli. *RNA*, *9*(5), 518–532. https://doi.org/10.1261/rna.2198203.

Lemay, D. G., Martin, W. F., Hinrichs, A. S., Rijnkels, M., German, J. B., Korf, I., et al. (2012). G-NEST: A gene neighborhood scoring tool to identify co-conserved, co-expressed genes. *BMC Bioinformatics*, *13*, 253. https://doi.org/10.1186/1471-2105-13-253.

Li, C., Tang, Z., Zhang, W., Ye, Z., & Liu, F. (2021). GEPIA2021: Integrating multiple deconvolution-based analysis into GEPIA. *Nucleic Acids Research*, *49*(W1), W242–W246. https://doi.org/10.1093/nar/gkab418.

Li, Y., & Ilie, L. (2017). SPRINT: Ultrafast protein-protein interaction prediction of the entire human interactome. *BMC Bioinformatics*, *18*(1), 485. https://doi.org/10.1186/s12859-017-1871-x.

Li, Z., Ivanov, A. A., Su, R., Gonzalez-Pecchi, V., Qi, Q., Liu, S., et al. (2017). The OncoPPi network of cancer-focused protein-protein interactions to inform biological insights and therapeutic strategies. *Nature Communications*, *8*, 14356. https://doi.org/10.1038/ncomms14356.

Lisowska, K., Olbryt, M., Dudaladava, V., Pamuła-Piłat, J., Kujawa, K., Grzybowska, E., et al. (2014). Gene expression analysis in ovarian cancer—Faults and hints from DNA microarray study. *Frontiers in Oncology*, *4*. https://www.frontiersin.org/article/10.3389/fonc.2014.00006.

Liu, X., Liu, B., Huang, Z., Shi, T., Chen, Y., & Zhang, J. (2012). SPPS: A sequence-based method for predicting probability of protein-protein interaction partners. *PLoS ONE*, 7(1), e30938. https://doi.org/10.1371/journal.pone.0030938.

Lockhart, D. J., Dong, H., Byrne, M. C., Follettie, M. T., Gallo, M. V., Chee, M. S., et al. (1996). Expression monitoring by hybridization to high-density oligonucleotide arrays. *Nature Biotechnology*, *14*(13), 1675–1680. https://doi.org/10.1038/nbt1296-1675.

Lonsdale, J., Thomas, J., Salvatore, M., Phillips, R., Lo, E., Shad, S., et al. (2013). The Genotype-Tissue Expression (GTEx) project. *Nature Genetics*, *45*(6), 580–585. https://doi.org/10.1038/ng.2653.

López, Y., Nakai, K., & Patil, A. (2015). HitPredict version 4: Comprehensive reliability scoring of physical protein–protein interactions from more than 100 species. *Database*, *2015*, bav117. https://doi.org/10.1093/database/bav117.

Luck, K., Kim, D.-K., Lambourne, L., Spirohn, K., Begg, B. E., Bian, W., et al. (2020). A reference map of the human binary protein interactome. *Nature*, *580*(7803), 402–408. https://doi.org/10.1038/s41586-020-2188-x.

Luo, Z., & Geschwind, D. H. (2001). Microarray applications in neuroscience. *Neurobiology of Disease*, *8*(2), 183–193. https://doi.org/10.1006/nbdi.2001.0387.

Madden, S. L., Wang, C. J., & Landes, G. (2000). Serial analysis of gene expression: From gene discovery to target identification. *Drug Discovery Today*, *5*(9), 415–425. https://doi.org/10.1016/s1359-6446(00)01544-0.

Mahalingam, R., & Fedoroff, N. (2001). Screening insertion libraries for mutations in many genes simultaneously using DNA microarrays. *Proceedings of the National Academy of Sciences*, *98*(13), 7420–7425. https://doi.org/10.1073/pnas.121189598.

Manyam, G., Birerdinc, A., & Baranova, A. (2015). KPP: KEGG pathway painter. *BMC Systems Biology*, *9*(Suppl 2), S3. https://doi.org/10.1186/1752-0509-9-S2-S3.

Marozzo, F., Talia, D., & Trunfio, P. (2013). Using clouds for scalable knowledge discovery applications. In I. Caragiannis, M. Alexander, R. M. Badia, M. Cannataro, A. Costan, M. Danelutto, … J. Weidendorfer (Eds.), *Vol. 7640. Euro-Par 2012: Parallel processing workshops* (pp. 220–227). Springer Berlin Heidelberg. https://doi.org/10.1007/978-3-642-36949-0_25.

Martin, S., Zhang, Z., Martino, A., & Faulon, J.-L. (2007). Boolean dynamics of genetic regulatory networks inferred from microarray time series data. *Bioinformatics*, *23*(7), 866–874. https://doi.org/10.1093/bioinformatics/btm021.

McDowall, M. D., Scott, M. S., & Barton, G. J. (2009). PIPs: Human protein-protein interaction prediction database. *Nucleic Acids Research*, *37*(Database), D651–D656. https://doi.org/10.1093/nar/gkn870.

McGeary, S. E., Lin, K. S., Shi, C. Y., Pham, T. M., Bisaria, N., Kelley, G. M., et al. (2019). The biochemical basis of microRNA targeting efficacy. *Science (New York, N.Y.)*, *366*(6472), eaav1741. https://doi.org/10.1126/science.aav1741.

McGuire, W. L. (1991). Breast cancer prognostic factors: Evaluation guidelines. *Journal of the National Cancer Institute*, *83*(3), 154–155. https://doi.org/10.1093/jnci/83.3.154.

Michael, K. L., Taylor, L. C., Schultz, S. L., & Walt, D. R. (1998). Randomly ordered addressable high-density optical sensor arrays. *Analytical Chemistry*, *70*(7), 1242–1248. https://doi.org/10.1021/ac971343r.

Michaut, M., Kerrien, S., Montecchi-Palazzi, L., Cassier-Chauvat, C., Chauvat, F., Aude, J.-C., et al. (2008). InteroPORC: An automated tool to predict highly conserved protein interaction networks. *BMC Bioinformatics*, *9*(10), P1. https://doi.org/10.1186/1471-2105-9-S10-P1.

Mischel, P. S., Cloughesy, T. F., & Nelson, S. F. (2004). DNA-microarray analysis of brain cancer: Molecular classification for therapy. *Nature Reviews Neuroscience*, *5*(10), 782–792. https://doi.org/10.1038/nrn1518.

Mishra, S., Shah, M. I., Udhaya Kumar, S., Thirumal Kumar, D., Gopalakrishnan, C., Al-Subaie, A. M., et al. (2021). Network analysis of transcriptomics data for the prediction and prioritization of membrane-associated biomarkers for idiopathic pulmonary fibrosis (IPF) by bioinformatics approach. In *Vol. 123. Advances in protein chemistry and structural biology* (pp. 241–273). Elsevier. https://doi.org/10.1016/bs.apcsb.2020.10.003.

Mitrea, C., Taghavi, Z., Bokanizad, B., Hanoudi, S., Tagett, R., Donato, M., et al. (2013). Methods and approaches in the topology-based analysis of biological pathways. *Frontiers in Physiology*, *4*. https://www.frontiersin.org/article/10.3389/fphys.2013.00278.

Mosca, R., Céol, A., & Aloy, P. (2013). Interactome3D: Adding structural details to protein networks. *Nature Methods*, *10*(1), 47–53. https://doi.org/10.1038/nmeth.2289.

Murakami, Y., & Mizuguchi, K. (2014). Homology-based prediction of interactions between proteins using averaged one-dependence estimators. *BMC Bioinformatics*, *15*, 213. https://doi.org/10.1186/1471-2105-15-213.

Muzio, G., O'Bray, L., & Borgwardt, K. (2021). Biological network analysis with deep learning. *Briefings in Bioinformatics*, *22*(2), 1515–1530. https://doi.org/10.1093/bib/bbaa257.

Nakanishi, T., Oka, T., & Akagi, T. (2001). Recent advances in DNA microarrays. *Acta Medica Okayama*, *55*(6), 319–328. https://doi.org/10.18926/AMO/32002.

Nannini, M., Pantaleo, M. A., Maleddu, A., Astolfi, A., Formica, S., & Biasco, G. (2009). Gene expression profiling in colorectal cancer using microarray technologies: Results and perspectives. *Cancer Treatment Reviews*, *35*(3), 201–209. https://doi.org/10.1016/j.ctrv.2008.10.006.

Negi, S. S., Schein, C. H., Oezguen, N., Power, T. D., & Braun, W. (2007). InterProSurf: A web server for predicting interacting sites on protein surfaces. *Bioinformatics*, *23*(24), 3397–3399. https://doi.org/10.1093/bioinformatics/btm474.

Novère, N. L., Hucka, M., Mi, H., Moodie, S., Schreiber, F., Sorokin, A., et al. (2009). The systems biology graphical notation. *Nature Biotechnology*, *27*(8), 735–741. https://doi.org/10.1038/nbt.1558.

Oba, S., Sato, M., Takemasa, I., Monden, M., Matsubara, K., & Ishii, S. (2003). A Bayesian missing value estimation method for gene expression profile data. *Bioinformatics*, *19*(16), 2088–2096. https://doi.org/10.1093/bioinformatics/btg287.

Ochoa, D., & Pazos, F. (2010). Studying the co-evolution of protein families with the Mirrortree web server. *Bioinformatics (Oxford, England)*, *26*(10), 1370–1371. https://doi.org/10.1093/bioinformatics/btq137.

Orchard, S., Ammari, M., Aranda, B., Breuza, L., Briganti, L., Broackes-Carter, F., et al., & Hermjakob, H. (2014). The MIntAct project—IntAct as a common curation platform for 11 molecular interaction databases. *Nucleic Acids Research*, *42*(D1), D358–D363. https://doi.org/10.1093/nar/gkt1115.
Orii, N., & Ganapathiraju, M. K. (2012). Wiki-Pi: A web-server of annotated human protein-protein interactions to aid in discovery of protein function. *PLoS One*, 7(11), e49029. https://doi.org/10.1371/journal.pone.0049029.
Ouyang, M., Welsh, W. J., & Georgopoulos, P. (2004). Gaussian mixture clustering and imputation of microarray data. *Bioinformatics*, *20*(6), 917–923. https://doi.org/10.1093/bioinformatics/bth007.
Pagel, P., Kovac, S., Oesterheld, M., Brauner, B., Dunger-Kaltenbach, I., Frishman, G., et al. (2005). The MIPS mammalian protein-protein interaction database. *Bioinformatics*, *21*(6), 832–834. https://doi.org/10.1093/bioinformatics/bti115.
Parkinson, H., Kapushesky, M., Shojatalab, M., Abeygunawardena, N., Coulson, R., Farne, A., et al. (2007). ArrayExpress—A public database of microarray experiments and gene expression profiles. *Nucleic Acids Research*, *35*(Database), D747–D750. https://doi.org/10.1093/nar/gkl995.
Parrish, M. L., Wei, N., Duenwald, S., Tokiwa, G. Y., Wang, Y., Holder, D., et al. (2004). A microarray platform comparison for neuroscience applications. *Journal of Neuroscience Methods*, *132*(1), 57–68. https://doi.org/10.1016/j.jneumeth.2003.09.013.
Patil, A. (2019). Protein–protein interaction databases. In *Encyclopedia of bioinformatics and computational biology* (pp. 849–855). Elsevier. https://doi.org/10.1016/B978-0-12-809633-8.20660-2.
Patil, A., Nakai, K., & Nakamura, H. (2011). HitPredict: A database of quality assessed protein–protein interactions in nine species. *Nucleic Acids Research*, *39*(Suppl. 1), D744–D749. https://doi.org/10.1093/nar/gkq897.
Pelizzola, M., Pavelka, N., Foti, M., & Ricciardi-Castagnoli, P. (2006). AMDA: An R package for the automated microarray data analysis. *BMC Bioinformatics*, 7, 335. https://doi.org/10.1186/1471-2105-7-335.
Perou, C. M., Sørlie, T., Eisen, M. B., van de Rijn, M., Jeffrey, S. S., Rees, C. A., et al. (2000). Molecular portraits of human breast tumours. *Nature*, *406*(6797), 747–752. https://doi.org/10.1038/35021093.
Perovic, V., Sumonja, N., Gemovic, B., Toska, E., Roberts, S. G., & Veljkovic, N. (2017). TRI_tool: A web-tool for prediction of protein-protein interactions in human transcriptional regulation. *Bioinformatics (Oxford, England)*, *33*(2), 289–291. https://doi.org/10.1093/bioinformatics/btw590.
Philipp, O., Osiewacz, H. D., & Koch, I. (2016). Path2PPI: An R package to predict protein-protein interaction networks for a set of proteins. *Bioinformatics (Oxford, England)*, *32*(9), 1427–1429. https://doi.org/10.1093/bioinformatics/btv765.
Piatetsky-Shapiro, G., & Tamayo, P. (2003). Microarray data mining: Facing the challenges. *ACM SIGKDD Explorations Newsletter*, *5*(2), 1–5. https://doi.org/10.1145/980972.980974.
Planas-Iglesias, J., Marin-Lopez, M. A., Bonet, J., Garcia-Garcia, J., & Oliva, B. (2013). iLoops: A protein–protein interaction prediction server based on structural features. *Bioinformatics*, *29*(18), 2360–2362. https://doi.org/10.1093/bioinformatics/btt401.
Polychemotherapy for early breast cancer. (1998). An overview of the randomised trials. *The Lancet*, *352*(9132), 930–942. https://doi.org/10.1016/S0140-6736(98)03301-7.
Porollo, A., & Meller, J. (2007). Prediction-based fingerprints of protein–protein interactions. *Proteins: Structure, Function, and Bioinformatics*, *66*(3), 630–645. https://doi.org/10.1002/prot.21248.
Prasad, T. S. K., Kandasamy, K., & Pandey, A. (2009). Human protein reference database and human proteinpedia as discovery tools for systems biology. *Methods in Molecular Biology (Clifton, N.J.)*, *577*, 67–79. https://doi.org/10.1007/978-1-60761-232-2_6.

Pritsker, M., Liu, Y.-C., Beer, M. A., & Tavazoie, S. (2004). Whole-genome discovery of transcription factor binding sites by network-level conservation. *Genome Research*, *14*(1), 99–108. https://doi.org/10.1101/gr.1739204.

Qi, Y., Tastan, O., Carbonell, J. G., Weston, J., & Klein-Seetharaman, J. (2010). Semi-supervised multi-task learning for predicting interactions between HIV-1 and human proteins. In *Vol. 26. Bioinformatics* (p. 1645). Oxford University Press. https://doi.org/10.1093/bioinformatics/btq394. PreJuSER-13095.

Qin, S., & Zhou, H.-X. (2007). meta-PPISP: A meta web server for protein-protein interaction site prediction. *Bioinformatics (Oxford, England)*, *23*(24), 3386–3387. https://doi.org/10.1093/bioinformatics/btm434.

Quackenbush, J. (2001). Computational analysis of microarray data. *Nature Reviews Genetics*, *2*(6), 418–427. https://doi.org/10.1038/35076576.

Ravine, D. (1999). Automated mutation analysis. *Journal of Inherited Metabolic Disease*, *22*(4), 503–518. https://doi.org/10.1023/A:1005508324048.

Rhodes, D. R., Kalyana-Sundaram, S., Mahavisno, V., Varambally, R., Yu, J., Briggs, B. B., et al. (2007). Oncomine 3.0: Genes, pathways, and networks in a collection of 18,000 cancer gene expression profiles. *Neoplasia*, *9*(2), 166–180. https://doi.org/10.1593/neo.07112.

Rhodes, D. R., Yu, J., Shanker, K., Deshpande, N., Varambally, R., Ghosh, D., et al. (2004). ONCOMINE: A cancer microarray database and integrated data-mining platform. *Neoplasia (New York, N.Y.)*, *6*(1), 1–6.

Ritchie, M. E., Phipson, B., Wu, D., Hu, Y., Law, C. W., Shi, W., et al. (2015). limma powers differential expression analyses for RNA-sequencing and microarray studies. *Nucleic Acids Research*, *43*(7), e47. https://doi.org/10.1093/nar/gkv007.

Rong, Z., Chen, H., Zhang, Z., Zhang, Y., Ge, L., Lv, Z., et al. (2022). Identification of cardiomyopathy-related core genes through human metabolic networks and expression data. *BMC Genomics*, *23*(1), 47. https://doi.org/10.1186/s12864-021-08271-0.

Rosenwald, A., Wright, G., Chan, W. C., Connors, J. M., Campo, E., Fisher, R. I., et al., & Lymphoma/Leukemia Molecular Profiling Project. (2002). The use of molecular profiling to predict survival after chemotherapy for diffuse large-B-cell lymphoma. *The New England Journal of Medicine*, *346*(25), 1937–1947. https://doi.org/10.1056/NEJMoa012914.

Ruepp, A., Waegele, B., Lechner, M., Brauner, B., Dunger-Kaltenbach, I., Fobo, G., et al. (2010). CORUM: The comprehensive resource of mammalian protein complexes—2009. *Nucleic Acids Research*, *38*(Suppl. 1), D497–D501. https://doi.org/10.1093/nar/gkp914.

Salwinski, L. (2004). The database of interacting proteins: 2004 update. *Nucleic Acids Research*, *32*(90001), 449D–451. https://doi.org/10.1093/nar/gkh086.

Sanfilippo, C., & Di Rosa, M. (2021). Chapter 6—Brain aging and microarray analysis. In C. R. Martin, V. R. Preedy, & R. Rajendram (Eds.), *Factors affecting neurological aging* (pp. 59–70). Academic Press. https://doi.org/10.1016/B978-0-12-817990-1.00006-8.

Schaefer, M. H., Fontaine, J.-F., Vinayagam, A., Porras, P., Wanker, E. E., & Andrade-Navarro, M. A. (2012). HIPPIE: Integrating protein interaction networks with experiment based quality scores. *PLoS One*, 7(2), e31826. https://doi.org/10.1371/journal.pone.0031826.

Schena, M. (1996). Genome analysis with gene expression microarrays. *BioEssays*, *18*(5), 427–431. https://doi.org/10.1002/bies.950180513.

Schena, M. (1998). Microarrays: Biotechnology's discovery platform for functional genomics. *Trends in Biotechnology*, *16*(7), 301–306. https://doi.org/10.1016/S0167-7799(98)01219-0.

Schena, M. (1999). *DNA microarrays: A practical approach*. Oxford University Press.

Schena, M., Shalon, D., Davis, R. W., & Brown, P. O. (1995). Quantitative monitoring of gene expression patterns with a complementary DNA microarray. *Science (New York, N.Y.)*, *270*(5235), 467–470. https://doi.org/10.1126/science.270.5235.467.

Schulze, A., & Downward, J. (2000). Analysis of gene expression by microarrays: Cell biologist's gold mine or minefield? *Journal of Cell Science*, *113*(23), 4151–4156. https://doi.org/10.1242/jcs.113.23.4151.

Schwenk, J. M., Gry, M., Rimini, R., Uhlén, M., & Nilsson, P. (2008). Antibody suspension bead arrays within serum proteomics. *Journal of Proteome Research*, 7(8), 3168–3179. https://doi.org/10.1021/pr700890b.

Selvaraj, S., & Natarajan, J. (2011). Microarray data analysis and mining tools. *Bioinformation*, *6*(3), 95–99. https://doi.org/10.6026/97320630006095.

Shang, J., Xiao, C., Ma, T., Li, H., & Sun, J. (2019). GAMENet: Graph Augmented MEmory Networks for recommending medication combination. *Proceedings of the AAAI Conference on Artificial Intelligence*, *33*, 1126–1133. https://doi.org/10.1609/aaai.v33i01.33011126.

Sharma, A., Virk, R., Khurana, M., & Kaur, R. (2018). Protein interaction databases: A review. *Research Journal of Life Sciences, Bioinformatics, Pharmaceutical and Chemical Sciences*.

Shen, R., Fan, J.-B., Campbell, D., Chang, W., Chen, J., Doucet, D., et al. (2005). High-throughput SNP genotyping on universal bead arrays. *Mutation Research*, *573*(1–2), 70–82. https://doi.org/10.1016/j.mrfmmm.2004.07.022.

Shoemaker, B. A., & Panchenko, A. R. (2007). Deciphering protein–protein interactions. Part II. Computational methods to predict protein and domain interaction partners. *PLoS Computational Biology*, *3*(4), e43. https://doi.org/10.1371/journal.pcbi.0030043.

Shulman-Peleg, A., Shatsky, M., Nussinov, R., & Wolfson, H. J. (2008). MultiBind and MAPPIS: Webservers for multiple alignment of protein 3D-binding sites and their interactions. *Nucleic Acids Research*, *36*(Web Server issue), W260–W264. https://doi.org/10.1093/nar/gkn185.

Si, W., Shen, J., Zheng, H., & Fan, W. (2019). The role and mechanisms of action of microRNAs in cancer drug resistance. *Clinical Epigenetics*, *11*(1), 25. https://doi.org/10.1186/s13148-018-0587-8.

Silva, T. C., Colaprico, A., Olsen, C., D'Angelo, F., Bontempi, G., Ceccarelli, M., et al. (2016). *TCGA Workflow: Analyze cancer genomics and epigenomics data using Bioconductor packages*. *F1000Research*, *5*, 1542. https://doi.org/10.12688/f1000research.8923.2.

Simon, R., Lam, A., Li, M.-C., Ngan, M., Menenzes, S., & Zhao, Y. (2007). Analysis of gene expression data using BRB-array tools. *Cancer Informatics*, *3*, 117693510700300020. https://doi.org/10.1177/117693510700300022.

Singh, R., Park, D., Xu, J., Hosur, R., & Berger, B. (2010). Struct2Net: A web service to predict protein-protein interactions using a structure-based approach. *Nucleic Acids Research*, *38*(Web Server issue), W508–W515. https://doi.org/10.1093/nar/gkq481.

Slonim, D. K., & Yanai, I. (2009). Getting started in gene expression microarray analysis. *PLoS Computational Biology*, *5*(10), e1000543. https://doi.org/10.1371/journal.pcbi.1000543.

Smith, M. L., Baggerly, K. A., Bengtsson, H., Ritchie, M. E., & Hansen, K. D. (2013). illuminaio: An open source IDAT parsing tool for Illumina microarrays. *F1000 Research*, *2*, 264. https://doi.org/10.12688/f1000research.2-264.v1.

Smith, A., & Farrah, K. (2019). *Gene Expression profiling tests for breast cancer: A rapid qualitative review*. Canadian Agency for Drugs and Technologies in Health. http://www.ncbi.nlm.nih.gov/books/NBK545134/.

Sosnowski, R. G., Tu, E., Butler, W. F., O'Connell, J. P., & Heller, M. J. (1997). Rapid determination of single base mismatch mutations in DNA hybrids by direct electric field control. *Proceedings of the National Academy of Sciences of the United States of America*, *94*(4), 1119–1123. https://doi.org/10.1073/pnas.94.4.1119.

Southern, E. M. (2001). DNA microarrays: History and overview. In J. B. Rampal (Ed.), *Vol. 170. DNA arrays* (pp. 1–15). Humana Press. https://doi.org/10.1385/1-59259-234-1:1.
Statnikov, A., Tsamardinos, I., Dosbayev, Y., & Aliferis, C. F. (2005). GEMS: A system for automated cancer diagnosis and biomarker discovery from microarray gene expression data. *International Journal of Medical Informatics*, *74*(7–8), 491–503. https://doi.org/10.1016/j.ijmedinf.2005.05.002.
Steemers, F. J., Ferguson, J. A., & Walt, D. R. (2000). Screening unlabeled DNA targets with randomly ordered fiber-optic gene arrays. *Nature Biotechnology*, *18*(1), 91–94. https://doi.org/10.1038/72006.
Steemers, F. J., & Gunderson, K. L. (2005). Illumina, Inc. *Pharmacogenomics*, *6*(7), 777–782. https://doi.org/10.2217/14622416.6.7.777.
Stowe-Evans, E. L., Ford, J., & Kehoe, D. M. (2004). Genomic DNA microarray analysis: Identification of new genes regulated by light color in the cyanobacterium Fremyella diplosiphon. *Journal of Bacteriology*, *186*(13), 4338–4349. https://doi.org/10.1128/JB.186.13.4338-4349.2004.
Sturn, A., Quackenbush, J., & Trajanoski, Z. (2002). Genesis: Cluster analysis of microarray data. *Bioinformatics (Oxford, England)*, *18*(1), 207–208. https://doi.org/10.1093/bioinformatics/18.1.207.
Subramanian, A., Tamayo, P., Mootha, V. K., Mukherjee, S., Ebert, B. L., Gillette, M. A., et al. (2005). Gene set enrichment analysis: A knowledge-based approach for interpreting genome-wide expression profiles. *Proceedings of the National Academy of Sciences*, *102*(43), 15545–15550.
Subramaniyan, V., Fuloria, S., Gupta, G., Kumar, D. H., Sekar, M., Sathasivam, K. V., et al. (2022). A review on epidermal growth factor receptor's role in breast and non-small cell lung cancer. *Chemico-Biological Interactions*, *351*, 109735. https://doi.org/10.1016/j.cbi.2021.109735.
Sukhwal, A., & Sowdhamini, R. (2015). PPCheck: A webserver for the quantitative analysis of protein-protein interfaces and prediction of residue hotspots. *Bioinformatics and Biology Insights*, *9*, 141–151. https://doi.org/10.4137/BBI.S25928.
Svrakic, N. M. (2003). Statistical approach to DNA chip analysis. *Recent Progress in Hormone Research*, *58*(1), 75–93. https://doi.org/10.1210/rp.58.1.75.
Szklarczyk, D., Franceschini, A., Wyder, S., Forslund, K., Heller, D., Huerta-Cepas, J., et al. (2015). STRING v10: Protein–protein interaction networks, integrated over the tree of life. *Nucleic Acids Research*, *43*(D1), D447–D452. https://doi.org/10.1093/nar/gku1003.
Szklarczyk, D., Morris, J. H., Cook, H., Kuhn, M., Wyder, S., Simonovic, M., et al. (2017). The STRING database in 2017: Quality-controlled protein–protein association networks, made broadly accessible. *Nucleic Acids Research*, *45*(D1), D362–D368. https://doi.org/10.1093/nar/gkw937.
Tang, Y., Li, M., Wang, J., Pan, Y., & Wu, F. X. (2015). CytoNCA: A cytoscape plugin for centrality analysis and evaluation of protein interaction networks. *Bio Systems*, *127*, 67–72. https://doi.org/10.1016/j.biosystems.2014.11.005.
Tang, Z., Li, C., Kang, B., Gao, G., Li, C., & Zhang, Z. (2017). GEPIA: A web server for cancer and normal gene expression profiling and interactive analyses. *Nucleic Acids Research*, *45*(W1), W98–W102. https://doi.org/10.1093/nar/gkx247.
Tatusov, R. L., Koonin, E. V., & Lipman, D. J. (1997). A genomic perspective on protein families. *Science (New York, N.Y.)*, *278*(5338), 631–637. https://doi.org/10.1126/science.278.5338.631.
Therneau, T. M., & Ballman, K. V. (2008). What does PLIER really do? *Cancer Informatics*, *6*, 423–431.
Thirumal Kumar, D., Shree Devi, M. S., Udhaya Kumar, S., Sherlin, A., Mathew, A., Lakshmipriya, M., et al. (2022). Chapter nine—Understanding the activating mechanism of the immune system against COVID-19 by Traditional Indian Medicine: Network

pharmacology approach. In R. Donev (Ed.), *Vol. 129. Advances in protein chemistry and structural biology* (pp. 275–379). Academic Press. https://doi.org/10.1016/bs.apcsb.2021.11.007.

Tian, L., Greenberg, S. A., Kong, S. W., Altschuler, J., Kohane, I. S., & Park, P. J. (2005). Discovering statistically significant pathways in expression profiling studies. *Proceedings of the National Academy of Sciences of the United States of America, 102*(38), 13544–13549. https://doi.org/10.1073/pnas.0506577102.

Tina, K. G., Bhadra, R., & Srinivasan, N. (2007). PIC: Protein interactions calculator. *Nucleic Acids Research, 35*(Web Server issue), W473–W476. https://doi.org/10.1093/nar/gkm423.

Triche, T. J., Schofield, D., & Buckley, J. (2001). DNA microarrays in pediatric cancer. *Cancer Journal (Sudbury, Mass.)*, 7(1), 2–15.

Troyanskaya, O., Cantor, M., Sherlock, G., Brown, P., Hastie, T., Tibshirani, R., et al. (2001). Missing value estimation methods for DNA microarrays. *Bioinformatics, 17*(6), 520–525. https://doi.org/10.1093/bioinformatics/17.6.520.

Troyanskaya, O. G., Garber, M. E., Brown, P. O., Botstein, D., & Altman, R. B. (2002). Nonparametric methods for identifying differentially expressed genes in microarray data. *Bioinformatics, 18*(11), 1454–1461. https://doi.org/10.1093/bioinformatics/18.11.1454.

Tseng, G. C., Ghosh, D., & Feingold, E. (2012). Comprehensive literature review and statistical considerations for microarray meta-analysis. *Nucleic Acids Research, 40*(9), 3785–3799. https://doi.org/10.1093/nar/gkr1265.

Tuikkala, J., Elo, L., Nevalainen, O. S., & Aittokallio, T. (2006). Improving missing value estimation in microarray data with gene ontology. *Bioinformatics, 22*(5), 566–572. https://doi.org/10.1093/bioinformatics/btk019.

Tuncbag, N., Keskin, O., & Gursoy, A. (2010). HotPoint: Hot spot prediction server for protein interfaces. *Nucleic Acids Research, 38*(Web Server issue), W402–W406. https://doi.org/10.1093/nar/gkq323.

Tusher, V. G., Tibshirani, R., & Chu, G. (2001). Significance analysis of microarrays applied to the ionizing radiation response. *Proceedings of the National Academy of Sciences, 98*(9), 5116–5121. https://doi.org/10.1073/pnas.091062498.

Udhaya Kumar, S., Madhana Priya, N., Thirumal Kumar, D., Anu Preethi, V., Kumar, V., Nagarajan, D., et al. (2021). An integrative analysis to distinguish between emphysema (EML) and alpha-1 antitrypsin deficiency-related emphysema (ADL)—A systems biology approach. In *Vol. 127. Advances in protein chemistry and structural biology* (pp. 315–342). Elsevier. https://doi.org/10.1016/bs.apcsb.2021.02.004.

Udhaya Kumar, S., Saleem, A., Thirumal Kumar, D., Anu Preethi, V., Younes, S., Zayed, H., et al. (2021). A systemic approach to explore the mechanisms of drug resistance and altered signaling cascades in extensively drug-resistant tuberculosis. In *Vol. 127. Advances in protein chemistry and structural biology* (pp. 343–364). Elsevier. https://doi.org/10.1016/bs.apcsb.2021.02.002.

Udhaya Kumar, S., Thirumal Kumar, D., Bithia, R., Sankar, S., Magesh, R., Sidenna, M., et al. (2020). Analysis of differentially expressed genes and molecular pathways in familial hypercholesterolemia involved in atherosclerosis: A systematic and bioinformatics approach. *Frontiers in Genetics, 11*, 734. https://doi.org/10.3389/fgene.2020.00734.

Udhaya Kumar, S., Thirumal Kumar, D., Siva, R., George Priya Doss, C., Younes, S., Younes, N., et al. (2020). Dysregulation of signaling pathways due to differentially expressed genes from the B-cell transcriptomes of systemic lupus erythematosus patients—A bioinformatics approach. *Frontiers in Bioengineering and Biotechnology, 8*. https://www.frontiersin.org/article/10.3389/fbioe.2020.00276.

Vaida, M., & Purcell, K. (2019). Hypergraph Link Prediction: Learning Drug Interaction Networks Embeddings. In *2019 18th IEEE International Conference On Machine Learning And Applications (ICMLA)* (pp. 1860–1865). https://doi.org/10.1109/ICMLA.2019.00299.

van Hal, N. L. W., Vorst, O., van Houwelingen, A. M. M. L., Kok, E. J., Peijnenburg, A., Aharoni, A., et al. (2000). The application of DNA microarrays in gene expression analysis. *Journal of Biotechnology*, *78*(3), 271–280. https://doi.org/10.1016/S0168-1656(00)00204-2.

von Mering, C., Krause, R., Snel, B., Cornell, M., Oliver, S. G., Fields, S., et al. (2002). Comparative assessment of large-scale data sets of protein–protein interactions. *Nature*, *417*(6887), 399–403. https://doi.org/10.1038/nature750.

Wallraff, G., Labadie, J., Brock, P., DiPietro, R., Nguyen, T., Huynh, T., et al. (1997). DNA sequencing on a chip. *ChemTech*, *27*(2). https://www.osti.gov/biblio/458263.

Walt, D. R. (2000). Techview: Molecular biology. Bead-based fiber-optic arrays. *Science (New York, N.Y.)*, *287*(5452), 451–452. https://doi.org/10.1126/science.287.5452.451.

Wang, S., Ren, P., Chen, Z., Ren, Z., Ma, J., & de Rijke, M. (2019). Order-free medicine combination prediction with graph convolutional reinforcement learning. In *Proceedings of the 28th ACM international conference on information and knowledge management* (pp. 1623–1632). https://doi.org/10.1145/3357384.3357965.

Wang, D., Urisman, A., Liu, Y.-T., Springer, M., Ksiazek, T. G., Erdman, D. D., et al. (2003). Viral discovery and sequence recovery using DNA microarrays. *PLoS Biology*, *1*(2), E2. https://doi.org/10.1371/journal.pbio.0000002.

Wang, D. D., Zheng, Y., Toledo, E., Razquin, C., Ruiz-Canela, M., Guasch-Ferré, M., et al. (2018). Lipid metabolic networks, Mediterranean diet and cardiovascular disease in the PREDIMED trial. *International Journal of Epidemiology*, *47*(6), 1830–1845. https://doi.org/10.1093/ije/dyy198.

Warde-Farley, D., Donaldson, S. L., Comes, O., Zuberi, K., Badrawi, R., Chao, P., et al. (2010). The GeneMANIA prediction server: Biological network integration for gene prioritization and predicting gene function. *Nucleic Acids Research*, *38*(Web Server issue), W214–W220. https://doi.org/10.1093/nar/gkq537.

Weeraratna, A. T., & Taub, D. D. (2007). Microarray data analysis. In M. J. Korenberg (Ed.), *Vol. 377. Microarray data analysis* (pp. 1–16). Humana Press. https://doi.org/10.1007/978-1-59745-390-5_1.

Wei, Q., La, D., & Kihara, D. (2017). BindML/BindML+: Detecting protein-protein interaction interface propensity from amino acid substitution patterns. *Methods in Molecular Biology (Clifton, N.J.)*, *1529*, 279–289. https://doi.org/10.1007/978-1-4939-6637-0_14.

Wei, C., Li, J., & Bumgarner, R. E. (2004). Sample size for detecting differentially expressed genes in microarray experiments. *BMC Genomics*, *5*, 87. https://doi.org/10.1186/1471-2164-5-87.

Weir, B. A., Woo, M. S., Getz, G., Perner, S., Ding, L., Beroukhim, R., et al. (2007). Characterizing the cancer genome in lung adenocarcinoma. *Nature*, *450*(7171), 893–898. https://doi.org/10.1038/nature06358.

Whetzel, P. L., Parkinson, H., Causton, H. C., Fan, L., Fostel, J., Fragoso, G., et al. (2006). The MGED ontology: A resource for semantics-based description of microarray experiments. *Bioinformatics*, *22*(7), 866–873. https://doi.org/10.1093/bioinformatics/btl005.

Wu, Z., Pan, S., Chen, F., Long, G., Zhang, C., & Yu, P. S. (2021). A comprehensive survey on graph neural networks. *IEEE Transactions on Neural Networks and Learning Systems*, *32*(1), 4–24. https://doi.org/10.1109/TNNLS.2020.2978386.

Wu, J., Vallenius, T., Ovaska, K., Westermarck, J., Mäkelä, T. P., & Hautaniemi, S. (2009). Integrated network analysis platform for protein-protein interactions. *Nature Methods*, *6*(1), 75–77. https://doi.org/10.1038/nmeth.1282.

Xenarios, I., Rice, D. W., Salwinski, L., Baron, M. K., Marcotte, E. M., & Eisenberg, D. (2000). DIP: The database of interacting proteins. *Nucleic Acids Research*, *28*(1), 289–291.

Xia, J., Benner, M. J., & Hancock, R. E. W. (2014). NetworkAnalyst—Integrative approaches for protein–protein interaction network analysis and visual exploration. *Nucleic Acids Research*, *42*(Web Server issue), W167–W174. https://doi.org/10.1093/nar/gku443.

Xia, J., Gill, E. E., & Hancock, R. E. (2015). NetworkAnalyst for statistical, visual and network-based meta-analysis of gene expression data. *Nature Protocols*, *10*(6), 823–844. https://doi.org/10.1038/nprot.2015.052.

Xia, X., McClelland, M., & Wang, Y. (2005). WebArray: An online platform for microarray data analysis. *BMC Bioinformatics*, *6*(1), 306. https://doi.org/10.1186/1471-2105-6-306.

Xiang, Q., Dai, X., Deng, Y., He, C., Wang, J., Feng, J., et al. (2008). Missing value imputation for microarray gene expression data using histone acetylation information. *BMC Bioinformatics*, *9*(1), 252. https://doi.org/10.1186/1471-2105-9-252.

Yang, J.-S., Campagna, A., Delgado, J., Vanhee, P., Serrano, L., & Kiel, C. (2012). SAPIN: A framework for the structural analysis of protein interaction networks. *Bioinformatics*, *28*(22), 2998–2999. https://doi.org/10.1093/bioinformatics/bts539.

Yuan, Q., Chen, J., Zhao, H., Zhou, Y., & Yang, Y. (2021). Structure-aware protein-protein interaction site prediction using deep graph convolutional network. *Bioinformatics (Oxford, England)*, btab643. https://doi.org/10.1093/bioinformatics/btab643.

Yuan, Q., Chen, J., Zhao, H., Zhou, Y., & Yang, Y. (2022). Structure-aware protein–protein interaction site prediction using deep graph convolutional network. *Bioinformatics*, *38*(1), 125–132. https://doi.org/10.1093/bioinformatics/btab643.

Zhang, B., Tian, Y., & Zhang, Z. (2014). Network biology in medicine and beyond. *Circulation. Cardiovascular Genetics*, 7(4), 536–547. https://doi.org/10.1161/CIRCGENETICS.113.000123.

Zhang, H., Yu, C. Y., Singer, B., & Xiong, M. (2001). Recursive partitioning for tumor classification with gene expression microarray data. *Proceedings of the National Academy of Sciences of the United States of America*, *98*(12), 6730–6735. https://doi.org/10.1073/pnas.111153698.

Zhang, L., Zhang, Y., (Derek), Zhao, P., & Huang, S.-M. (2009). Predicting drug–drug interactions: An FDA perspective. *The AAPS Journal*, *11*(2), 300–306. https://doi.org/10.1208/s12248-009-9106-3.

Zhang, Q. C., Petrey, D., Garzón, J. I., Deng, L., & Honig, B. (2013). PrePPI: A structure-informed database of protein-protein interactions. *Nucleic Acids Research*, *41*(Database issue), D828–D833. https://doi.org/10.1093/nar/gks1231.

Zhong, H., Kim, S., Zhi, D., & Cui, X. (2019). Predicting gene expression using DNA methylation in three human populations. *PeerJ*, 7, e6757. https://doi.org/10.7717/peerj.6757.

Zhou, G., Soufan, O., Ewald, J., Hancock, R. E. W., Basu, N., & Xia, J. (2019). NetworkAnalyst 3.0: A visual analytics platform for comprehensive gene expression profiling and meta-analysis. *Nucleic Acids Research*, *47*(W1), W234–W241. https://doi.org/10.1093/nar/gkz240.

Ziauddin, J., & Sabatini, D. M. (2001). Microarrays of cells expressing defined cDNAs. *Nature*, *411*(6833), 107–110. https://doi.org/10.1038/35075114.

Zitnik, M., Agrawal, M., & Leskovec, J. (2018). Modeling polypharmacy side effects with graph convolutional networks. *Bioinformatics (Oxford, England)*, *34*(13), i457–i466. https://doi.org/10.1093/bioinformatics/bty294.

Zou, M., & Conzen, S. D. (2005). A new dynamic Bayesian network (DBN) approach for identifying gene regulatory networks from time course microarray data. *Bioinformatics*, *21*(1), 71–79. https://doi.org/10.1093/bioinformatics/bth463.

Further reading

Adriaens, M. E., Jaillard, M., Waagmeester, A., Coort, S. L. M., Pico, A. R., & Evelo, C. T. A. (2008). The public road to high-quality curated biological pathways. *Drug Discovery Today*, *13*, 856–862. https://doi.org/10.1016/j.drudis.2008.06.013.

Backes, C., Keller, A., Kuentzer, J., Kneissl, B., Comtesse, N., Elnakady, Y. A., et al. (2007). GeneTrail—Advanced gene set enrichment analysis. *Nucleic Acids Research*, *35*(Suppl. 2), W186–W192.

Bengtsson, H., Simpson, K., Bullard, J., & Hansen, K. (2008). *aroma.affymetrix: A generic framework in R for analyzing small to very large Affymetrix data sets in bounded memory.* Report 745, Department of Statistics, University of California, Berkeley.

Bisognin, A., Coppe, A., Ferrari, F., Risso, D., Romualdi, C., Bicciato, S., et al. (2009). A-MADMAN: Annotation-based microarray data meta-analysis tool. *BMC Bioinformatics*, *10*(1), 201. https://doi.org/10.1186/1471-2105-10-201.

Boegel, S. (Ed.). (2020). *Vol. 2120. Bioinformatics for cancer immunotherapy: Methods and protocols* Springer US. https://doi.org/10.1007/978-1-0716-0327-7.

Cahan, P., Ahmad, A. M., Burke, H., Fu, S., Lai, Y., Florea, L., et al. (2005). List of lists-annotated (LOLA): A database for annotation and comparison of published microarray gene lists. *Gene*, *360*(1), 78–82. https://doi.org/10.1016/j.gene.2005.07.008.

Caspi, R., Billington, R., Keseler, I. M., Kothari, A., Krummenacker, M., Midford, P. E., et al. (2020). The MetaCyc database of metabolic pathways and enzymes—A 2019 update. *Nucleic Acids Research*, *48*(D1), D445–D453. https://doi.org/10.1093/nar/gkz862.

Cerami, E. G., Bader, G. D., Gross, B. E., & Sander, C. (2006). cPath: Open source software for collecting, storing, and querying biological pathways. *BMC Bioinformatics*, 7(1), 497. https://doi.org/10.1186/1471-2105-7-497.

Chen, J. Y., Mamidipalli, S., & Huan, T. (2009). HAPPI: An online database of comprehensive human annotated and predicted protein interactions. *BMC Genomics*, *10*(1), S16. https://doi.org/10.1186/1471-2164-10-S1-S16.

Chen, K., & Rajewsky, N. (2006). Natural selection on human microRNA binding sites inferred from SNP data. *Nature Genetics*, *38*(12), 1452–1456. https://doi.org/10.1038/ng1910.

Chen, E. Y., Tan, C. M., Kou, Y., Duan, Q., Wang, Z., Meirelles, G. V., et al. (2013). Enrichr: Interactive and collaborative HTML5 gene list enrichment analysis tool. *BMC Bioinformatics*, *14*(1), 128. https://doi.org/10.1186/1471-2105-14-128.

Cheng, W.-C., Tsai, M.-L., Chang, C.-W., Huang, C.-L., Chen, C.-R., Shu, W.-Y., et al. (2010). Microarray meta-analysis database (M2DB): A uniformly pre-processed, quality controlled, and manually curated human clinical microarray database. *BMC Bioinformatics*, *11*, 421. https://doi.org/10.1186/1471-2105-11-421.

Chu, V. T., Gottardo, R., Raftery, A. E., Bumgarner, R. E., & Yeung, K. (2008). MeV+R: Using MeV as a graphical user interface for Bioconductor applications in microarray analysis. *Genome Biology*, *9*(7), R118. https://doi.org/10.1186/gb-2008-9-7-r118.

Chung, H.-J., Kim, M., Park, C. H., Kim, J., & Kim, J. H. (2004). ArrayXPath: Mapping and visualizing microarray gene-expression data with integrated biological pathway resources using Scalable Vector Graphics. *Nucleic Acids Research*, *32*(Suppl. 2), W460–W464. https://doi.org/10.1093/nar/gkh476.

Clough, E., & Barrett, T. (2016). The gene expression omnibus database. In E. Mathé, & S. Davis (Eds.), *1418. Statistical genomics* (pp. 93–110). New York: Springer. https://doi.org/10.1007/978-1-4939-3578-9_5.

Culhane, A. C., Schröder, M. S., Sultana, R., Picard, S. C., Martinelli, E. N., Kelly, C., et al. (2012). GeneSigDB: A manually curated database and resource for analysis of gene expression signatures. *Nucleic Acids Research*, *40*(Database issue), D1060–D1066. https://doi.org/10.1093/nar/gkr901.

Dietrich, S., Wiegand, S., & Liesegang, H. (2014). TraV: A genome context sensitive transcriptome browser. *PLoS One*, *9*(4), e93677. https://doi.org/10.1371/journal.pone.0093677.

Draghici, S., Khatri, P., Bhavsar, P., Shah, A., Krawetz, S. A., & Tainsky, M. A. (2003). Onto-Tools, the toolkit of the modern biologist: Onto-express, onto-compare, onto-design and onto-translate. *Nucleic Acids Research*, *31*(13), 3775–3781. https://doi.org/10.1093/nar/gkg624.

Ekins, S., Bugrim, A., Brovold, L., Kirillov, E., Nikolsky, Y., Rakhmatulin, E., et al. (2006). Algorithms for network analysis in systems-ADME/Tox using the MetaCore and MetaDrug platforms. *Xenobiotica*, *36*(10–11), 877–901. https://doi.org/10.1080/00498250600861660.

Fang, H., Harris, S. C., Su, Z., Chen, M., Qian, F., Shi, L., et al. (2009). ArrayTrack: An FDA and public genomic tool. *Methods in Molecular Biology (Clifton, N.J.)*, *563*, 379–398. https://doi.org/10.1007/978-1-60761-175-2_20.

Frolkis, A., Knox, C., Lim, E., Jewison, T., Law, V., Hau, D. D., et al. (2010). SMPDB: The small molecule pathway database. *Nucleic Acids Research*, *38*(Database issue), D480–D487. https://doi.org/10.1093/nar/gkp1002.

Gene Ontology Consortium. (2004). The Gene Ontology (GO) database and informatics resource. *Nucleic Acids Research*, *32*(90001), 258D–261. https://doi.org/10.1093/nar/gkh036.

Goffard, N., & Weiller, G. (2007). PathExpress: A web-based tool to identify relevant pathways in gene expression data. *Nucleic Acids Research*, *35*(Web Server issue), W176–W181. https://doi.org/10.1093/nar/gkm261.

Heiskanen, M. A., Lorenz, J., Bian, X., Madhavan, S., Gustafson, S., Muju, S., et al. (2005). Cancer microarray informatics (caArray) data management and analysis tools at the National Cancer Institute (NCI) Center for Bioinformatics. *Cancer Research*, *65*(Suppl. 9), 7.

Hijikata, A., Kitamura, H., Kimura, Y., Yokoyama, R., Aiba, Y., Bao, Y., et al. (2007). Construction of an open-access database that integrates cross-reference information from the transcriptome and proteome of immune cells. *Bioinformatics*, *23*(21), 2934–2941. https://doi.org/10.1093/bioinformatics/btm430.

Hong, F., Breitling, R., McEntee, C. W., Wittner, B. S., Nemhauser, J. L., & Chory, J. (2006). RankProd: A bioconductor package for detecting differentially expressed genes in meta-analysis. *Bioinformatics*, *22*(22), 2825–2827. https://doi.org/10.1093/bioinformatics/btl476.

Hong, S.-E., Rho, S.-H., Yeom, Y. I., & Kim, D. H. (2006). HCNet: A database of heart and calcium functional network. *Bioinformatics (Oxford, England)*, *22*(16), 2053–2054. https://doi.org/10.1093/bioinformatics/btl331.

Hu, Z., Ng, D. M., Yamada, T., Chen, C., Kawashima, S., Mellor, J., et al. (2007). VisANT 3.0: New modules for pathway visualization, editing, prediction and construction. *Nucleic Acids Research*, *35*(Suppl. 2), W625–W632. https://doi.org/10.1093/nar/gkm295.

Ivliev, A. E., Hoen, P. A. C., 't Villerius, M. P., den Dunnen, J. T., & Brandt, B. W. (2008). Microarray retriever: A web-based tool for searching and large scale retrieval of public microarray data. *Nucleic Acids Research*, *36*(Web Server issue), W327–W331. https://doi.org/10.1093/nar/gkn213.

Kanehisa, M., Goto, S., Sato, Y., Kawashima, M., Furumichi, M., & Tanabe, M. (2014). Data, information, knowledge and principle: Back to metabolism in KEGG. *Nucleic Acids Research*, *42*(Database issue), D199–D205. https://doi.org/10.1093/nar/gkt1076.

Kapushesky, M., Kemmeren, P., Culhane, A. C., Durinck, S., Ihmels, J., Körner, C., et al. (2004). Expression Profiler: Next generation—An online platform for analysis of microarray data. *Nucleic Acids Research, 32*(Web Server issue), W465–W470. https://doi.org/10.1093/nar/gkh470.

Karp, P. D., Paley, S. M., Midford, P. E., Krummenacker, M., Billington, R., Kothari, A., et al. (2020). *Pathway tools version 24.0: Integrated software for pathway/genome informatics and systems biology*. ArXiv:1510.03964 [q-Bio] http://arxiv.org/abs/1510.03964.

Kayala, M. A., & Baldi, P. (2012). Cyber-T web server: Differential analysis of high-throughput data. *Nucleic Acids Research, 40*(Web Server issue), W553–W559. https://doi.org/10.1093/nar/gks420.

Kim, S.-B., Yang, S., Kim, S.-K., Kim, S. C., Woo, H. G., Volsky, D. J., et al. (2007). GAzer: Gene set analyzer. *Bioinformatics, 23*(13), 1697–1699. https://doi.org/10.1093/bioinformatics/btm144.

Kodama, Y., Mashima, J., Kosuge, T., & Ogasawara, O. (2019). DDBJ update: The Genomic Expression Archive (GEA) for functional genomics data. *Nucleic Acids Research, 47*(D1), D69–D73. https://doi.org/10.1093/nar/gky1002.

Kolpakov, F., Poroikov, V., Sharipov, R., Kondrakhin, Y., Zakharov, A., Lagunin, A., et al. (2007). CYCLONET—An integrated database on cell cycle regulation and carcinogenesis. *Nucleic Acids Research, 35*(Database issue), D550–D556. https://doi.org/10.1093/nar/gkl912.

Kupershmidt, I., Su, Q. J., Grewal, A., Sundaresh, S., Halperin, I., Flynn, J., et al. (2010). Ontology-based meta-analysis of global collections of high-throughput public data. *PLoS One, 5*(9), e13066. https://doi.org/10.1371/journal.pone.0013066.

Lakiotaki, K., Vorniotakis, N., Tsagris, M., Georgakopoulos, G., & Tsamardinos, I. (2018). BioDataome: A collection of uniformly preprocessed and automatically annotated datasets for data-driven biology. *Database, 2018*. https://doi.org/10.1093/database/bay011.

Li, C., Donizelli, M., Rodriguez, N., Dharuri, H., Endler, L., Chelliah, V., et al. (2010). BioModels Database: An enhanced, curated and annotated resource for published quantitative kinetic models. *BMC Systems Biology, 4*(1), 92. https://doi.org/10.1186/1752-0509-4-92.

Liao, Y., Wang, J., Jaehnig, E. J., Shi, Z., & Zhang, B. (2019). WebGestalt 2019: Gene set analysis toolkit with revamped UIs and APIs. *Nucleic Acids Research, 47*(W1), W199–W205. https://doi.org/10.1093/nar/gkz401.

Liberzon, A., Birger, C., Thorvaldsdóttir, H., Ghandi, M., Mesirov, J. P., & Tamayo, P. (2015). The Molecular Signatures Database (MSigDB) hallmark gene set collection. *Cell Systems, 1*(6), 417–425. https://doi.org/10.1016/j.cels.2015.12.004.

Lottaz, C., Yang, X., Scheid, S., & Spang, R. (2006). OrderedList—A bioconductor package for detecting similarity in ordered gene lists. *Bioinformatics, 22*(18), 2315–2316. https://doi.org/10.1093/bioinformatics/btl385.

Manduchi, E., Grant, G. R., He, H., Liu, J., Mailman, M. D., Pizarro, A. D., et al. (2004). RAD and the RAD Study-Annotator: An approach to collection, organization and exchange of all relevant information for high-throughput gene expression studies. *Bioinformatics (Oxford, England), 20*(4), 452–459. https://doi.org/10.1093/bioinformatics/btg428.

Manduchi, E., Grant, G. R., McKenzie, S. E., Overton, G. C., Surrey, S., & Stoeckert, C. J. (2000). Generation of patterns from gene expression data by assigning confidence to differentially expressed genes. *Bioinformatics (Oxford, England), 16*(8), 685–698. https://doi.org/10.1093/bioinformatics/16.8.685.

Marot, G., & Bruyere, R. (2015). *Using metaMA for differential gene expression analysis from multiple studies* (p. 7). Retrieved from. http://cran.r-project.org/web/packages/metaMA/metaMA.pdf.

Martens, M., Ammar, A., Riutta, A., Waagmeester, A., Slenter, D. N., Hanspers, K., et al. (2021). WikiPathways: Connecting communities. *Nucleic Acids Research*, *49*(D1), D613–D621. https://doi.org/10.1093/nar/gkaa1024.

Marzec, J., Dayem Ullah, A. Z., Pirrò, S., Gadaleta, E., Crnogorac-Jurcevic, T., Lemoine, N. R., et al. (2018). The pancreatic expression database: 2018 update. *Nucleic Acids Research*, *46*(D1), D1107–D1110. https://doi.org/10.1093/nar/gkx955.

Merico, D., Isserlin, R., Stueker, O., Emili, A., & Bader, G. D. (2010). Enrichment map: A network-based method for gene-set enrichment visualization and interpretation. *PLoS One*, *5*(11), e13984. https://doi.org/10.1371/journal.pone.0013984.

Mi, H., & Thomas, P. (2009). PANTHER pathway: An ontology-based pathway database coupled with data analysis tools. *Methods in Molecular Biology (Clifton, NJ)*, *563*, 123–140. https://doi.org/10.1007/978-1-60761-175-2_7.

Mizuno, H., Kitada, K., Nakai, K., & Sarai, A. (2009). PrognoScan: A new database for meta-analysis of the prognostic value of genes. *BMC Medical Genomics*, *2*, 18. https://doi.org/10.1186/1755-8794-2-18.

Nagasaki, M., Saito, A., Doi, A., Matsuno, H., & Miyano, S. (2009). Pathway databases. In A. Dress, M. Vingron, G. Myers, R. Giegerich, W. Fitch, & P. A. Pevzner (Eds.), *Vol. 13. Foundations of systems biology* (pp. 5–18). London: Springer. https://doi.org/10.1007/978-1-84882-023-4_2.

Nikitin, A., Egorov, S., Daraselia, N., & Mazo, I. (2003). Pathway studio—The analysis and navigation of molecular networks. *Bioinformatics (Oxford, England)*, *19*(16), 2155–2157. https://doi.org/10.1093/bioinformatics/btg290.

Numata, K., Yoshida, R., Nagasaki, M., Saito, A., Imoto, S., & Miyano, S. (2008). ExonMiner: Web service for analysis of GeneChip Exon array data. *BMC Bioinformatics*, *9*(1), 494. https://doi.org/10.1186/1471-2105-9-494.

Obayashi, T., Kagaya, Y., Aoki, Y., Tadaka, S., & Kinoshita, K. (2019). COXPRESdb v7: A gene coexpression database for 11 animal species supported by 23 coexpression platforms for technical evaluation and evolutionary inference. *Nucleic Acids Research*, *47*(D1), D55–D62. https://doi.org/10.1093/nar/gky1155.

Pan, F., Chiu, C.-H., Pulapura, S., Mehan, M. R., Nunez-Iglesias, J., Zhang, K., et al. (2007). Gene Aging Nexus: A web database and data mining platform for microarray data on aging. *Nucleic Acids Research*, *35*(Database issue), D756–D759. https://doi.org/10.1093/nar/gkl798.

Pan, F., Kamath, K., Zhang, K., Pulapura, S., Achar, A., Nunez-Iglesias, J., et al. (2006). Integrative array analyzer: A software package for analysis of cross-platform and cross-species microarray data. *Bioinformatics*, *22*(13), 1665–1667. https://doi.org/10.1093/bioinformatics/btl163.

Pandey, R., Guru, R. K., & Mount, D. W. (2004). Pathway Miner: Extracting gene association networks from molecular pathways for predicting the biological significance of gene expression microarray data. *Bioinformatics (Oxford, England)*, *20*(13), 2156–2158. https://doi.org/10.1093/bioinformatics/bth215.

Paszkowski-Rogacz, M., Slabicki, M., Pisabarro, M. T., & Buchholz, F. (2010). PhenoFam-gene set enrichment analysis through protein structural information. *BMC Bioinformatics*, *11*(1), 254. https://doi.org/10.1186/1471-2105-11-254.

Pihur, V., Datta, S., & Datta, S. (2009). RankAggreg, an R package for weighted rank aggregation. *BMC Bioinformatics*, *10*(1), 62. https://doi.org/10.1186/1471-2105-10-62.

Pomaznoy, M., Ha, B., & Peters, B. (2018). GOnet: A tool for interactive Gene Ontology analysis. *BMC Bioinformatics*, *19*(1), 470. https://doi.org/10.1186/s12859-018-2533-3.

Ruschhaupt, L. L., Lusa, I., & Gentleman, R. (2017). GeneMeta. *Bioconductor*. https://doi.org/10.18129/B9.BIOC.GENEMETA.

Santos-Zavaleta, A., Salgado, H., Gama-Castro, S., Sánchez-Pérez, M., Gómez-Romero, L., Ledezma-Tejeida, D., et al. (2019). RegulonDB v 10.5: Tackling challenges to unify

classic and high throughput knowledge of gene regulation in E. coli K-12. *Nucleic Acids Research*, *47*(D1), D212–D220. https://doi.org/10.1093/nar/gky1077.
Sartor, M., Mahavisno, V., Keshamouni, V., Cavalcoli, J., Wright, Z., Karnovsky, A., et al. (2009). ConceptGen: A gene set enrichment and gene set relation mapping tool. *Bioinformatics (Oxford, England)*, *26*, 456–463. https://doi.org/10.1093/bioinformatics/btp683.
Schaefer, C. F., Anthony, K., Krupa, S., Buchoff, J., Day, M., Hannay, T., et al. (2009). PID: The pathway interaction database. *Nucleic Acids Research*, *37*(Database issue), D674–D679. https://doi.org/10.1093/nar/gkn653.
Shamir, R., Maron-Katz, A., Tanay, A., Linhart, C., Steinfeld, I., Sharan, R., et al. (2005). EXPANDER—An integrative program suite for microarray data analysis. *BMC Bioinformatics*, *6*(1), 1–12.
Shin, G., Kang, T.-W., Yang, S., Baek, S.-J., Jeong, Y.-S., & Kim, S.-Y. (2011). GENT: Gene Expression Database of Normal and Tumor Tissues. *Cancer Informatics*, *10*, CIN.S7226. https://doi.org/10.4137/CIN.S7226.
Stülke, J., Lassak, J., & Karp, P. D. (2021). The EcoCyc database in 2021. *Frontiers in Microbiology*, *12*, 10.
Subramanian, A., Kuehn, H., Gould, J., Tamayo, P., & Mesirov, J. P. (2007). GSEA-P: A desktop application for Gene Set Enrichment Analysis. *Bioinformatics (Oxford, England)*, *23*(23), 3251–3253. https://doi.org/10.1093/bioinformatics/btm369.
Tomczak, K., Czerwińska, P., & Wiznerowicz, M. (2015). The Cancer Genome Atlas (TCGA): An immeasurable source of knowledge. *Contemporary Oncology*, *19*(1A), A68–A77. https://doi.org/10.5114/wo.2014.47136.
van Schothorst, E. M., & Ost, M. (2016, September 5). Insights in working with complex datasets: Benefits and limitations using the bioinformatics tool MetaCore. In *Nugoweek: Phenotypes and prevention - the interplay of genes, life-style factors and gut environment*. https://research.wur.nl/en/publications/insights-in-working-with-complex-datasets-benefits-and-limitation.
Vastrik, I., D'Eustachio, P., Schmidt, E., Joshi-Tope, G., Gopinath, G., Croft, D., et al. (2007). Reactome: A knowledge base of biologic pathways and processes. *Genome Biology*, *8*(3), R39. https://doi.org/10.1186/gb-2007-8-3-r39.
Viswanathan, G. A., Nudelman, G., Patil, S., & Sealfon, S. C. (2007). BioPP: A tool for web-publication of biological networks. *BMC Bioinformatics*, *8*(1), 168. https://doi.org/10.1186/1471-2105-8-168.
Vollrath, A. L., Smith, A. A., Craven, M., & Bradfield, C. A. (2009). EDGE3: A web-based solution for management and analysis of Agilent two color microarray experiments. *BMC Bioinformatics*, *10*(1), 280. https://doi.org/10.1186/1471-2105-10-280.
Wang, G., Oh, D.-H., & Dassanayake, M. (2020). GOMCL: A toolkit to cluster, evaluate, and extract non-redundant associations of Gene Ontology-based functions. *BMC Bioinformatics*, *21*(1), 139. https://doi.org/10.1186/s12859-020-3447-4.
Xia, Y., Campen, A., Rigsby, D., Guo, Y., Feng, X., Su, E. W., et al. (2007). DGEM—A microarray gene expression database for primary human disease tissues. *Molecular Diagnosis & Therapy*, *11*(3), 145–149. https://doi.org/10.1007/BF03256235.
Xu, J., Kelly, R., Fang, H., et al. (2010). ArrayTrack: A free FDA bioinformatics tool to support emerging biomedical research—An update. *Human Genomics*, *4*, 428. https://doi.org/10.1186/1479-7364-4-6-428.
Yang, Y., Adelstein, S. J., & Kassis, A. I. (2012). Target discovery from data mining approaches. *Drug Discovery Today*, *17*, S16–S23. https://doi.org/10.1016/j.drudis.2011.12.006.
Yu, J., Gu, X., & Yi, S. (2016). Ingenuity pathway analysis of gene expression profiles in distal nerve stump following nerve injury: Insights into Wallerian degeneration. *Frontiers in Cellular Neuroscience*, *10*. https://doi.org/10.3389/fncel.2016.00274.

Zeeberg, B. R., Feng, W., Wang, G., Wang, M. D., Fojo, A. T., Sunshine, M., et al. (2003). GoMiner: A resource for biological interpretation of genomic and proteomic data. *Genome Biology*, *4*(4), R28. https://doi.org/10.1186/gb-2003-4-4-r28.

Zhang, Z., & Fenstermacher, D. (2005). An introduction to MAMA (Meta-Analysis of MicroArray data) system. In *2005 IEEE engineering in medicine and biology 27th annual conference* (pp. 7730–7733). https://doi.org/10.1109/IEMBS.2005.1616304.

Zhang, J., Finney, R. P., Rowe, W., Edmonson, M., Yang, S. H., Dracheva, T., et al. (2007). Systematic analysis of genetic alterations in tumors using Cancer Genome WorkBench (CGWB). *Genome Research*, *17*(7), 1111–1117. https://doi.org/10.1101/gr.5963407.

Zhou, Y., Zhou, B., Pache, L., Chang, M., Khodabakhshi, A. H., Tanaseichuk, O., et al. (2019). Metascape provides a biologist-oriented resource for the analysis of systems-level datasets. *Nature Communications*, *10*(1), 1523. https://doi.org/10.1038/s41467-019-09234-6.

Zoubarev, A., Hamer, K. M., Keshav, K. D., McCarthy, E. L., Santos, J. R. C., Van Rossum, T., et al. (2012). Gemma: A resource for the reuse, sharing and meta-analysis of expression profiling data. *Bioinformatics*, *28*(17), 2272–2273. https://doi.org/10.1093/bioinformatics/bts430.

CHAPTER FOUR

Fusion proteins mediate alternation of protein interaction networks in cancers

Sunanda Biswas Mukherjee, Sumit Mukherjee, and Milana Frenkel-Morgenstern*
Cancer Genomics and BioComputing of Complex Diseases Lab, Azrieli Faculty of Medicine, Bar-Ilan University, Safed, Israel
*Corresponding author: e-mail address: milana.morgenstern@biu.ac.il

Contents

Abstract

Fusions of two different genes could lead to the production of chimeric RNAs, which could be translated into novel fusion (or chimeric) proteins. Fusion proteins often act as oncoproteins and drive cancer development, particularly in leukemia and lymphomas. Fusion proteins modify the existing protein-protein interaction (PPI) networks, which could eliminate some PPIs by removing protein domains in such fusions. This alternation of protein interaction networks could impact the signaling pathways and switch on the cancer-promoting activity that could drive the generation of cancer phenotypes and/or loss of controlled apoptosis. Thus, knowledge of the fusion proteins and their protein interaction networks could facilitate a deeper molecular understanding of cancer development, which could help to design new approaches for cancer therapies. Here, we discuss the structural features of fusion proteins and how they impact the PPI networks in cancers. Further, we discuss how to analyze the fusion protein-mediated alternation of PPI networks in cancers.

Advances in Protein Chemistry and Structural Biology, Volume 131
ISSN 1876-1623
https://doi.org/10.1016/bs.apcsb.2022.05.007

Abbreviations

ACC adenoid cystic carcinomas
APL acute promyelocytic leukemia
ChiPPI Chimeric protein-protein interactions
CML chronic myeloid leukemia

1. Introduction

The fusion of two or more genes results in the formation of chimeric RNA, which plays a significant role in the generation of new functions and regulatory mechanisms in the cell (Jia, Xie, & Li, 2016; Li, Wang, Ma, & Sklar, 2009; Mertens, Johansson, Fioretos, & Mitelman, 2015; Mukherjee, Detroja, et al., 2021; Sun & Li, 2022). Some chimeric RNAs can encode new proteins known as fusion or chimeric proteins, which play an important role in generating new disease phenotypes (Chwalenia, Facemire, & Li, 2017; Mukherjee & Frenkel-Morgenstern, 2021; Zheng, 2013). Gene fusions are one of the characteristic features of cancers, where fusion proteins are produced by chromosomal translocations or trans-splicing and have a role in cancer development, cell differentiation, cancer heterogeneity, and drug resistance (Mukherjee, Heng, & Frenkel-Morgenstern, 2021; Wu, Li, & Li, 2019). The recurrent fusions observed in various cancer patients are mostly produced by chromosomal translocations, which are the essential driver for oncogenesis (Zheng, 2013). In the last decades, several fusion proteins have been detected in different cancer types that are associated with cancer development. BCR-ABL has been described as one such classic example of gene fusion in chronic myeloid leukemia (CML) that occurs upon translocation of the q arms of chromosomes 9 and 22 and is denoted as t(9;22) (Bedi et al., 1995; Shtivelman, Lifshitz, Gale, & Canaani, 1985). The fusion protein produced by this BCR-ABL fusion alters constitutively active ABL1 genes and activates a variety of signaling pathways that promote the development of CML. Another example of chromosomal translocation t(15;17) was detected in acute promyelocytic leukemia (APL), leading to the formation of promyelocytic leukemia–retinoic acid receptor (PML-RARα) fusion protein, which interacts with retinoic X receptors (RXR) and promotes epigenetic changes (Saeed, Logie, Stunnenberg, & Martens, 2011). In follicular thyroid carcinomas and some follicular variants of papillary thyroid carcinomas, the translocation of t(2;3)(q13;p25) was found to produce the PAX8-PPAR fusion protein, which could act as a dominant-negative inhibitor of wild-type PPARγ and can activate or repress PAX8-responsive genes (Kroll et al., 2000). In adenoid cystic carcinomas (ACC), the recurrent t(6;9)(q22–23;p23–24) translocation involves the fusion of two transcription factors, MYB and NFIB (Persson et al., 2009; Wysocki et al., 2016). Fusion proteins of MYB and NFIB promote the upregulation of MYB protein, which can lead to the development of ACC. This emerging evidence suggests that the appearance of the fusions proteins by chromosomal translocations could promote cancer development. Fusion proteins that arise in cancers have altered protein domains and could carry diverse structural signatures, which could differentiate them from their parental proteins (Latysheva & Babu, 2019). When they are recruited to the protein interaction network, such different structural signatures could impact the network's global changes, which could activate cell growth pathways and switch various cellular pathways (Frenkel-Morgenstern et al., 2017; Latysheva et al., 2016). Therefore, understanding the structural features of fusion proteins and their mechanism to alter the global PPI networks in the cell could decipher the molecular insights into the fusion proteins mediated functional dysregulation and cancer development.

2. Structural features of fusion proteins in cancers

Fusion proteins generated by the novel chimeric RNAs have unique structural features that differ from their parental proteins, which could help them gain new functional features (Frenkel-Morgenstern et al., 2012). Several studies attempted to understand the structural principles of fusion proteins by analyzing their translocation breakpoint positions, domain architectures, posttranslational modification, and the role of intrinsic disorder regions. Particularly, some recent studies have demonstrated that translocation breakpoints tend to occur in intrinsic disorder regions and maintain the reading frame (Hegyi, Buday, & Tompa, 2009; Latysheva & Babu, 2016). Further, translocation breakpoints have been observed to avoid the splitting of domains. This trend could highlight a selection of fusions to preserve viability and evade degradation pathways. Analysis of posttranslational modification on fusion proteins highlights that phosphorylation and methylation modification are enriched in the disorder areas, suggesting that these posttranslational modifications could impact the generation of different functionality from their parental proteins by epigenetic markers (Wang et al., 2015). A study demonstrated that fusion proteins that originated from chromosomal translocations tend to have fewer domains than non-translocated proteins (Hegyi et al., 2009). However, another study demonstrated that fusion proteins contain complete protein domains significantly more often than random datasets (Frenkel-Morgenstern & Valencia, 2012). The complete functional domains present in fusion proteins could help them compete with parental proteins and generate a dominant-negative effect (Frenkel-Morgenstern & Valencia, 2012). For example, in their dataset (Frenkel-Morgenstern & Valencia, 2012), DNA binding domains were found to be prevalent, but transcriptional activation domains were uncommon, which highlights a known dominant-negative mechanism utilized by some oncogenic fusion proteins (Suzuki et al., 2010). In principle, fusions involving transcriptional activators or repressors could utilize dominant-negative effects for their parental proteins. One of the best examples to explain this mechanism is the RUNX1-ETO fusion protein was detected in acute myeloid leukemia (Lam & Zhang, 2012; Ptasinska et al., 2012). RUNX1-ETO employs a dominant-negative impact over RUNX1, a key regulator of hematopoietic stem cell differentiation, by interfering with normal RUNX1 function and preventing differentiation. The RUNX1-ETO fusion proteins act as constitutive transcriptional repressors. They retain the DNA-binding Runt homology domain from the RUNX1 transcription factor and inherit the ability to bind to RUNX1 target genes, and integrate almost all of the transcriptional repressor ETO protein (Latysheva & Babu, 2016). The domain architecture analysis in fusion proteins reveals that the same domain architecture is reused in different gene fusion events (Ortiz de Mendíbil, Vizmanos, & Novo, 2009). The most prevalently reused domain architectures in fusion proteins involve tyrosine kinases, Runt domain, AT-hook DNA-binding domains, NUP repeats, and coiled-coil domains (Frenkel-Morgenstern & Valencia, 2012). Most of these domains are DNA-binding domains, indicating that such signature of domain combinations was subject to potential functional constraints. Fusion proteins could generate novel domain combinations which are not found in their parental proteins, such as HLH (Helix–Loop–Helix), GTP-binding domain, Hydrolase_3 domain, coiled_coil domain and ZnF_C2C2 (Frenkel-Morgenstern & Valencia, 2012). Further, a study demonstrated that fusion proteins enriched of transmembrane domains and signal peptides which suggests that fusion protein functionality could be modulated by changing the cellular localization

(Frenkel-Morgenstern et al., 2012). Altogether, we could say all the above mentioned unique structural features of fusion proteins lead them to recruit in the normal protein interaction network resulting global change of PPIs in the cells and activation of the oncogenic pathways.

3. How do fusion proteins alter the protein interaction network?

Fusions played a significant role in alternating cellular PPI networks in various cancers (Frenkel-Morgenstern et al., 2017). Some principles for fusion proteins mediated alternation of PPI networks in cancers are designed (Latysheva et al., 2016). Fusion proteins mediated network rewiring is appeared to have one of the common mechanisms of oncogenesis (Fig. 1A). For example, the simultaneous rewiring of signaling pathways could be vital for BCR-ABL1 mediated transformation (Pawson & Warner, 2007). Therefore, targeting the interacting partners or downstream signaling of fusion proteins could be useful for the development of therapeutic targets. Fusion proteins impact the loss of parental interaction-mediating features and regulatory sites. A recent study used a fusion network approach to show that the parental proteins of fusions dominate central positions in PPI networks, losing their functional activities after the recruitment of fusion proteins in the PPI networks in cancers (Latysheva et al., 2016). Further, these fusions were found to escape regulation by losing post-translational modification sites. Therefore, whenever there is a fusion event in cancer, the parental genes tend to lose their original functionality, including the regulatory mechanism and improper connection of the newly formed fusion protein within the protein interacting network, which could switch the cancer cell's functional landscape. This mechanism facilitates fusion proteins to function independently of cellular regulation and to rewire protein interaction networks in a constitutive fashion. Proteins reliant on the function of different molecular features (such as the interface residues and nuclear import/export signal motifs in nucleophosmin; Fig. 1B), also proteins sensitive to changes in post-translational modifications content (such as EWSR1; Fig. 1C), could be specifically disrupted by fusion events. Altogether, it can be suggested that fusion proteins could alter the topology of cellular signaling pathways and promote cancer by deregulating central interaction-prone proteins in the PPI networks. Therefore, targeting the interacting partners or downstream signaling of fusion proteins could be useful for development of personalized therapy in cancers.

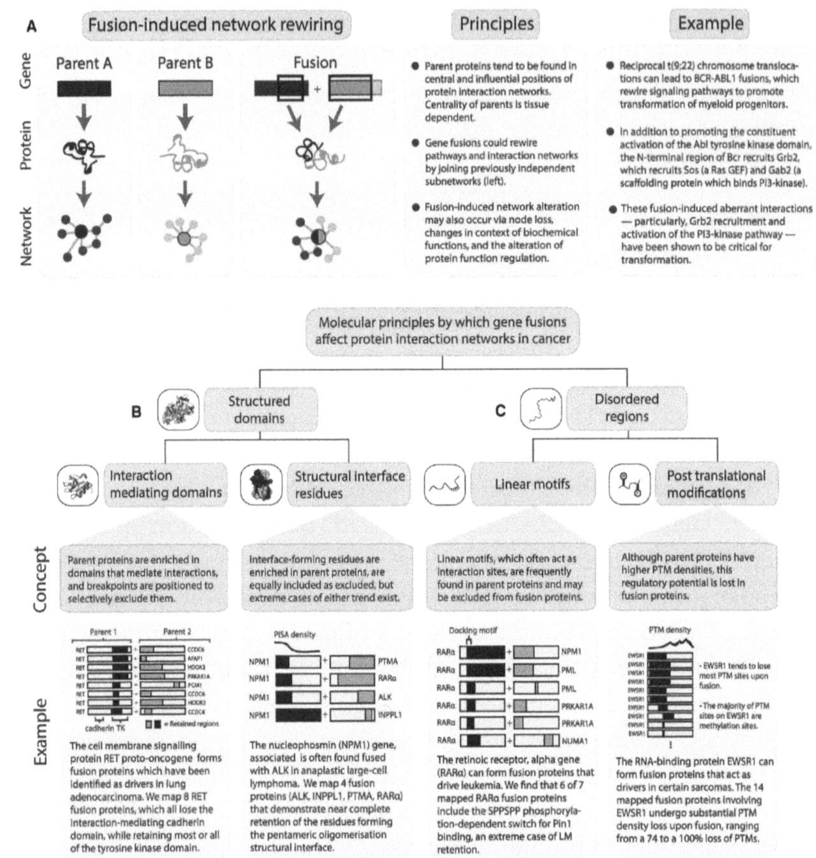

Fig. 1 Principles of fusion protein-mediated alternation of protein interaction networks in cancer (A) Fusion tends to involve highly central proteins in interaction networks and can alter networks by several mechanisms. (B) principles of fusion protein-mediated alterations of molecular interactions of proteins by shuffling interaction-prone regions within ordered protein segment (C) principles of fusion protein-mediated alterations of molecular interactions of proteins by shuffling interaction-prone regions within ordered protein segment disordered protein segments. *Figure adapted from Latysheva, N. S., Oates, M. E., Maddox, L., Flock, T., Gough, J., Buljan, M., et al. (2016). Molecular principles of gene fusion mediated rewiring of protein interaction networks in cancer.* Molecular Cell, 63*(4). doi:10.1016/j.molcel.2016.07.008.*

4. How to analyze the fusion proteins mediated alternation of PPI networks?

Human interactome can be largely defined as networks that comprise physical and functional interactions (Vidal, Cusick, & Barabási, 2011). Physical interactions typically involve the biophysical contact between participating proteins in the network. The PPI networks might be altered in various diseases, including cancers (Gonzalez & Kann, 2012; Yeger-Lotem & Sharan, 2015). Therefore, network-based analysis becomes an essential new tool for understanding the mechanistic insights into disease development. Fusion proteins can alter PPI networks, and PPI network analyses have effectively identified cancer risk factors (Li et al., 2017). At present, only a few methods exist that are useful to analyze the fusion proteins mediated alternation of PPIs. The most popular method is ChiPPI, which uses the domain-domain co-occurrence scores to identify the potential interactors of fusion proteins (Frenkel-Morgenstern et al., 2017). This method uses a domain–domain co-occurrence (DDCOS) score to calculate PPI likelihood. The DDCOS score was calculated based on the previous observations about the preference for domain-domain interactions and the co-occurrence methods (Apic, Gough, & Teichmann, 2001; Itzhaki, Akiva, Altuvia, & Margalit, 2006; Memišević, Wallqvist, & Reifman, 2013). The final score for a fusion protein and its interactors combines all the domains of its two parental proteins. Finally, this method predicts the alternations in networks of the fusion protein in the context of the protected domains of its two parental proteins (Fig. 2). ChiPPI offers a comprehensive

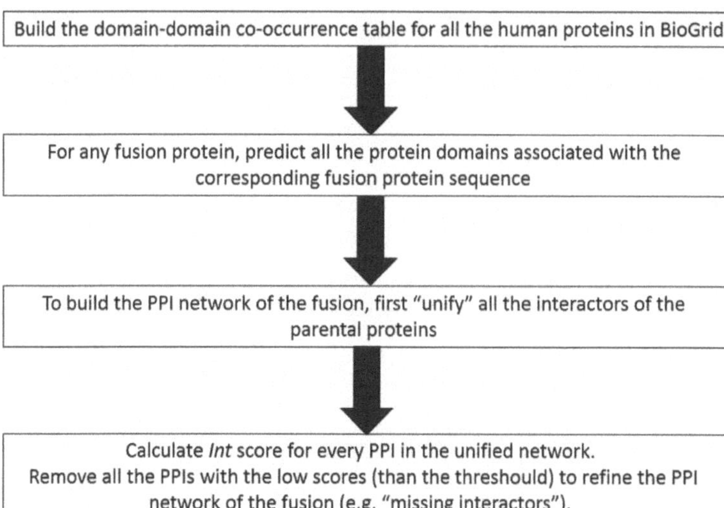

Fig. 2 Generalized scheme to generate a proteome-wide scoring system for all the ChiPPI interactions using discrete protein domains. *Figure adapted from Frenkel-Morgenstern, M., Gorohovski, A., Tagore, S., Sekar, V., Vazquez, M., Valencia, A. (2017). ChiPPI: A novel method for mapping chimeric protein-protein interactions uncovers selection principles of protein fusion events in cancer.* Nucleic Acids Research. *doi:10.1093/nar/gkx423.*

method for analyzing the fusion protein networks and can be used to discover the alternations in the cancer cellular PPI networks. Recently, another method, FusionPathway was developed, which could be promising for predicting both protein-protein and protein-DNA interactions of fusions (Wu, Beird, Zhang, & Futreal, 2018). Therefore, this method could be able to decipher the molecular mechanisms of those fusions that act as transcription factors to deregulate pathways via protein-DNA interactions. Further, this method can identify the pathways which are associated with the individual fusion proteins. Therefore, both ChiPPI and FusionPathway are important tools for analyzing the fusion proteins mediated PPI network alternations and evaluating their functional impact.

5. Identifying altered fusion PPI networks and targeting fusion proteins in cancers could help to develop the precision medicine platform

Recurrent chimeric RNAs could generate the fusion proteins which are found to drive cancer development (Mertens et al., 2015). However, most of the chimeric RNAs detected from the cancer genomes and transcriptomes are not recurrent and are patient and/or sample-specific. Recent studies have shown that patient/sample-specific chimeric RNAs could promote cancer heterogeneity (Mukherjee, Heng, et al., 2021). Further, chimeric RNAs are found to be generated at the later stage of cancers after the drug treatment (Mukherjee, Heng, et al., 2021). These patients and/or sample-specific chimeric RNAs may cause drug resistance cancer relapse. Recent studies suggest that cancer cells change their transcriptional dynamics to adapt and survive with high-stress levels (Kotsantis et al., 2016; Moreno-Smith, Lutgendorf, & Sood, 2010). Different cancer patients have different cancer-related stress and tumor burden; therefore, every patient responds differently to cancer treatments. The fusion proteins translated from chimeric RNA, which can generate after getting a chemotherapy treatment at the later stage of cancers, can alter the functionality such that they can help the cells survive in the face of drug-induced stress (Mukherjee, Heng, et al., 2021). Therefore, identifying the chimeric RNAs and their protein products might be the first step towards understanding the patient-specific cancer heterogeneity. Then, understanding the patient-specific fusion protein-mediated alternation of PPI could help understand the functional significance of the fusion proteins, which could help design a more effective personalized treatment strategy. Because the newly generated fusion proteins could act on the PPI networks, it can change the global alternation of PPIs in the cell, altering the cell functionality. It has been known that gene-gene fusions that are associated with cancers have the potential to serve as sources of high immunogenic neoantigens (Weber et al., 2022). Neoantigens are cancer-specific antigens that are only expressed in tumor cells (Zhang et al., 2021). So, the fusion proteins generated in tumor cells are unique and may serve as neoantigen. A targeted immune response to the tumor can be obtained by identifying the patient-specific neoantigens (Brennick, George, Corwin, Srivastava, & Ebrahimi-Nik, 2017). Hence, fusion proteins could be used as these are tumor-specific and target these proteins for the personalized drug response. Therefore, identifying the unique fusion proteins from the cancer patients with a detailed understanding of the functional importance of these proteins by analyzing the PPI networks could be a new strategy for developing effective precision medicine.

6. Concluding remarks

In summary, we highlighted that fusion proteins have the potential to alter the PPI networks in the cell, thereby altering the cell functionality; it causes dysregulation in the system and finally can lead to cancer development. Recent studies sequenced many genomes and RNA sequences of different tumors across various cancers and uncovered a large cohort of patient-specific chimeric RNAs, which may be translated into fusions proteins. Fusion proteins could be both drivers and passengers in cancers. If they recruit into the PPI networks, they can disrupt the functions of several tumor suppressors and activate the cancer-related pathways. So, it becomes crucial to identify and analyze the patient-specific tumor-related fusion proteins and understand their functional impact on the alternation of PPI networks in cancers, including gained and missing interactors. This could uncover how these fusion proteins are associated with cancer-related processes in a particular patient that could be further useful in designing personalized treatments. Thus, the patient-specific unique and highly expressed fusion proteins, which potentially alter the PPI networks in cancers, could be potential targets for developing diagnostic and personalized treatment.

Acknowledgments

We thank the members of the Cancer Genomics and BioComputing of Complex Diseases laboratory for the useful discussions. We thank the Israel Council for Higher Education (CHE) for the PBC Fellowship awarded to Dr. Sumit Mukherjee under the category of outstanding postdoctoral researchers from China and India (2019–21).

References

Apic, G., Gough, J., & Teichmann, S. A. (2001). Domain combinations in archaeal, eubacterial and eukaryotic proteomes. *Journal of Molecular Biology*, *310*(2), 311–325. https://doi.org/10.1006/jmbi.2001.4776.

Bedi, A., Barber, J. P., Bedi, G. C., El-Deiry, W. S., Sidransky, D., Vala, M. S., et al. (1995). BCR-ABL-mediated inhibition of apoptosis with delay of G2/M transition after DNA damage: A mechanism of resistance to multiple anticancer agents. *Blood*, *86*(3), 1148–1158. https://doi.org/10.1182/blood.v86.3.1148.1148.

Brennick, C. A., George, M. M., Corwin, W. L., Srivastava, P. K., & Ebrahimi-Nik, H. (2017). Neoepitopes as cancer immunotherapy targets: Key challenges and opportunities. *Immunotherapy*, *9*(4), 361–371. https://doi.org/10.2217/imt-2016-0146.

Chwalenia, K., Facemire, L., & Li, H. (2017). Chimeric RNAs in cancer and normal physiology. *Wiley Interdisciplinary Reviews: RNA*, *8*(6), e1427. https://doi.org/10.1002/wrna.1427.

Frenkel-Morgenstern, M., Gorohovski, A., Tagore, S., Sekar, V., Vazquez, M., & Valencia, A. (2017). ChiPPI: A novel method for mapping chimeric protein-protein interactions uncovers selection principles of protein fusion events in cancer. *Nucleic Acids Research*, *45*(12), 7094–7105. https://doi.org/10.1093/nar/gkx423.

Frenkel-Morgenstern, M., Lacroix, V., Ezkurdia, I., Levin, Y., Gabashvili, A., Prilusky, J., et al. (2012). Chimeras taking shape: Potential functions of proteins encoded by chimeric RNA transcripts. *Genome Research*, *22*(7), 1231–1242. https://doi.org/10.1101/gr.130062.111.

Frenkel-Morgenstern, M., & Valencia, A. (2012). Novel domain combinations in proteins encoded by chimeric transcripts. *Bioinformatics*, *28*(12), i67–i74. https://doi.org/10.1093/bioinformatics/bts216.

Gonzalez, M. W., & Kann, M. G. (2012). Chapter 4: Protein interactions and disease. *PLoS Computational Biology*, *8*(12). https://doi.org/10.1371/journal.pcbi.1002819.

Hegyi, H., Buday, L., & Tompa, P. (2009). Intrinsic structural disorder confers cellular viability on oncogenic fusion proteins. *PLoS Computational Biology*, *5*(10). https://doi.org/10.1371/journal.pcbi.1000552.

Itzhaki, Z., Akiva, E., Altuvia, Y., & Margalit, H. (2006). Evolutionary conservation of domain-domain interactions. *Genome Biology*, 7(12), 1–12. https://doi.org/10.1186/gb-2006-7-12-r125.

Jia, Y., Xie, Z., & Li, H. (2016). Intergenically spliced chimeric RNAs in cancer. *Trends in Cancer*, *2*(9), 475–484. https://doi.org/10.1016/j.trecan.2016.07.006.

Kotsantis, P., Silva, L. M., Irmscher, S., Jones, R. M., Folkes, L., Gromak, N., et al. (2016). Increased global transcription activity as a mechanism of replication stress in cancer. *Nature Communications*, 7, 1–13. https://doi.org/10.1038/ncomms13087.

Kroll, T. G., Sarraf, P., Pecciarini, L., Chen, C. J., Mueller, E., Spiegelman, B. M., et al. (2000). PAX8-PPARγ1 fusion in oncogene human thyroid carcinoma. *Science*, *289*(5483), 1357–1360. https://doi.org/10.1126/science.289.5483.1357.

Lam, K., & Zhang, D. E. (2012). RUNX1 and RUNX1-ETO: Roles in hematopoiesis and leukemogenesis. *Frontiers in Bioscience*, *17*(3). https://doi.org/10.2741/3977.

Latysheva, N. S., & Babu, M. M. (2016). Discovering and understanding oncogenic gene fusions through data intensive computational approaches. *Nucleic Acids Research*, *44*(10), 4487–4503. https://doi.org/10.1093/nar/gkw282.

Latysheva, N. S., & Babu, M. M. (2019). Molecular signatures of fusion proteins in cancer. *ACS Pharmacology & Translational Science*, *2*(2), 122–133. https://doi.org/10.1021/acsptsci.9b00019.

Latysheva, N. S., Oates, M. E., Maddox, L., Flock, T., Gough, J., Buljan, M., et al. (2016). Molecular principles of gene fusion mediated rewiring of protein interaction networks in Cancer. *Molecular Cell*, *63*(4), 579–592. https://doi.org/10.1016/j.molcel.2016.07.008.

Li, Z., Ivanov, A. A., Su, R., Gonzalez-Pecchi, V., Qi, Q., Liu, S., et al. (2017). The OncoPPi network of cancer-focused protein-protein interactions to inform biological insights and therapeutic strategies. *Nature Communications*, *8*(1), 1–14. https://doi.org/10.1038/ncomms14356.

Li, H., Wang, J., Ma, X., & Sklar, J. (2009). Gene fusions and RNA trans-splicing in normal and neoplastic human cells. *Cell Cycle*, *8*(2), 218–222. https://doi.org/10.4161/cc.8.2.7358.

Memišević, V., Wallqvist, A., & Reifman, J. (2013). Reconstituting protein interaction networks using parameter-dependent domain-domain interactions. *BMC Bioinformatics*, *14*(1), 1–15. https://doi.org/10.1186/1471-2105-14-154.

Mertens, F., Johansson, B., Fioretos, T., & Mitelman, F. (2015). The emerging complexity of gene fusions in cancer. *Nature Reviews Cancer*, *15*(6), 371–381. https://doi.org/10.1038/nrc3947.

Moreno-Smith, M., Lutgendorf, S. K., & Sood, A. K. (2010). Impact of stress on cancer metastasis. *Future Oncology*, *6*(12). https://doi.org/10.2217/fon.10.142.

Mukherjee, S., Detroja, R., Balamurali, D., Matveishina, E., Medvedeva, Y. A., Valencia, A., et al. (2021). Computational analysis of sense-antisense chimeric transcripts reveals their potential regulatory features and the landscape of expression in human cells. *NAR Genomics and Bioinformatics*, *3*(3), lqab074. https://doi.org/10.1093/nargab/lqab074.

Mukherjee, S., & Frenkel-Morgenstern, M. (2021). Evolutionary impact of chimeric RNAs on generating phenotypic plasticity in human cells. *Trends in Genetics*, *38*(1), 4–7. https://doi.org/10.1016/j.tig.2021.08.015.

Mukherjee, S., Heng, H. H., & Frenkel-Morgenstern, M. (2021). Emerging role of chimeric RNAs in cell plasticity and adaptive evolution of cancer cells. *Cancers*, *13*(17), 4328. https://doi.org/10.3390/CANCERS13174328.

Ortiz de Mendíbil, I., Vizmanos, J. L., & Novo, F. J. (2009). Signatures of selection in fusion transcripts resulting from chromosomal translocations in human cancer. *PLoS One*, *4*(3). https://doi.org/10.1371/journal.pone.0004805.

Pawson, T., & Warner, N. (2007). Oncogenic re-wiring of cellular signaling pathways. *Oncogene*, *26*(9), 1268–1275. https://doi.org/10.1038/sj.onc.1210255.

Persson, M., Andrén, Y., Mark, J., Horlings, H. M., Persson, F., & Stenman, G. (2009). Recurrent fusion of MYB and NFIB transcription factor genes in carcinomas of the breast and head and neck. *Proceedings of the National Academy of Sciences of the United States of America*, *106*(44), 18740–18744. https://doi.org/10.1073/pnas.0909114106.

Ptasinska, A., Assi, S. A., Mannari, D., James, S. R., Williamson, D., Dunne, J., et al. (2012). Depletion of RUNX1/ETO in t(8;21) AML cells leads to genome-wide changes in chromatin structure and transcription factor binding. *Leukemia*, *26*(8). https://doi.org/10.1038/leu.2012.49.

Saeed, S., Logie, C., Stunnenberg, H. G., & Martens, J. H. A. (2011). Genome-wide functions of PML-RARα in acute promyelocytic leukaemia. *British Journal of Cancer*, *104*(4), 554–558. https://doi.org/10.1038/sj.bjc.6606095.

Shtivelman, E., Lifshitz, B., Gale, R. P., & Canaani, E. (1985). Fused transcript of abl and bcr genes in chronic myelogenous leukaemia. *Nature*, *315*(6020), 550–554. https://doi.org/10.1038/315550a0.

Sun, Y., & Li, H. (2022). Chimeric RNAs discovered by RNA sequencing and their roles in Cancer and rare genetic diseases. *Genes*, *13*(5), 741. https://doi.org/10.3390/GENES13050741.

Suzuki, S., Sasaki, S., Morita, H., Oki, Y., Turiya, D., Ito, T., et al. (2010). The role of the amino-terminal domain in the interaction of unliganded peroxisome proliferator-activated receptor γ-2 with nuclear receptor co-repressor. *Journal of Molecular Endocrinology*, *45*(3), 133–145. https://doi.org/10.1677/JME-10-0007.

Vidal, M., Cusick, M. E., & Barabási, A. L. (2011). Interactome networks and human disease. *Cell*, *144*(6), 986–998. https://doi.org/10.1016/j.cell.2011.02.016.

Wang, D., Li, D., Qin, G., Zhang, W., Ouyang, J., Zhang, M., et al. (2015). The structural characterization of tumor fusion genes and proteins. *Computational and Mathematical Methods in Medicine*, *2015*. https://doi.org/10.1155/2015/912742.

Weber, D., Ibn-Salem, J., Sorn, P., Suchan, M., Holtsträter, C., Lahrmann, U., et al. (2022). Accurate detection of tumor-specific gene fusions reveals strongly immunogenic personal neo-antigens. *Nature Biotechnology*, *2022*, 1–9. https://doi.org/10.1038/s41587-022-01247-9.

Wu, C. C., Beird, H. C., Zhang, J., & Futreal, P. A. (2018). FusionPathway: Prediction of pathways and therapeutic targets associated with gene fusions in cancer. *PLoS Computational Biology*, *14*(7). https://doi.org/10.1371/journal.pcbi.1006266.

Wu, H., Li, X., & Li, H. (2019). Gene fusions and chimeric RNAs, and their implications in cancer. *Genes and Diseases*, *6*(4), 385–390. https://doi.org/10.1016/j.gendis.2019.08.002.

Wysocki, P. T., Izumchenko, E., Meir, J., Ha, P. K., Sidransky, D., & Brait, M. (2016). Adenoid cystic carcinoma: Emerging role of translocations and gene fusions. *Oncotarget*, 7(40), 66239. https://doi.org/10.18632/oncotarget.11288.

Yeger-Lotem, E., & Sharan, R. (2015). Human protein interaction networks across tissues and diseases. *Frontiers in Genetics*, *6*, 257. https://doi.org/10.3389/fgene.2015.00257.

Zhang, Z., Lu, M., Qin, Y., Gao, W., Tao, L., Su, W., et al. (2021). Neoantigen: A new breakthrough in tumor immunotherapy. *Frontiers in Immunology*, *12*, 1297. https://doi.org/10.3389/fimmu.2021.672356.

Zheng, J. (2013). Oncogenic chromosomal translocations and human cancer. *Oncology Reports*, *30*(5), 2011–2019. https://doi.org/10.3892/or.2013.2677.

CHAPTER FIVE

Integrative ontology and pathway-based approach identifies distinct molecular signatures in transcriptomes of esophageal squamous cell carcinoma

S. Udhaya Kumar[a], Ambritha Balasundaram[a], V. Anu Preethi[b], Sayoni Chatterjee[a], G.V. Kameshwari Gollakota[a], Manoj Kumar Kashyap[c], C. George Priya Doss[a,*], and Hatem Zayed[d,*]
[a]Laboratory of Integrative Genomics, Department of Integrative Biology, School of Biosciences and Technology, Vellore Institute of Technology, Vellore, India
[b]School of Computer Science and Engineering, Vellore Institute of Technology, Vellore, India
[c]Amity Stem Cell Institute, Amity Medical School, Amity University Haryana, Gurugram, India
[d]Department of Biomedical Sciences, College of Health and Sciences, Qatar University, QU Health, Doha, Qatar
*Corresponding authors: e-mail address: georgepriyadoss@vit.ac.in; hatem.zayed@qu.edu.qa

Contents

Advances in Protein Chemistry and Structural Biology, Volume 131
ISSN 1876-1623
https://doi.org/10.1016/bs.apcsb.2022.04.003

Abstract

Esophageal squamous cell carcinoma (ESCC) remains a serious concern globally due to many factors that including late diagnosis, lack of an ideal biomarker for diagnosis and prognosis, and high rate of mortality. In this study, we aimed to identify the essential dysregulated genes and molecular signatures associated with the progression and development of ESCC. The dataset with 15 ESCCs and the 15 adjacent normal tissue samples from the surrounding histopathologically tumor-free mucosa was selected. We applied bioinformatics pipelines including various topological parameters from MCODE, CytoNCA, and cytoHubba to prioritize the most significantly associated DEGs with ESCC. We performed functional enrichment annotation for the identified DEGs using DAVID and MetaCore™ GeneGo platforms. Furthermore, we validated the essential core genes in TCGA and GTEx datasets between the normal mucosa and ESCC for their expression levels. These DEGs were primarily enriched in positive regulation of transferase activity, negative regulation of organelle organization, cell cycle mitosis/S-phase transition, spindle organization/assembly, development, and regulation of angiogenesis. Subsequently, the DEGs were associated with the pathways such as oocyte meiosis, cell cycle, and DNA replication. Our study identified the eight-core genes (*AURKA*, *AURKB*, *MCM2*, *CDC20*, *TPX2*, *PLK1*, *FOXM1*, and *MCM7*) that are highly expressed among the ESCC, and TCGA dataset. The multigene comparison and principal component analysis resulted in elevated signals for the *AURKA*, *MCM2*, *CDC20*, *TPX2*, *PLK1*, and *FOXM1*. Overall, our study reported GO profiles and molecular signatures that might help researchers to grasp the pathological mechanisms underlying ESCC development and eventually provide novel therapeutic and diagnostic strategies.

Abbreviations

AURKA	aurora kinase A
AURKB	aurora kinase B
BP	biological process
CDC20	cell division cycle 20
CRC	colorectal cancer
DEGs	differentially expressed genes
EAC or EAD	esophageal adenocarcinoma
EC	esophageal carcinoma
EPC	edge percolated component
ESCC	esophageal squamous cell carcinoma
FDR	false discovery rate
FOXM1	forkhead box M1
GO	gene ontology
HNSCC	head and neck squamous cell carcinoma
KEGG	Kyoto Encyclopedia of Genes and Genomes
MCC	maximal clique centrality
MCM2	minichromosome maintenance complex component 2

MCM7	minichromosome maintenance complex component 7
MF	molecular function
MNC	maximum neighborhood component
PCA	principal component analysis
PLK1	polo-like kinase 1
PPIs	protein-protein interactions
TPX2	targeting protein for Xklp2
UMAP	uniform manifold approximation and projection

1. Introduction

Esophageal cancer (EC) remains a major threat to public health worldwide. Histologically, the two major subtypes of esophageal cancer are Esophageal Squamous Cell Carcinoma (ESCC) and Esophageal Adenocarcinoma (EAC or EAD) (Smyth, Lagergren, Fitzgerald, et al., 2017). ESCC and EAC are predominantly present in developing and developed countries, respectively. ESCC accounts for about 80% of all esophageal cancers (Ferlay, Soerjomataram, Dikshit, et al., 2015). Central and Eastern Asia and parts of the African subcontinent are major hotspots for the occurrence of ESCC (Abnet, Arnold, & Wei, 2018). The underlying cause for ESCC differs from region to region (Abnet et al., 2018; Yan, Cui, Zhou, et al., 2019). The commonly associated factors such as smoking and alcohol consumption increase the high rates of incidence in some areas. At the same time, it falls short of explaining the occurrence in other regions. Along with these etiological factors, consumption of hot beverages and chronic diseases like the chagasic megaesophagus, which is the late manifestation of Chagas' disease, can be attributed to the development of this deadly cancer (Munari, Cruvinel-Carloni, Lacerda, et al., 2018; Prabhu, Obi, & Rubenstein, 2014). There are predisposing factors that associated with ESCC including tylosis, Plummer-Vinson syndrome, lye structure, and history of Head and Neck Squamous Cell carcinoma (HNSCC). Several studies have been conducted to decode the multifactorial causation of ESCC. Comprehensive genomic analysis reveals that the highest mutation is found in the *TP53*, with more than 83% of ESCC harboring somatic mutations in the *TP53* gene (Song, Li, Ou, et al., 2014). *PIK3CA* mutations in ESCC have been demonstrated to be associated with chagasic megaesophagus (Prabhu et al., 2014). *PI3K/AKT* signaling pathway activation is proved to play a crucial role in the ESCC. Somatic aberrations have also been found in other notable tumor-associated pathways like *WNT* and *NOTCH* signaling (Ohashi, Miyamoto, Kikuchi, et al., 2015; Song et al., 2014). Unfortunately, late-stage diagnosis of ESCC often deprives most patients of the mainstream therapeutic options widely used, leading to a 5-year survival rate below 15% (Nicolau-Neto, Costa, et al., 2018). ESCC is one of the foremost causes of mortality related to cancer. This is due to the genetic complexity of this tumor type. Though many genetic alterations have been identified, diagnosis and prognosis related to ESCC and the clinical applicability of the genetic alterations still requires further research (Carneiro, Isinger, Karlsson, et al., 2008). The prognosis of ESCC is poor, and currently there is no targeted therapy/treatment options available for the ESCC patients (Smyth et al., 2017).

Assessment of the ESCC genetic profile by employing microarray-based gene expression profiling provides indispensable insights into molecular pathogenesis. Microarray technology has become a powerful and frequently availed tool for high throughput comprehensive analysis of gene expression, and it is also used to analyze differentially expressed genes (DEGs) in patients who have distinct conditions (Kumar, Kumar, Siva, et al., 2019; Mishra, Shah, Udhaya Kumar, et al., 2021; Udhaya Kumar, Rajan, Thirumal Kumar, et al., 2020; Udhaya Kumar, Thirumal Kumar, Siva, et al., 2020). Studies dealing with the analysis of DEGs in malignant and non-malignant types in ESCC are very few. Hence, this study aims at bridging the gap by employing a bioinformatics-based method to investigate DEGs between esophageal tumors and non-malignant surrounding mucosa in the microarray dataset (GSE75241). Identification of DEGs, gene networks, protein-protein interactions (PPIs) that are distinctive for ESCC, gene ontology, and enriched molecular pathway analyses were carried out to capture prospective targets. Discerning the core genes (potential targets) in ESCC will provide a diagnostic biomarker and promising therapeutic applications for treatment.

2. Materials and methods

2.1 Data accession and pre-processing

The NCBI's GEO database was utilized to obtain the gene expression profiling data from the microarray dataset with ID GSE75241 (NCBI Resource Coordinators, 2016; Nicolau-Neto et al., 2018). The freely accessible GEO database is an international public repository that allows the research community to submit high-throughput microarray and next-generation sequence functional genomic data sets. In order to obtain the gene expression profiling data with the existing datasets associated with ESCC from GEO, keywords such as "esophageal squamous cell carcinoma," "ESCC," and "gene expression" were used in combination with the boolean operators such as AND, OR, and NOT (Tungekar, Mandarthi, Mandaviya, et al., 2018). Overall, 556 hits were obtained for the keywords and GSE75241 dataset was selected due to the high number of samples in the control and ESCC group. The dataset was focused on finding the DEGs between the 15 tumor (esophageal) and 15 non-malignant samples (surrounding mucosa). In any genomics or transcriptomics study, it is always preferred to use paired samples over unpaired, the reason is simply that in paired samples the comparison can be made when the genetic background is the same, and that way you do not add bias in the analysis (Kashyap, Harsha, Renuse, et al., 2010). With the help of a robust multi-array standard model, the preliminary data from the microarray were put through quantile normalization, log transition, and correction of background. Pre-processing of data includes conversion to gene symbols from probe IDs. In order to achieve this, the gene ID converter Entrez was employed. The online statistical tool, GEO2R, which has Bioconductor/R and Limma package incorporated with it, was utilized to assess the raw expression of the genes (Davis & Meltzer, 2007; Smyth, 2005). In order to determine the DEGs, *P*-value and false discovery rate (FDR) were calculated. This was achieved by performing a *t*-test and Benjamini and Hochberg, which are inbuilt methods of GEO2R (Benjamini & Hochberg, 1995). The significant DEGs from the dataset were acquired by setting principal standard values of fold change $(FC) \geq 2$ and $P \leq 0.05$. Among the acquired DEGs, the ones with $FC \leq -2$ were considered downregulated, and the ones with $FC \geq 2$ were considered upregulated. A volcano plot was constructed using RStudio, and the subsequent DEGs that were attained were used for further investigation. The "bioMart" R package was used to fetch the data from the BioMart database (Durinck, Moreau, Kasprzyk, et al., 2005). The results obtained were further given as an input in the UCSC genome browser to obtain the DEG's chromosomal locations (Karolchik, Hinrichs, Furey, et al., 2004). The "ShinyCircos" software, a freely accessible web tool, was used to visualize the DEGs in their respective chromosome locations (Yu, Ouyang, & Yao, 2018).

2.2 Constructing protein network

To evaluate the relationship between the DEGs obtained from the dataset, a PPI network was established using the web-based tool STRING (Szklarczyk, Franceschini, Wyder, et al., 2015). A cut-off criterion of value 0.4 was set as a high confident interaction score in order to obtain only the significant nodes. The obtained PPI network was visualized by the Cytoscape software (Shannon, Markiel, Ozier, et al., 2003). We implemented the NetworkAnalyzer tool to perform the centrality and node degree distribution analysis for the identified DEGs (Saito, Smoot, Ono, et al., 2012). It helps in the identification of the nodes with the highest degree, but also will reveal the node numbers with a given degree that share a neighbor with another node (Assenov, Ramírez, Schelhorn, et al., 2008; Stelzl, Worm, Lalowski, et al., 2005). The PPI network was screened for the potential genes based on the topology scores (Network, Eigenvector, Betweenness, and Closeness) using CytoNCA from Cytoscape (Tang, Li, Wang, et al., 2015). Consequently, for ranking the nodes from the PPI network, we used the cytoHubba (Maximal Clique Centrality (MCC), Edge Percolated Component (EPC), Maximum Neighborhood Component (MNC), and Stress) from Cytoscape (Chin, Chen, Wu, et al., 2014). Based on the above analyses with significant degree scores and by comparing the overlapped genes, we selected the top eight crucial DEGs for further process.

2.3 DAVID and MetaCore™ GeneGo analysis of DEGs

We used the DAVID v6.85 web-based tool for functional GO annotation (Dennis, Sherman, Hosack, et al., 2003), and KEGG enrichment analysis (Huang, Sherman, & Lempicki, 2009). DAVID database is an important source for functional assessment of high-throughput expression profiles of genes. Further, the results from DAVID were imported into RStudio for topGO analysis (Alexa & Rahnenfuhrer, 2020). The topGO is a Bioconductor package in R, which is used to perform enrichment analysis for DEGs. An examination of curated enriched pathway and GO analysis were carried out using MetaCore™, a Cortellis Solution software. GeneGo enables speedy evaluation of molecular pathways, networks from biological proteins, and pathways from elevated clinical and molecular genetic experimental resources. A molecular interactions of DEGs are depicted graphically and were generated based on a significant $P \leq 0.05$ threshold from the groups of study. DEGs-ontological element intersection could be random, and it was estimated by calculating the hypergeometric *P*-value. To assess a considerable number of DEGs that made up a processor pathway, a criterion of FDR < 0.05 was used.

3. Principal component analysis (PCA) of hub genes

The PCA was performed for the identified hub genes to explore the gene expression in comparison to the ESCC normal, ESCC tumor, and esophageal mucosa. We used the GEPIA2 web-server that contains the datasets from TCGA and GTEx projects with clinical annotations. The violin plots were generated for the identified hub genes to validate their expression levels at different pathological stages of ESCC. In addition, the hub genes were compared to each other with their expression using multiple gene comparison methods between the healthy and cancer cohorts (Li, Tang, Zhang, et al., 2021).

4. Results

4.1 Identification DEGs from ESCC

The NCBI's GEO database was used to retrieve the expression profiles for GSE75241. The GSE75241 dataset was generated using the GPL5175 platform (Affymetrix Human Exon 1.0 ST Array). Table S1 in Supplementary File in the online version at https://doi.org/10.1016/bs.apcsb.2022.04.003 contains information on the samples used in this study. The GEO dataset GSE75241 possesses 15 paired ESCC samples with matched non-malignant mucosa. The dataset was analyzed using the GEO2R tool. We identified a total of 250 DEGs between ESCC and normal. Among those 35 were upregulated and 215 were downregulated (Table S2 in Supplementary File in the online version at https://doi.org/10.1016/bs.apcsb.2022.04.003). The UMAP was plotted to visualize the relationship between the samples with the help of the dimension reduction technique and the number of nearest neighbors utilized in the calculation (Fig. 1A). The Volcano plot was generated by visualizing the DEGs with $FC \geq 2$ or ≤ -2 and *P*-values ≤ 0.05 (Fig. 1B). With the help of the Circos plot, we were able to understand the DEGs that are distributed among all the chromosomes except for chromosome Y. The highest number of DEGs were found to be located on chromosome 1. We also characterized the genomic distributions of *P*-values and logFC values of DEGs between ESCC and controls using the Circos plot. The outermost first layer

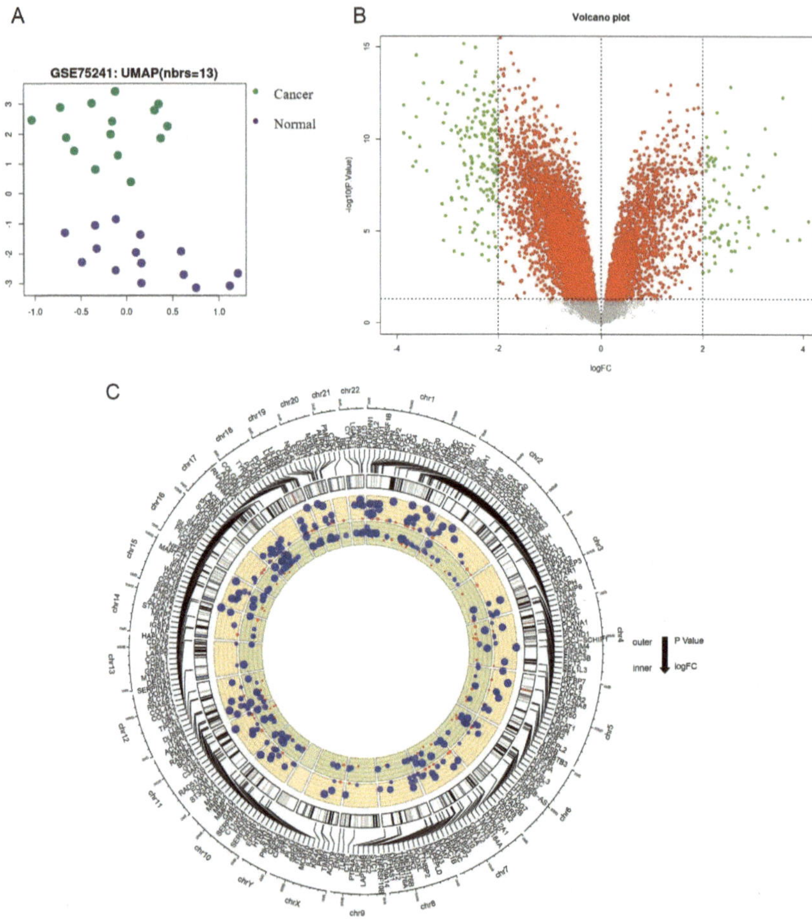

Fig. 1 (A) Uniform Manifold Approximation and Projection (UMAP) plot for the samples from GSE75241. (B) From the GSE75241 dataset, volcano plot of the DEGs in ESCC *versus* non-malignant surrounding mucosa. The plot's green color dots represent significantly expressed DEGs that map upregulated genes (logFC ≥ 2; $P \leq 0.05$) and downregulated genes (log FC≤ -2; $P \leq 0.05$). The plot's red color dots represent less significantly expressed DEGs that map upregulated genes (logFC 0 to <2; $P \leq 0.05$) and down-regulated genes (logFC 0 to > -2; $P \leq 0.05$). (C) Circos plot depicting the *P*-value and logFC of the DEGs. The outer circle represents the cytoband ideogram. The *P*-values of the DEGs are represented by the inner yellow circle, with blue indicating the lowest and red representing the greatest *P*-value. The innermost green circle represents the logFC values of the DEGs, with blue indicating the lowest logFC value and red indicating the greatest logFC value.

represents the chromosome position of the DEGs. Whereas the second outermost layer indicates genes in the barcode, it ranges from dark to light based on the expression of each gene in ESCC. The innermost first layer represents the logFC value of DEGs. The *P*-value of DEGs was displayed on the second innermost layer (Fig. 1C).

4.2 PPI network construction

The identified DEGs were imported into STRING to find PPI networks for up and downregulated DEGs. A high confidence score of 0.4 for the nodes indicated an interaction of PPI, and also eliminated the disconnected nodes. The resulting PPI network was then downloaded as a ".txt" and uploaded as a .csv into the Cytoscape for graphical depiction. The PPI network consists of a total of 197 nodes with 1061 edges, in which the node size indicates the node degree (A smaller size denotes a lower degree, whereas a larger size denotes a higher degree) (Fig. 2). The color ranges from red to green of the node is the node's degree score (red color node indicates a higher degree whereas green color node indicates a lower degree). MCODE helps to find the highly interconnected regions in the network. Using the MCODE plugin examines the entire network and selected three clusters. A total of 52 genes were found in the three clusters (Fig. 3). Cluster 1 consists of 23 nodes and 234 edges (Fig. 3A), Cluster 2 has 25 nodes and 100 edges (Fig. 3B), Cluster 3 consists of 24 nodes and 84 edges (Fig. 3C), with the MCODE score of 21.27, 8.33, and 7.30. The maximum number of significant scores was found in Cluster 1 (Table 1).

4.3 Hub gene identification

To discover the hub genes in the complex PPI furthermore, we employed the CytoNCA and cytoHubba plugins. CytoNCA and cytoHubba are two centrality analysis Cytoscape plugins that contain information on the major nodes or edges in the network. We used the CytoNCA application to retrieve data from four different calculating methods (Network, Eigenvector, Betweenness, and Closeness). The top 100 nodes were selected in each of these four methods (Table S3 in Supplementary File in the online version at https://doi.org/10.1016/bs.apcsb.2022.04.003). In this PPI, using the calculating methods such as Network, Eigenvector, Betweenness, and Closeness, we discovered 61 overlapping genes. In addition, four methods (MCC, MNC, EPC, and Stress) were utilized in the cytoHubba program, and the top 100 nodes were

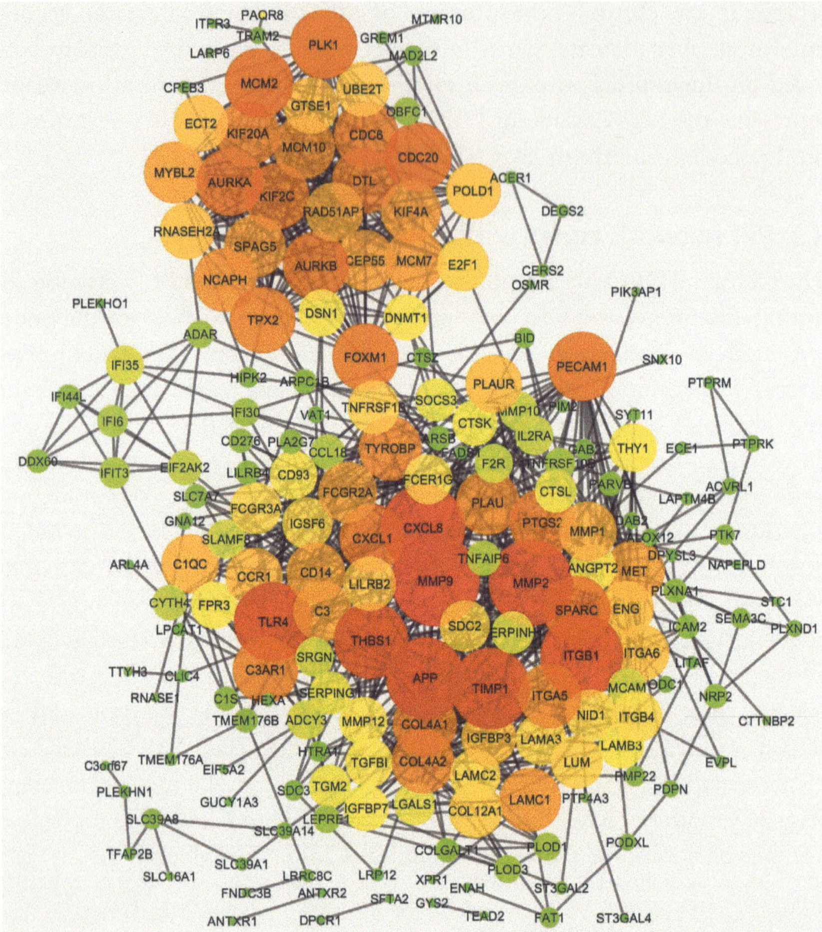

Fig. 2 The protein-protein interaction (PPI) network of ESCC DEGs was determined using Cytoscape. The nodes and edges are represented as spheres and lines, respectively. The size of the node indicated the degree of the node. A smaller size denotes a lower degree, whereas a larger size denotes a higher degree. The degree score of the node was also represented by its color, which ranges from red to green.

obtained based on these four ways (Table S4 in Supplementary File in the online version at https://doi.org/10.1016/bs.apcsb.2022.04.003). We found 76 overlapping genes in MCC, MNC, EPC, and stress. A Venn diagram was created to identify the significant DEGs shared by all the groups. CytoNCA and cytoHubba had 61 genes in common, and the output has been illustrated in Table S5 in Supplementary File in the online version at https://doi.org/10.1016/bs.apcsb.2022.04.003. Cluster 1 from

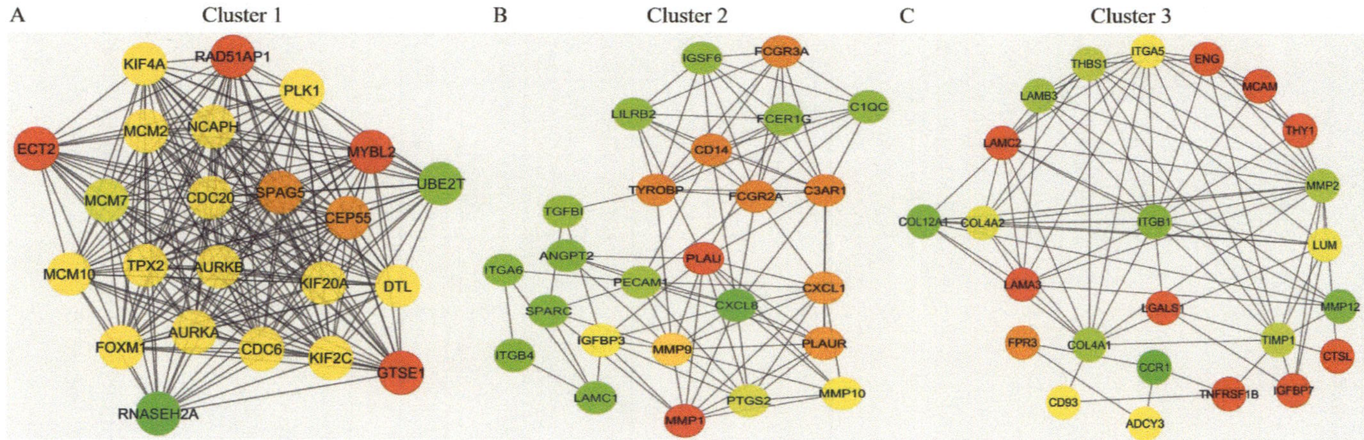

Fig. 3 Based on Cytoscape's MCODE, the most significant cluster from the complex PPI of ESCC DEGs was constructed. (A) The first and most significant cluster, with the highest MCODE score of 21.273. (B) The second and most significant cluster, with the MCODE score of 8.333. (C) The third and most significant cluster, with the MCODE score of 7.304.

Table 1 Identified clusters from the complex network of ESCC between the normal mucosa and esophageal tumor.

Clusters	Score (density*#nodes)	Nodes	Edges	Node IDs
1	21.27	23	234	*MCM7, UBE2T, DTL, RAD51AP1, MYBL2, GTSE1, MCM2, CEP55, ECT2, NCAPH, SPAG5, FOXM1, KIF4A, KIF20A, KIF2C, AURKB, PLK1, CDC20, AURKA, TPX2, RNASEH2A, CDC6, MCM10*
2	8.33	25	100	*MMP10, C3AR1, CXCL1, CXCL8, ITGB4, MMP9, PTGS2, PLAU, PECAM1, C1QC, SPARC, IGSF6, FCER1G, CD14, LAMC1, TYROBP, LILRB2, PLAUR, TGFBI, IGFBP3, FCGR2A, ANGPT2, FCGR3A, ITGA6, MMP1*
3	7.30	24	84	*LAMB3, ADCY3, THY1, CCR1, COL4A2, MMP12, FPR3, COL4A1, IGFBP7, TIMP1, MMP2, THBS1, MCAM, ITGA5, COL12A1, ITGB1, CTSL, LUM, LAMA3, ENG, CD93, LGALS1, LAMC2, TNFRSF1B*
4	4.4	11	22	*IFIT3, SERPING1, C3, TLR4, DDX60, IFI44L, SDC2, EIF2AK2, SRGN, IFI6, IFI35*
5	3.33	4	5	*PLOD1, PLOD3, LEPRE1, COLGALT1*
6	3	3	3	*CERS2, DEGS2, ACER1*
7	3	3	3	*SLC39A8, SLC39A1, SLC39A14*

the complex PPI network consists of 23 DEGs. As highly interconnected proteins aggregate in a cluster in the network, we have chosen only eight hub genes aurora kinase B (*AURKB*), polo-like kinase 1 (*PLK1*), minichromosome maintenance complex component 2 (*MCM2*), cell division cycle 20 (*CDC20*), Targeting protein for Xklp2 (*TPX2*), forkhead box M1 (*FOXM1*), minichromosome maintenance complex component 7 (*MCM7*), and aurora kinase A (*AURKA*) from cluster 1 was identified by overlapping analysis and visualized in a Venn diagram. All steps for hub gene selection are shown in Fig. 4.

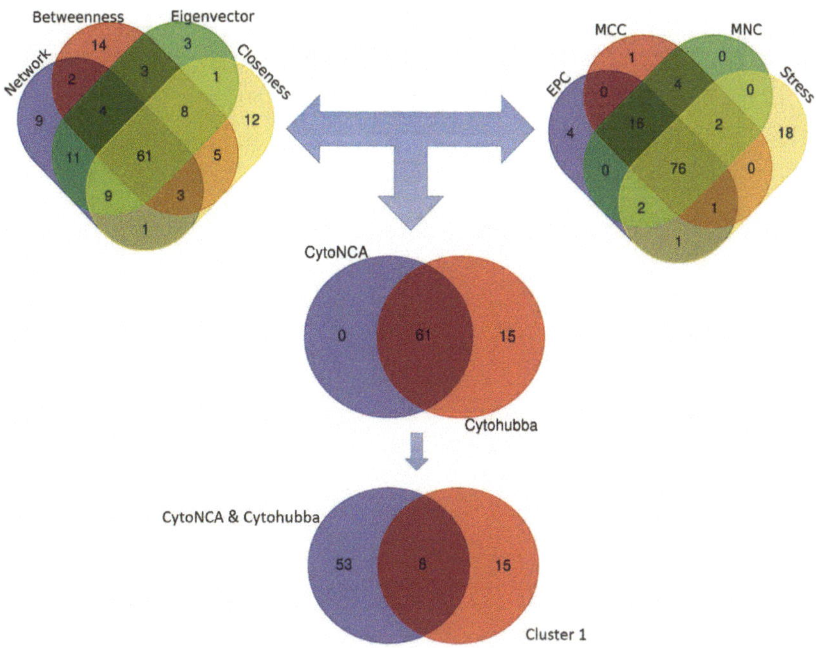

Fig. 4 CytoNCA and cytoHubba plug-in from Cytoscape was used to screen the significant DEGs in a PPI network. The right side panel shows a Venn diagram of 76 overlapping genes across four different cytoHubba methods; the left side panel shows a Venn diagram of 61 overlapping genes across four different CytoNCA methods. An overall Venn diagram of 61 overlapping genes was screened across eight different cytoHubba and CytoNCA calculating algorithms (middle panel). The overlapped CytoNCA and cytoHubba genes are further overlapped with the most significant Cluster 1 of the PPI network and found eight significant DEGs from them (lower panel).

4.4 GO and enrichment analysis

To investigate the role of the top DEGs in ESCC further, we utilized the DAVID tool to perform Gene Ontology (GO) categorization (biological process (BP) and molecular function (MF)) and KEGG pathway enrichment analysis. The results from the GO-BP assessment showed that the DEGs were mainly enriched in regulation of cell migration (GO:0030334), extracellular structure organization (GO:0043062), blood vessel morphogenesis (GO:0048514), angiogenesis (GO:0001525), regulation of locomotion (GO:0040012), regulation of cellular component organization (GO:0051128), regulation of response to external stimulus (GO:0032101), regulation of cell proliferation (GO:0042127) and vasculature development (GO:0001944) (Fig. 5A). For the GO-MF assessment, the DEGs

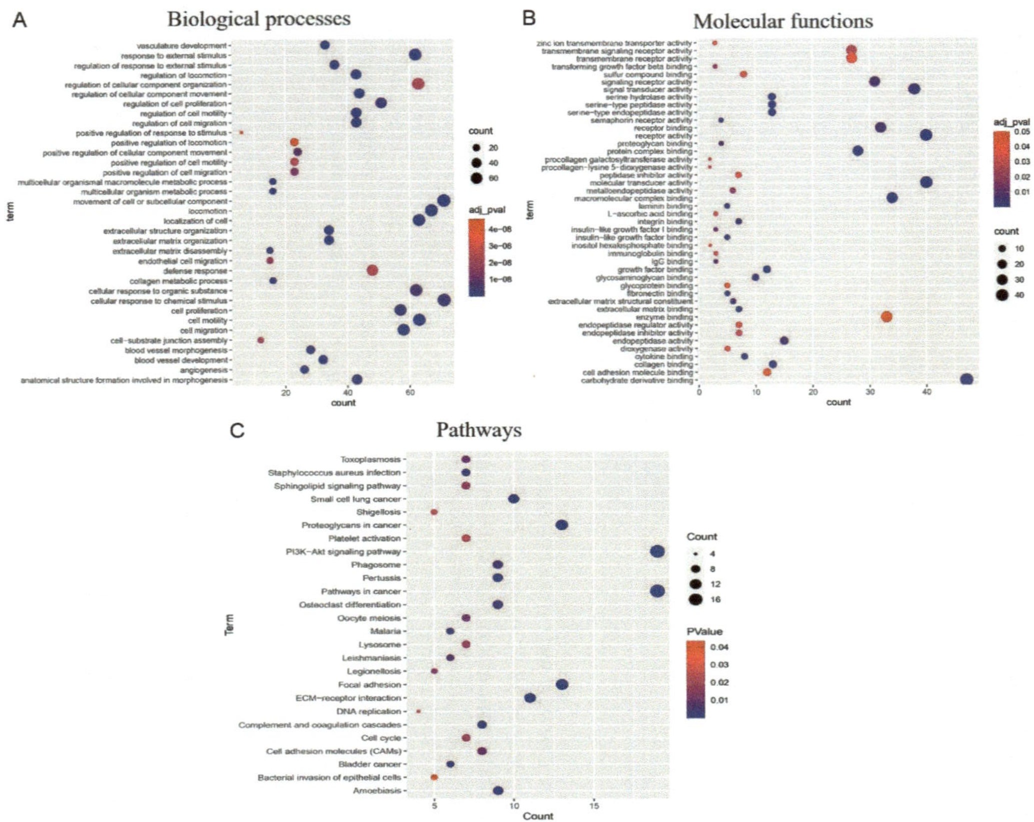

Fig. 5 Gene ontology analysis revealed the functional enrichment of DEGs from the ESCC dataset. The size of the sphere reflects the number of genes involved, and the color of the sphere denotes the *P*-value. (A) Top 30 GO-biological pathways of DEGs in ESCC; (B) Top 30 GO-molecular function of DEGs in ESCC; (C) Top 30 GO-KEGG pathways of DEGs in ESCC.

were highly enriched in the activity of signaling receptor (GO:0038023), semaphorin receptor binding (GO:0017154), sulfur compound binding (GO:1901681), activity of zinc ion transmembrane transporter (GO:0005385), activity of serine-type peptidase (GO:0008236), binding of proteoglycan (GO:0043394), metalloendopeptidase activity (GO:0004222), binding of enzyme (GO:0019899), and transmembrane signaling receptor activity (GO:0004888) (Fig. 5B). We discovered dysregulated KEGG pathways depending on the adjusted $P \leq 0.05$ and FDR ≤ 0.05. At last, the genes were mainly enriched in pathways including DNA replication (hsa03030), pathways in cancer (hsa05200), cell cycle (hsa04110), and cell adhesion molecule (hsa04514) (Fig. 5C).

4.5 Enrichment analysis using Metacore™

Furthermore, the Metacore™ was utilized to execute enrichment analysis to validate and identify the genes and pathways that were expressed differentially in the ESCC patient. Fig. 6 depicts the top 10 BPs, process networks, pathway maps, and disease biomarkers identified through relative functional analysis of the DEGs. Based on low *P*-value, these are the most highly relevant data for each category. The top 10 biological processes of DEGs in ESCC consist of regulation of cellular component organization (GO:0051128), organelle organization (GO:0006996), activity of positive regulation of transferase (GO:0051347), positive regulation of the cellular process (GO:0048522), positive regulation of intracellular transport (GO:0032388), positive regulation of the biological process (GO:0048518), activity of positive regulation of kinase (GO:0033674), cytoskeleton organization (GO:0007010), cellular localization (GO:0051641), and activity of transferase regulation (GO:0051338) (Fig. 6A). The top 10 process network were as follows: cell cycle S-phase, cytoskeleton actin filaments, cell cycle mitosis, cytoskeleton spindle microtubules, cell cycle core, development regulation of angiogenesis, proteolysis connective tissue degradation, reproduction progesterone signaling, transcription chromatin modification, and transport synaptic vesicle exocytosis (Fig. 6B). The top 10 enriched pathway maps were as follows: cell cycle role of *APC* in cell cycle regulation, G-protein signaling regulation of *CDC42* activity, cell cycle the metaphase checkpoint, cell cycle spindle assembly, and chromosome separation, cell cycle chromosome condensation in prometaphase, immune response Fc epsilon RI pathway, immune response antigen presentation by MHC class I: cross-presentation, n-6 polysaturated fatty acid

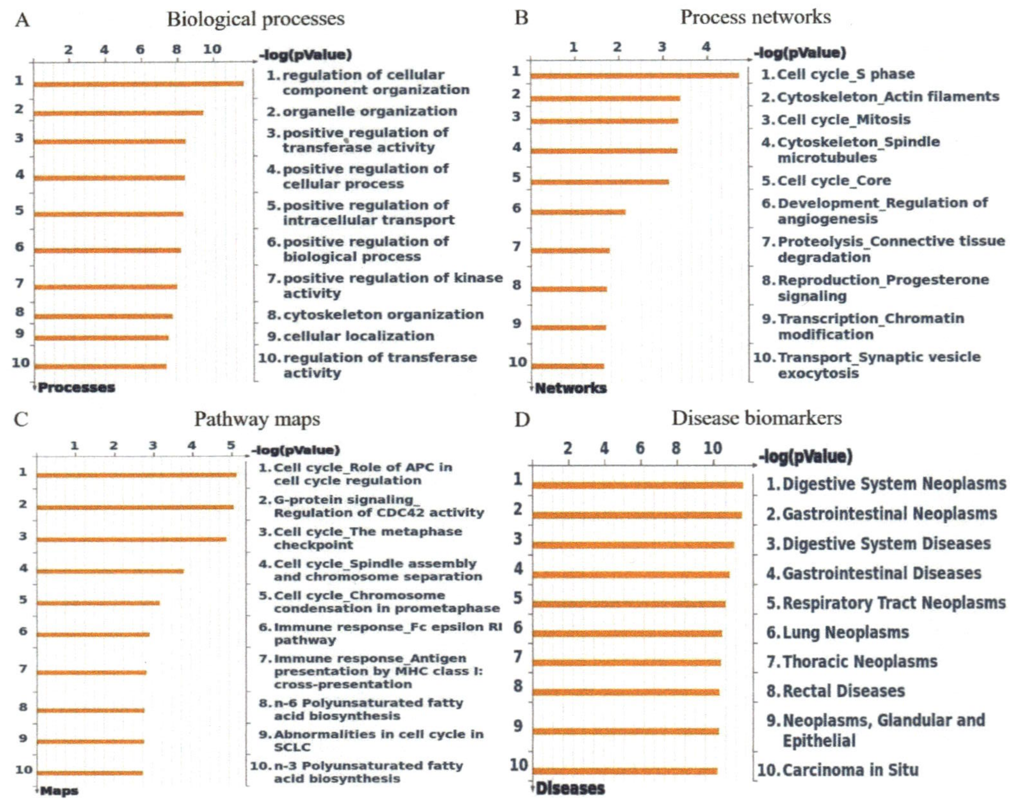

Fig. 6 Enrichment analysis of the DEGs from the ESCC dataset using MetaCore™ GeneGo software (Clarivate Analytics). (A) Top 10 biological processes involved in DEGs; (B) Top 10 process networks of DEGs; (C) Top 10 pathway maps of DEGs; and (D) Top 10 disease biomarkers of DEGs.

Cell cycle role of APC in cell cycle regulation

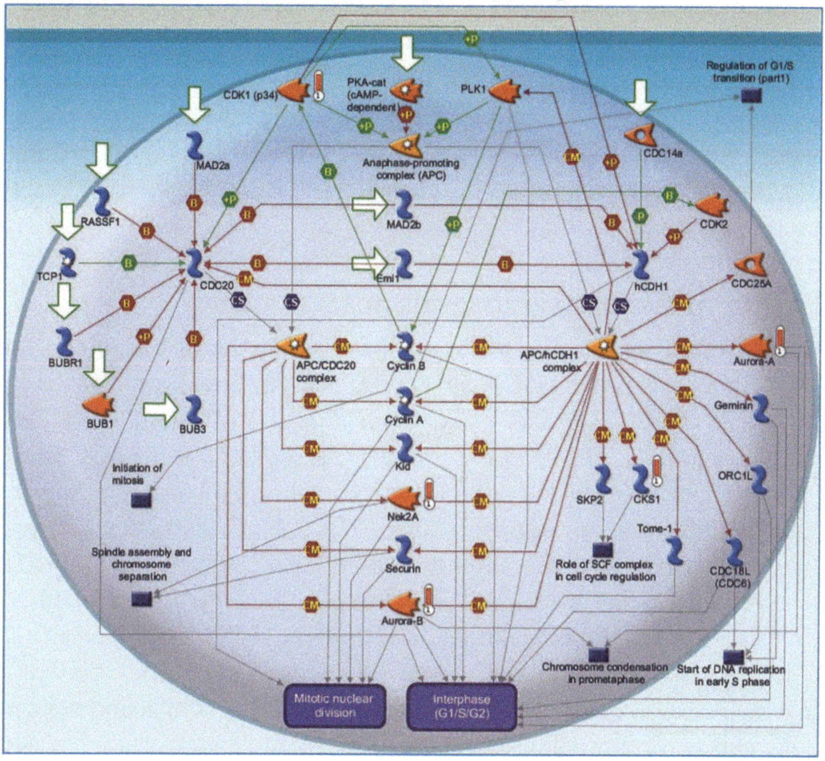

Fig. 7 The cell cycle role of *APC* in cell cycle control, a top-most pathway of ESCC DEGs generated by MetaCore™ software. Green lines in the pathway represent positive/activation interaction, red lines in the pathway represent negative/inhibition interaction, and gray lines in the pathway represent unspecified interaction between the proteins or protein complexes. The experimental data are shown as "thermometer-like" figures on the pathway map. Upward thermometers (red color) indicate upregulated gene transcripts. *AURKA* and *AURKB* genes are with an upward thermometer. *AURKB, CDC20, PLK1*, and *AURKA* are four of eight hub genes from ESCC that were involved in this pathway. Further information is available at https://portal.genego.com/help/MC_legend.pdf.

biosynthesis, abnormalities in the cell cycle in SCLC, and n-3 polyunsaturated fatty acid biosynthesis (Fig. 6C). The first top-most pathway, based on gene enrichment distribution was the cell cycle role of *APC* in cell cycle regulation, as illustrated in graphics (Fig. 7). When compared to the other top-listed pathways, this one has the lowest *P*-value. Four of the eight identified hub genes (*AURKB*, *CDC20*, *PLK1*, and *AURKA*) were found to be engaged in this pathway. These four hub genes are overexpressed in

ESCC patients as compared with paired non-malignant normal mucosa. Furthermore, we also analyzed biomarkers by disease of DEGs using MetaCore™ software, which resulted in digestive system neoplasms, digestive system diseases, gastrointestinal disease, respiratory tract neoplasms, lung neoplasms, thoracic neoplasms, rectal diseases, neoplasms, glandular and epithelial, and carcinoma *in situ* (Fig. 6D). The finalized DEGs were primarily associated with the regulation of the cell cycle, sister chromatid cohesion, spindle organization, and DNA replication.

4.6 Hub gene expression levels in ESCC stages

The GEPIA2 tool was used to investigate the expression levels of the eight ESCC hub genes in the TCGA and GTEx data. The boxplot in Fig. 8 shows the gene expression profiles of the eight hub genes (*AURKA*, *AURKB*, *MCM2*, *CDC20*, *TPX2*, *PLK1*, *FOXM1*, and *MCM7*) in ESCC tumor ($N=182$) and normal ($N=286$) tissues. Each hub gene had a mean value of log2 (TPM + 1) as the signature score. The red box shows the samples of the tumor, while the gray one shows the normal tissues. Gene expression levels based on the signature score for hub genes (*AURKA*, *MCM2*, *CDC20*, *TPX2*, *PLK1*, and *FOXM1*) were considerably greater in ESCC tumors than in normal tissues. A key identification of the ESCC development was based on the pathological stage. The connection between eight hub gene expression levels and ESCC pathological stages has been established in violin plots using GEPIA2 (Fig. 9). According to cancer stage correlation analysis, *AURKB* (Pr = 0.0424) and *CD20* (Pr = 0.018) were found to be significantly different between the pathological stages of ESCC ($P < 0.05$). At the same time, *AURKA* (Pr = 0.244), *FOXM1* (Pr = 0.25), *MCM2* (Pr = 0.504), *MCM7* (Pr = 0.189), *PLK* (Pr = 0.179), and *TPX2* (Pr = 0.14) were discovered to be associated with ESCC, but with no significant differences between ESCC pathological stages.

4.7 Hub genes comparison and principal component analysis

GEPIA was used for multiple gene comparison analysis in eight hub genes of ESCC, using the Match TCGA normal and GTEx data. Six hub genes (*AURKA*, *MCM2*, *CDC20*, *TPX2*, *PLK1*, and *FOXM1*) showed the highest expression levels in ESCC than normal among the eight hub genes analyzed. The expression level of two hub genes (*AURKB* and *MCM7*) in ESCC and normal had similar expression patterns (Fig. 10A). In GEPIA2, we also conducted principal component analysis (PCA) to

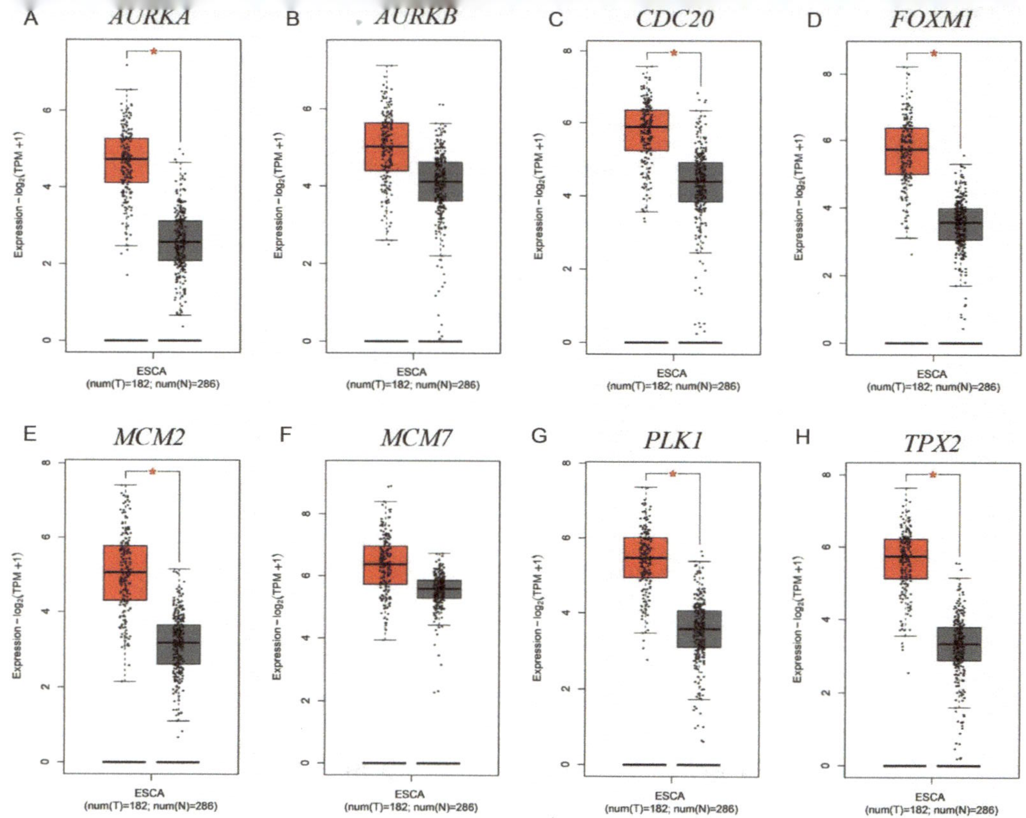

Fig. 8 Box plot representation of eight hub genes expression levels from ESCC (TCGA project) relative to the corresponding normal tissues (GTEx database) using GEPIA2. The red box indicates tumor samples (ESCC), while the gray box represents normal samples. Where * denotes $P < 0.05$ and are significantly showed higher expression among 182 ESCC compared to 286 normal samples. (A) *AURKA* exhibited higher expression in ESCC compared to normal; (B) *AURKB* exhibited more or less similar expression in ESCC and normal; (C) *CDC20* exhibited higher expression in ESCC compared to normal; (D) *FOXM1* exhibited higher expression in ESCC compared to normal; (E) *MCM2* exhibited higher expression in ESCC compared to normal; (F) *MCM7* exhibited more or less similar expression in ESCC and normal; (G) *PLK1* exhibited higher expression in ESCC compared to normal; and (H) *TPX2* exhibited higher expression in ESCC compared to normal.

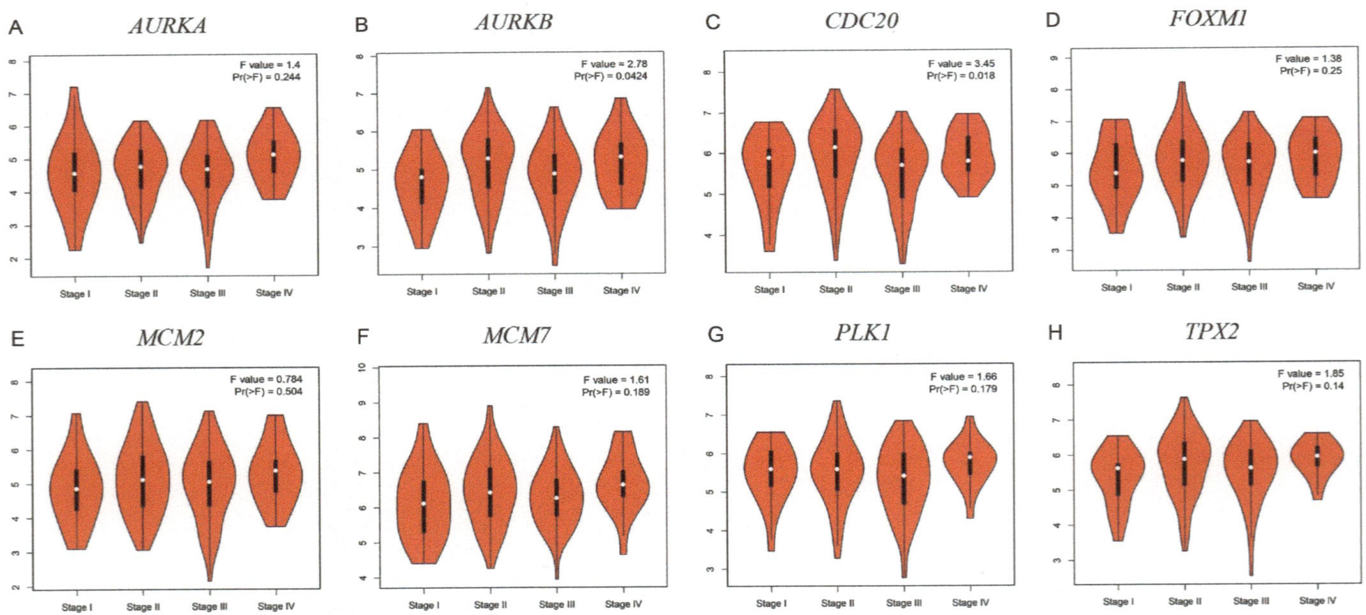

Fig. 9 GEPIA2 provides a stage-dependent expression level of ESCC in violin plots. ESCC pathological stages such as stage I, stage II, stage III, and stage IV were assessed and compared with TCGA data. The expression of eight hub genes (A) *AURKA*; (B) *AURKB*; (C) *CDC20*; (D) *FOXM1*; (E) *MCM2*; (F) *MCM7*; (G) *PLK1*; and (H) *TPX2* in each pathological stage are illustrated.

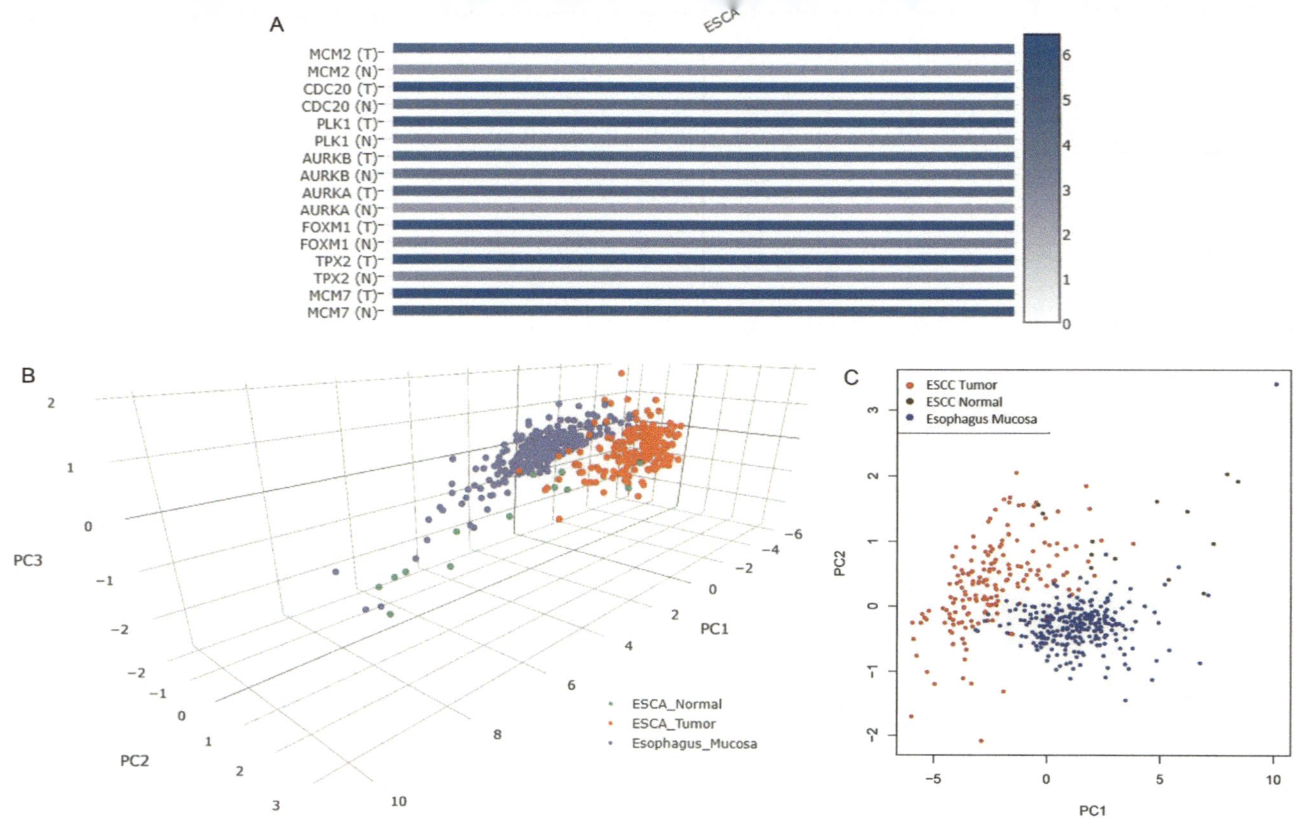

Fig. 10 (A) The multiple gene comparison analysis of eight hub genes in ESCC using GEPIA2. For each gene, it correlates with tumor (ESCC) and normal. Gene expression is measured on a scale of 0 to 6 (white to dark blue color), where 0 (white color) indicates no or less expression and 6 (dark blue colors) indicates higher expression. (B) In three dimensions, scatter plots of eight hub genes of tumor (ESCC), esophageal mucosa, and normal tissue variance distributions were displayed. Green dots represent ESCC normal (TCGA), orange dots represent ESCC tumor (TCGA), and blue dots represent esophageal mucosa (GTEx database). (C) Two-dimensional scatter plots of eight hub genes showing tumor (ESCC) (red dots), esophageal mucosa (brown dots), and normal (blue) tissue variance distributions.

validate the eight hub genes for biomarker possibilities in the diagnosis of ESCC using TCGA tumor data, TCGA normal data, and GTEx data (esophagus mucosa). Based on the correlation analysis, 3D (Fig. 10B) and 2D (Fig. 10C) scatter plots of ESCC, esophagus mucosa, and normal tissue variance distributions were exhibited. The percentage of variances for PCA dimensionality reduction is 90%, 10%, and 8% for PC1, PC2, and PC3, respectively and are represented as bar plots (Fig. S1 in Supplementary File in the online version at https://doi.org/10.1016/bs.apcsb.2022.04.003). We found that the eight hub genes were effectively able to discriminate between ESCC tumor and esophageal mucosa (normal), suggesting the potential of these genes as the probable candidate for ESCC biomarkers.

5. Discussion

Although the plethora of information provided on the ESCC's differential expression, the molecular signature of the cancer progression remains unclear. A global epidemiological survey estimated that 442,000 individuals had been diagnosed with ESCC (Edgren, Adami, Weiderpass, & Nyrén, 2013; Smyth et al., 2017). At the initial stages of the ESCC, there are no distinct clinical symptoms, but warning symptoms such as dysphagia and hoarseness are seen (Smyth et al., 2017). Melena is another symptom in patients with ESCC, and all the features of a tumor are usually documented during endoscopy and bronchoscopy (Lordick, Mariette, Haustermans, et al., 2016). The North American and European patients with EC were diagnosed at the advanced stage or metastasis stage and not responsive to therapeutic intervention. Therefore, it is essential to find novel gene markers that indicate dysregulated pathways, their expression, ESCC recurrence, and disease-free survival, and thereafter help in segregating patients into low or high-risk cohorts for improved therapeutic response. Our present study focused on exploring the DEGs that are responsible for the dysregulated signaling in patients with ESCC compared to surrounding normal mucosa. Overall, 9505 DEGs (based on significant P-value) were present in the dataset; among them, 35 are upregulated, and 215 are downregulated (shortlisted upon FC values). Based on the network parameters, we found significant DEGs with higher node degree distribution (Fig. 2). The topological parameters from the NetworkAnalyzer help to differentiate and comprehend the nodes and edges within the network for understanding the complex biological networks (Assenov et al., 2008). From the complex PPI, we observed three top clusters, which are strongly connected regions that are defined by high scores (Fig. 3). The gene clusters are screened to find out the vital genes, their interaction with each other, and a possible link to the progression and development of ESCC. The MCODE clusters often help in identifying the mechanism of the network and extensive comprehension of biological network features (Krogan, Cagney, Yu, et al., 2006; Rahman, Islam, Banik, et al., 2013; Udhaya Kumar, Saleem, Thirumal Kumar, et al., 2021; Udhaya Kumar, Thirumal Kumar, Bithia, et al., 2020).

Furthermore, the PPI network was evaluated for the identification of hub genes with the help of CytoNCA and cytoHubba. We combined both the methods and compared them with cluster 1 (highest MCODE rank) to identify the most significant DEGs that may help in capturing the hub genes, which are crucial in the ESCC progression. Sixty-one DEGs were shared between the CytoNCA and cytoHubba, and eight DEGs that were shared between the CytoNCA, cytoHubba, and cluster 1 respectively. Taken together, we found the DEGs such as *AURKA*, *AURKB*, *CDC20*, *FOXM1*, *MCM2*, *MCM7*, *PLK1*, and *TPX2* as hub genes and intrinsically linked to the ESCC development. To investigate the role of DEGs in BP, KEGG pathway, and MF of ESCC, we used the DAVID platform. The results from the DAVID were further processed using the TopGo from Rstudio with the help of an in-house R-script. In addition, the Metacore™ was utilized to delineate the extensive BPs and pathway maps involved in the DEGs. By comparing both the enrichment models and hub genes, we found that the DEGs were primarily involved in the positive regulation of transferase activity, negative regulation of organelle organization, cell cycle mitosis/S-phase transition, spindle organization/assembly, development, and regulation of angiogenesis. The GO-MF analysis showed that the DEGs enriched in enzyme binding, receptor binding/activity, and protein serine/threonine kinase activity. The pathways from KEGG analysis such as oocyte meiosis, cell cycle, and DNA replication showed significance for the identified hub genes. To support the above analysis, a study conducted by Chuyong Lin, Song, Gong, et al. (2013) found that the ESCC angiogenesis occurred upon downregulation of NKX2–8 and upregulation of VEGF-C (Lin et al., 2013). Moreover, the regulation of transferase activity in several biological processes was reported in a previous study conducted in Indian patients with ESCC (Chattopadhyay, Kapur, Purkayastha, et al., 2007). Recently, Scott, Li, Jammula, et al. (2021) reported that the inconsistency in the number of proteins in the EAC might implicate the adhesion of chromosomes to microtubules (spindle) (Scott et al., 2021). In this context, our results from the GO enrichment captured these MF and BP that are associated with cancer-causing pathway networks. The involvement of such networks and pathways in the development of breast and prostate cancer has been identified (Dashti, Taheri, & Ghafouri-Fard, 2020; Fan, Liang, Gao, et al., 2018; Udhaya Kumar, Madhana Priya, Thirumal Kumar, et al., 2021; Zayed, 2020).

Aurora-A overexpression is associated with the differentiation rates of the tumor and with malignant status (Tong, Zhong, Kong, et al., 2004). Furthermore, a study reported that *AURKA* overexpression could contribute to the suppression of human ESCC cell apoptosis and contribute to ESCC cells' resistance to anti-carcinogenic medications (Wang, Liu, Jin, et al., 2006). Previous research reported that *AURKA* overexpression was involved in upregulating β-catenin expression in ESCC cells using GST pull-down and Co-IP assays. AURKB protein expression was discovered to be upregulated in the early stages of ESCC formation, suggesting that it plays a role in cancer cell-cell interactions with neighboring stromal cells (Kwon, Lee, Koh, et al., 2009). High levels of the *AURKB* protein have been linked to a poor overall survival rate in patients with ESCC. *AURKB* depletion reduced the malignant phenotypes in ESCC cells (Yu, Liang, Liu, et al., 2017). A recent study discovered that inhibiting AURKA or AURKB expression inhibited EC cell proliferation. APIO-EE-9 has anticancer action by targeting AURKA and AURKB, suggesting a potential treatment for EC (Jin, Yao, Guo, et al., 2017). Similarly, our study found that *AURKA* and *AURKB* have been differently expressed in the ESCC patients and in pathological stages (Figs. 8A and B and 9A and B). In line with this, our findings suggest that the elevated signals of *AURKA* and *AURKB* in ESCC result in the dysregulated signaling of mitotic nuclear division and interphase (G1/S/G2); therefore, increases the ESCC development.

At metaphase, *CDC20* is essential for *APC/C* activation and participates in mitotic exit. Increased *CDC20* levels were associated with colorectal cancer (CRC) carcinogenesis and a shorter estimated overall survival time. As a result, it was reported that *CDC20* was an independent prognostic factor (Wu, Hu, Wang, et al., 2013). *CDC20* overexpression was associated with aneuploidy in oral cancer by disrupting the timing of the anaphase-promoting complex (APC/C) and encouraging premature anaphase (Mondal, Sengupta, Panda, et al., 2007). *FOXM1* mRNA and protein levels were higher in ESCC tissues as compared with the corresponding para-cancerous tissues. *FOXM1* knockdown inhibited the growth of ESCC cells (Song, Wang, & Feng, 2018). A previous study revealed that early-stage

ESCC is associated with low cytoplasmic *FOXM1* levels, while nuclear *FOXM1* expression was identified in young ESCC patients. *FOXM1* has been suggested as a biomarker for the late-stage ESCC (Hui, Chan, Luk, et al., 2012). Overall, these previous findings support our computational exploration of the *CDC20* and *FOXM1,* where they exhibited increased expression in the TCGA dataset and were validated by box and violin plots (Figs. 8 and 9).

The overexpression of MCM proteins have been discovered as proliferation markers and diagnostic indicators in various cancer (Yu, Wang, Shi, et al., 2020). *MCM2* overexpression was associated with a poor outcome, and it's an independent ESCC prognostic factor. *MCM7* stimulates tumor cell proliferation, colony formation, and migration of ESCC *via* stimulating the AKT1/mTOR signaling pathway (Qiu, Wang, Zhang, et al., 2017). *PLK1* was engaged in spindle formation and chromosomal segregation during mitosis in the process of cell cycle regulation. A previous study found that the *PLK1* expression has been higher in 69.6% of the ESCC tissues and linked to differentiation grades and invasiveness in ESCC (Zhao, Gong, Li, & Chen, 2010). *PLK1* overexpression was found to be strongly and independently associated with a bad prognosis. *PLK1* deficiency caused cell apoptosis through the mitochondrial pathway (Feng, Lin, Shi, et al., 2009). *TPX2* is necessary for *AURKA* targeting the spindle apparatus (Kufer, Silljé, Körner, et al., 2002). The *TPX2* stability regulation by APC/C is a biological mechanism for controlling the cell cycle's spindle assembly, reported that overexpression of a *TPX2* variant (stable) induces prometaphase arrest and an aberrant spindle (Stewart & Fang, 2005). *TPX2* expression was required for the progression and invasiveness of colon cancer. Inhibiting *TPX2* expression deactivates the PI3K/Akt signaling pathway, reducing colon cancer cell tumorigenicity. It also causes *MMP2* to be downregulated, which reduces metastasis (Wei, Zhang, Xu, et al., 2013). The elevated signals for the hub genes (*MCM2, TPX2,* and *PLK1*) were validated by box and violin plots in comparison with ESCC and normal samples (Figs. 8 and 9). Altogether, our bioinformatic analysis captured the differential expression of the above-discussed genes through the multigene comparison method with ESCC and normal samples (Fig. 10A). Here, the *MCM7* exhibited similar expression in both ESCC and normal samples, but our dataset showed increased expression of *MCM7* in ESCC and decreased expression in the surrounding mucosa. *MCM7*'s interacting partner is *CDC6,* we found the increased expression of *CDC6* in the ESCC, interestingly, a previous study found higher expression of *CDC6* and *MCM7* in the pre-cancerous lesion and oral squamous cell carcinoma (Feng, Li, Li, et al., 2008). These findings support that the increased expression of *MCM7* and *CDC6* in our dataset and associated with each other. Moreover, the PCA analysis showed that eight genes could adequately differentiate between the ESCC tumors, ESCC normal, and esophagus mucosa (Fig. 10B and C), referring to the development of a gene biomarker for diagnosis of ESCC and therapeutic interventions.

6. Conclusion

Overall, our combined bioinformatic methods identified the crucial role of DEGs that have the ability to increase ESCC oncogenesis. The hub genes (*AURKA*, *MCM2*, *CDC20*, *TPX2*, *PLK1*, and *FOXM1*) were well-correlated with previous research and exhibited higher expression levels in the dataset. All the hub genes are significantly associated with the pathological stages and prognosis of the ESCC. The identification of hug genes can help in developing gene-specific targeted therapy for ESCC. Personalized medications could be implemented based on contrasting but aberrant molecular signature that are responsible for ESCC tumorigenesis. Overall, we speculate that the identified hub genes could help us in improving the ESCC diagnosis efficiency and assessing future therapeutics.

Acknowledgments

The authors would like to take this opportunity to thank the management of Vellore Institute of Technology (VIT), Vellore, India, Qatar University, Doha, Qatar, and Amity University, Haryana, Gurugram for providing the necessary facilities and encouragement to carry out this work.

Author contributions

SUK, HZ, and CGPD were involved in the design of the study. SUK, AB, and VAP were involved in the data collection and conducted the experiment. SUK, AB, VAP, MKK, and CGPD were involved in the acquisition, analysis, and interpretation of the results. SUK and AB drafted the manuscript. CGPD and HZ supervised the entire study and were involved in study design, the acquisition, analysis, understanding of the data, and critically reviewed the manuscript. All authors edited and approved the submitted version of the article.

Conflict of interest

The authors have declared that no conflicts of interest exist.

Role of funding source

No funding agency was involved in the present study. MKK is recipient of the TARE fellowship (Grant # TAR/2018/001054) from the Science and Engineering Research Board (SERB), Department of Science and Technology, Government of India.

References

Abnet, C. C., Arnold, M., & Wei, W.-Q. (2018). Epidemiology of esophageal squamous cell carcinoma. *Gastroenterology, 154*, 360–373. https://doi.org/10.1053/j.gastro.2017.08.023.

Alexa, A., & Rahnenfuhrer, J. (2020). *topGO: Enrichment analysis for gene ontology version 2.42.0 from Bioconductor*. https://rdrr.io/bioc/topGO/. (Accessed 31 May 2021).

Assenov, Y., Ramírez, F., Schelhorn, S.-E., et al. (2008). Computing topological parameters of biological networks. *Bioinformatics, 24*, 282–284. https://doi.org/10.1093/bioinformatics/btm554.

Benjamini, Y., & Hochberg, Y. (1995). Controlling the false discovery rate: A practical and powerful approach to multiple testing. *Journal of the Royal Statistical Society: Series B: Methodological, 57*, 289–300.

Carneiro, A., Isinger, A., Karlsson, A., et al. (2008). Prognostic impact of array-based genomic profiles in esophageal squamous cell cancer. *BMC Cancer, 8*, 98. https://doi.org/10.1186/1471-2407-8-98.

Chattopadhyay, I., Kapur, S., Purkayastha, J., et al. (2007). Gene expression profile of esophageal cancer in north East India by cDNA microarray analysis. *World Journal of Gastroenterology, 13*, 1438–1444. https://doi.org/10.3748/wjg.v13.i9.1438.

Chin, C.-H., Chen, S.-H., Wu, H.-H., et al. (2014). cytoHubba: Identifying hub objects and sub-networks from complex interactome. *BMC Systems Biology, 8*, S11. https://doi.org/10.1186/1752-0509-8-S4-S11.

Dashti, S., Taheri, M., & Ghafouri-Fard, S. (2020). An in-silico method leads to recognition of hub genes and crucial pathways in survival of patients with breast cancer. *Scientific Reports, 10*, 18770. https://doi.org/10.1038/s41598-020-76024-2.

Davis, S., & Meltzer, P. S. (2007). GEOquery: A bridge between the gene expression omnibus (GEO) and BioConductor. *Bioinformatics, 23*, 1846–1847. https://doi.org/10.1093/bioinformatics/btm254.

Dennis, G., Jr., Sherman, B. T., Hosack, D. A., et al. (2003). DAVID: Database for Annotation, Visualization, and Integrated Discovery. *Genome Biology, 4*(5), P3.

Durinck, S., Moreau, Y., Kasprzyk, A., et al. (2005). BioMart and bioconductor: A powerful link between biological databases and microarray data analysis. *Bioinformatics, 21*, 3439–3440. https://doi.org/10.1093/bioinformatics/bti525.

Edgren, G., Adami, H.-O., Weiderpass, E., & Nyrén, O. (2013). A global assessment of the oesophageal adenocarcinoma epidemic. *Gut, 62*, 1406–1414. https://doi.org/10.1136/gutjnl-2012-302412.

Fan, S., Liang, Z., Gao, Z., et al. (2018). Identification of the key genes and pathways in prostate cancer. *Oncology Letters, 16*, 6663–6669. https://doi.org/10.3892/ol.2018.9491.

Feng, C.-J., Li, H.-J., Li, J.-N., et al. (2008). Expression of Mcm7 and Cdc6 in oral squamous cell carcinoma and precancerous lesions. *Anticancer Research, 28*, 3763–3769.

Feng, Y.-B., Lin, D.-C., Shi, Z.-Z., et al. (2009). Overexpression of PLK1 is associated with poor survival by inhibiting apoptosis via enhancement of survivin level in esophageal squamous cell carcinoma. *International Journal of Cancer, 124*, 578–588. https://doi.org/10.1002/ijc.23990.

Ferlay, J., Soerjomataram, I., Dikshit, R., et al. (2015). Cancer incidence and mortality worldwide: Sources, methods and major patterns in GLOBOCAN 2012. *International Journal of Cancer, 136*, E359–E386. https://doi.org/10.1002/ijc.29210.

Huang, D. W., Sherman, B. T., & Lempicki, R. A. (2009). Bioinformatics enrichment tools: Paths toward the comprehensive functional analysis of large gene lists. *Nucleic Acids Research, 37*, 1–13. https://doi.org/10.1093/nar/gkn923.

Hui, M. K. C., Chan, K. W., Luk, J. M., et al. (2012). Cytoplasmic Forkhead box M1 (FoxM1) in esophageal squamous cell carcinoma significantly correlates with pathological disease stage. *World Journal of Surgery, 36*, 90–97. https://doi.org/10.1007/s00268-011-1302-5.

Jin, G., Yao, K., Guo, Z., et al. (2017). APIO-EE-9 is a novel Aurora A and B antagonist that suppresses esophageal cancer growth in a PDX mouse model. *Oncotarget, 8*, 53387–53404. https://doi.org/10.18632/oncotarget.18508.

Karolchik, D., Hinrichs, A. S., Furey, T. S., et al. (2004). The UCSC table browser data retrieval tool. *Nucleic Acids Research, 32*, D493–D496. https://doi.org/10.1093/nar/gkh103.

Kashyap, M. K., Harsha, H. C., Renuse, S., et al. (2010). SILAC-based quantitative proteomic approach to identify potential biomarkers from the esophageal squamous cell carcinoma secretome. *Cancer Biology & Therapy, 10*, 796–810. https://doi.org/10.4161/cbt.10.8.12914.

Krogan, N. J., Cagney, G., Yu, H., et al. (2006). Global landscape of protein complexes in the yeast Saccharomyces cerevisiae. *Nature, 440*, 637–643. https://doi.org/10.1038/nature04670.

Kufer, T. A., Silljé, H. H. W., Körner, R., et al. (2002). Human TPX2 is required for targeting Aurora-A kinase to the spindle. *Journal of Cell Biology, 158*, 617–623. https://doi.org/10.1083/jcb.200204155.

Kumar, S. U., Kumar, D. T., Siva, R., et al. (2019). Integrative bioinformatics approaches to map potential novel genes and pathways involved in ovarian cancer. *Frontiers in Bioengineering and Biotechnology*, 7, 391. https://doi.org/10.3389/fbioe.2019.00391.

Kwon, Y.-J., Lee, S. J., Koh, J. S., et al. (2009). Expression patterns of aurora kinase B, heat shock protein 47, and periostin in esophageal squamous cell carcinoma. *Oncology Research, 18*, 141–151. https://doi.org/10.3727/096504009790217407.

Li, C., Tang, Z., Zhang, W., et al. (2021). GEPIA2021: Integrating multiple deconvolution-based analysis into GEPIA. *Nucleic Acids Research, 49*, W242–W246. https://doi.org/10.1093/nar/gkab418.

Lin, C., Song, L., Gong, H., et al. (2013). Nkx2-8 downregulation promotes angiogenesis and activates NF-κB in esophageal cancer. *Cancer Research, 73*, 3638–3648. https://doi.org/10.1158/0008-5472.CAN-12-4028.

Lordick, F., Mariette, C., Haustermans, K., et al. (2016). Oesophageal cancer: ESMO clinical practice guidelines for diagnosis, treatment and follow-up. *Annals of Oncology, 27*, v50–v57. https://doi.org/10.1093/annonc/mdw329.

Mishra, S., Shah, M. I., Udhaya Kumar, S., et al. (2021). Chapter eleven—Network analysis of transcriptomics data for the prediction and prioritization of membrane-associated biomarkers for idiopathic pulmonary fibrosis (IPF) by bioinformatics approach. In R. Donev (Ed.), *Advances in protein chemistry and structural biology* (pp. 241–273). Academic Press. https://doi.org/10.1016/bs.apcsb.2020.10.003.

Mondal, G., Sengupta, S., Panda, C. K., et al. (2007). Overexpression of Cdc20 leads to impairment of the spindle assembly checkpoint and aneuploidization in oral cancer. *Carcinogenesis, 28*, 81–92. https://doi.org/10.1093/carcin/bgl100.

Munari, F. F., Cruvinel-Carloni, A., Lacerda, C. F., et al. (2018). PIK3CA mutations are frequent in esophageal squamous cell carcinoma associated with chagasic megaesophagus and are associated with a worse patient outcome. *Infectious Agents and Cancer, 13*, 43. https://doi.org/10.1186/s13027-018-0216-3.

NCBI Resource Coordinators. (2016). Database resources of the National Center for Biotechnology Information. *Nucleic Acids Research*, *44*, D7–D19. https://doi.org/10.1093/nar/gkv1290.

Nicolau-Neto, P., Costa, N. M. D., de Santos, P. T. S., et al. (2018). Esophageal squamous cell carcinoma transcriptome reveals the effect of FOXM1 on patient outcome through novel PIK3R3 mediated activation of PI3K signaling pathway. *Oncotarget*, *9*, 16634–16647. https://doi.org/10.18632/oncotarget.24621.

Ohashi, S., Miyamoto, S., Kikuchi, O., et al. (2015). Recent advances from basic and clinical studies of esophageal squamous cell carcinoma. *Gastroenterology*, *149*, 1700–1715. https://doi.org/10.1053/j.gastro.2015.08.054.

Prabhu, A., Obi, K. O., & Rubenstein, J. H. (2014). The synergistic effects of alcohol and tobacco consumption on the risk of esophageal squamous cell carcinoma: A meta-analysis. *Official Journal of the American College of Gastroenterology | ACG*, *109*, 822–827. https://doi.org/10.1038/ajg.2014.71.

Qiu, Y.-T., Wang, W.-J., Zhang, B., et al. (2017). MCM7 amplification and overexpression promote cell proliferation, colony formation and migration in esophageal squamous cell carcinoma by activating the AKT1/mTOR signaling pathway. *Oncology Reports*, *37*, 3590–3596. https://doi.org/10.3892/or.2017.5614.

Rahman, K. M. T., Islam, F., Banik, R. S., et al. (2013). Changes in protein interaction networks between normal and cancer conditions: Total chaos or ordered disorder? *Network Biology*, *3*, 14.

Saito, R., Smoot, M. E., Ono, K., et al. (2012). A travel guide to Cytoscape plugins. *Nature Methods*, *9*, 1069–1076. https://doi.org/10.1038/nmeth.2212.

Scott, S. J., Li, X., Jammula, S., et al. (2021). Evidence that polyploidy in esophageal adenocarcinoma originates from mitotic slippage caused by defective chromosome attachments. *Cell Death and Differentiation*, *28*, 2179–2193. https://doi.org/10.1038/s41418-021-00745-8.

Shannon, P., Markiel, A., Ozier, O., et al. (2003). Cytoscape: A software environment for integrated models of biomolecular interaction networks. *Genome Research*, *13*, 2498–2504. https://doi.org/10.1101/gr.1239303.

Smyth, G. K. (2005). Limma: Linear models for microarray data. In R. Gentleman, V. J. Carey, W. Huber, et al. (Eds.), *Bioinformatics and computational biology solutions using R and bioconductor* (pp. 397–420). New York, NY: Springer.

Smyth, E. C., Lagergren, J., Fitzgerald, R. C., et al. (2017). Oesophageal cancer. *Nature Reviews. Disease Primers*, *3*, 1–21. https://doi.org/10.1038/nrdp.2017.48.

Song, Y., Li, L., Ou, Y., et al. (2014). Identification of genomic alterations in oesophageal squamous cell cancer. *Nature*, *509*, 91–95. https://doi.org/10.1038/nature13176.

Song, L., Wang, X., & Feng, Z. (2018). Overexpression of FOXM1 as a target for malignant progression of esophageal squamous cell carcinoma. *Oncology Letters*, *15*, 5910–5914. https://doi.org/10.3892/ol.2018.8035.

Stelzl, U., Worm, U., Lalowski, M., et al. (2005). A human protein-protein interaction network: A resource for annotating the proteome. *Cell*, *122*, 957–968. https://doi.org/10.1016/j.cell.2005.08.029.

Stewart, S., & Fang, G. (2005). Anaphase-promoting complex/cyclosome controls the stability of TPX2 during mitotic exit. *Molecular and Cellular Biology*, *25*, 10516–10527. https://doi.org/10.1128/MCB.25.23.10516-10527.2005.

Szklarczyk, D., Franceschini, A., Wyder, S., et al. (2015). STRING v10: Protein-protein interaction networks, integrated over the tree of life. *Nucleic Acids Research*, *43*, D447–D452. https://doi.org/10.1093/nar/gku1003.

Tang, Y., Li, M., Wang, J., et al. (2015). CytoNCA: A cytoscape plugin for centrality analysis and evaluation of protein interaction networks. *Biosystems*, *127*, 67–72. https://doi.org/10.1016/j.biosystems.2014.11.005.

Tong, T., Zhong, Y., Kong, J., et al. (2004). Overexpression of Aurora-a contributes to malignant development of human esophageal squamous cell carcinoma. *Clinical Cancer Research*, *10*, 7304–7310. https://doi.org/10.1158/1078-0432.CCR-04-0806.

Tungekar, A., Mandarthi, S., Mandaviya, P. R., et al. (2018). ESCC ATLAS: A population wide compendium of biomarkers for esophageal squamous cell carcinoma. *Scientific Reports*, *8*, 12715. https://doi.org/10.1038/s41598-018-30579-3.

Udhaya Kumar, S., Madhana Priya, N., Thirumal Kumar, D., et al. (2021). Chapter Ten—An integrative analysis to distinguish between emphysema (EML) and alpha-1 antitrypsin deficiency-related emphysema (ADL)—A systems biology approach. In R. Donev, & T. Karabencheva-Christova (Eds.), *Advances in protein chemistry and structural biology* (pp. 315–342). Academic Press. https://doi.org/10.1016/bs.apcsb.2021.02.004.

Udhaya Kumar, S., Rajan, B., Thirumal Kumar, D., et al. (2020). Involvement of essential signaling cascades and analysis of gene networks in diabesity. *Genes*, *11*, 1256. https://doi.org/10.3390/genes11111256.

Udhaya Kumar, S., Saleem, A., Thirumal Kumar, D., et al., & Donev, R. (2021). Chapter Eleven—A systemic approach to explore the mechanisms of drug resistance and altered signaling cascades in extensively drug-resistant tuberculosis. In T. Karabencheva-Christova (Ed.), *Advances in protein chemistry and structural biology* (pp. 343–364). Academic Press. https://doi.org/10.1016/bs.apcsb.2021.02.002.

Udhaya Kumar, S., Thirumal Kumar, D., Bithia, R., et al. (2020). Analysis of differentially expressed genes and molecular pathways in familial hypercholesterolemia involved in atherosclerosis: A systematic and bioinformatics approach. *Frontiers in Genetics*, *11*, 734. https://doi.org/10.3389/fgene.2020.00734.

Udhaya Kumar, S., Thirumal Kumar, D., Siva, R., et al. (2020). Dysregulation of signaling pathways due to differentially expressed genes from the B-cell transcriptomes of systemic lupus erythematosus patients—A bioinformatics approach. *Frontiers in Bioengineering and Biotechnology*, *8*, 276. https://doi.org/10.3389/fbioe.2020.00276.

Wang, X. X., Liu, R., Jin, S. Q., et al. (2006). Overexpression of Aurora-A kinase promotes tumor cell proliferation and inhibits apoptosis in esophageal squamous cell carcinoma cell line. *Cell Research*, *16*, 356–366. https://doi.org/10.1038/sj.cr.7310046.

Wei, P., Zhang, N., Xu, Y., et al. (2013). TPX2 is a novel prognostic marker for the growth and metastasis of colon cancer. *Journal of Translational Medicine*, *11*, 313. https://doi.org/10.1186/1479-5876-11-313.

Wu, W., Hu, K., Wang, D., et al. (2013). CDC20 overexpression predicts a poor prognosis for patients with colorectal cancer. *Journal of Translational Medicine*, *11*, 142. https://doi.org/10.1186/1479-5876-11-142.

Yan, T., Cui, H., Zhou, Y., et al. (2019). Multi-region sequencing unveils novel actionable targets and spatial heterogeneity in esophageal squamous cell carcinoma. *Nature Communications*, *10*, 1670. https://doi.org/10.1038/s41467-019-09255-1.

Yu, X., Liang, Q., Liu, W., et al. (2017). Deguelin, an Aurora B kinase inhibitor, exhibits potent anti-tumor effect in human esophageal squamous cell carcinoma. *eBioMedicine*, *26*, 100–111. https://doi.org/10.1016/j.ebiom.2017.10.030.

Yu, Y., Ouyang, Y., & Yao, W. (2018). shinyCircos: An R/shiny application for interactive creation of Circos plot. *Bioinformatics*, *34*, 1229–1231. https://doi.org/10.1093/bioinformatics/btx763.

Yu, S., Wang, G., Shi, Y., et al. (2020). MCMs in cancer: Prognostic potential and mechanisms. *Analytical Cellular Pathology*, *2020*, e3750294. https://doi.org/10.1155/2020/3750294.

Zayed, H. (2020). The identification of highly upregulated genes in claudin-low breast cancer through an integrative bioinformatics approach. *Computers in Biology and Medicine*, *127*, 103806. https://doi.org/10.1016/j.compbiomed.2020.103806.

Zhao, C., Gong, L., Li, W., & Chen, L. (2010). Overexpression of Plk1 promotes malignant progress in human esophageal squamous cell carcinoma. *Journal of Cancer Research and Clinical Oncology*, *136*, 9–16. https://doi.org/10.1007/s00432-009-0630-4.

CHAPTER SIX

Protein interaction networks of the mammalian core clock proteins

Ibrahim Halil Kavakli[a,b,*], Nuri Ozturk[c], and Ibrahim Baris[b]
[a]Department of Chemical and Biological Engineering, Koc University, Istanbul, Turkey
[b]Department of Molecular Biology and Genetics, Koc University, Istanbul, Turkey
[c]Department of Molecular Biology and Genetics, Gebze Technical University, Kocaeli, Turkey
*Corresponding author: e-mail address: hkavakli@ku.edu.tr

Contents

Abstract

Circadian rhythm is a 24-h cycle that regulates the biochemical and behavioral changes of organisms. It controls a wide range of functions, from gene expression to behavior, allowing organisms to anticipate daily changes in their environment. In mammals, circadian rhythm is generated by a complex transcriptional and translational feedback loop mechanism. The binding of CLOCK/BMAL1 heterodimer to the E-box of DNA located within the promoter region initiates transcription of clock control genes including the transcription of the other two core clock genes of *Periods (Pers)* and *Cryptochromes (Crys)*. Then PERs and CRYs along with casein kinase 1ε/Δ translocate into the nucleus where they suppress CLOCK/BMAL1 transactivation and, in turn, clock-regulated gene expression. Various clock components must be operational to aid in their stabilization and period extension in circadian rhythm. In this review, we have highlighted the recent progress for the core clock interacting proteins to maintain and to stabilize circadian rhythm in mammals.

Advances in Protein Chemistry and Structural Biology, Volume 131
ISSN 1876-1623
https://doi.org/10.1016/bs.apcsb.2022.04.001

1. Introduction

Circadian rhythms are endogenous daily rhythms generated by the molecular circadian clock. These rhythms allow organisms to adapt their physiological and behavioral activities to the 24-h cyclic changes in the environment and are found in nearly all organisms ranging from cyanobacteria to mammals. Studies with different organisms revealed that a circadian rhythm has three key characteristics (Kavakli & Sancar, 2002; Pittendrigh, 1993). First, circadian rhythm is endogenous and can be generated in the absence of any environmental cues. Second, the period of a circadian rhythm is temperature compensated. In other words, the period of circadian clock is not affected over a broad range of physiological temperatures. Finally, circadian rhythms can be entrained/reset by external time cues, called Zeitgebers, such as light.

Having a circadian rhythm for an organism is thought to provide selective advantages to organisms in a given environment. For example, a study with cyanobacteria indicates that wild type cyanobacteria have better reproductive fitness than mutant cyanobacteria with different circadian periods (Johnson & Golden, 1999).

Several epidemiological and genetic studies show that a variety of physiological functions and behaviors are controlled by circadian rhythm in mammals. Sleep–wake cycle, body temperature, heart rate, blood pressure are all clock-controlled (Eckel-Mahan & Storm, 2009; Franken & Dijk, 2009).

2. Circadian rhythm in mammals

The mammalian circadian system consists of three basic components: (i) input signals (environmental cues such as light, temperature, feeding, social interactions, etc.), (ii) a circadian oscillator as an intrinsic rhythm generator, and (iii) output rhythms (as gene expression, neuronal or hormonal signals).

The circadian rhythm system is mainly composed of a central pacemaker and a peripheral clock in the peripheral tissues in mammals (Stephan & Zucker, 1972). The suprachiasmatic nucleus (SCN) in the hypothalamus is the central circadian oscillator. The SCN consists of approximately 20,000 neurons which can generate autonomous circadian oscillations (Mieda, 2019). It receives photic information from the eyes via the retinohypothalamic tract to synchronize the circadian timing system with environmental time (Yamazaki et al., 2000). Therefore, environmental light–dark cues do not generate circadian rhythms per se, but rather they serve as Zeitgebers of the period and phase of the endogenous molecular clock. The resetting information from SCN is transmitted to other regions of the brain and peripheral cells via humoral, systemic, and neuronal signals (Kramer et al., 2001; Mieda, 2019). Peripheral clocks then regulate behavioral and physiological variables such as body temperature, bowel movement, blood pressure in a circadian-dependent manner (Mohawk, Green, & Takahashi, 2012). The effector element of the circadian oscillator is the output component. All circadian changes in physiology, metabolism, and behavior are most likely linked to rhythmic gene expression in some way. Microarray studies revealed that about 10% of expressed genes exhibit circadian rhythm in peripheral cells (Panda et al., 2002; Storch et al., 2002; Washietl, Kellis, & Garber, 2014). A high-resolution transcriptomic study revealed that 43% of all protein-coding genes exhibit circadian rhythms in transcription in at least one organ (Zhang, Lahens, Ballance, Hughes, & Hogenesch, 2014). Additionally, metabolomic studies suggest that 50% of the detectable metabolites are regulated by the circadian clock (Krishnaiah et al., 2017). However, how the circadian clock differentially controls the transcription of the genes and metabolites in tissue specific manner is not yet fully known. As discussed below, several studies suggest that post-translational/transcriptional regulations and tissues specific clock factors may enable cells to regulate clock output in tissue-specific manner.

2.1 Molecular mechanism of circadian rhythm in mammals

The circadian clock is the cell-autonomous and composed of interlocking core and auxiliary transcriptional/translational feedback loops (Takahashi, 2017). In the core loop, the CLOCK (Circadian Locomotor Output Cycles Kaput, and its paralogue NPAS2) interacts with BMAL1 (Brain and Muscle ARNT-Like 1) and they bind *E*-boxes region on the DNA, where they initiate the transcription of the clock-controlled genes including *Period* (*Per1*, *Per2*, and *Per3*) and *Cryptochrome* (*Cry1* and *Cry2*) genes in mammals (Allada & Bass, 2021; Kavakli et al., 2017; Takahashi, 2017). After accumulation of the PERs and CRYs in the cytosol, they translocate into the nucleus along with Casein kinase Iε/Δ (CKIε/Δ), where they

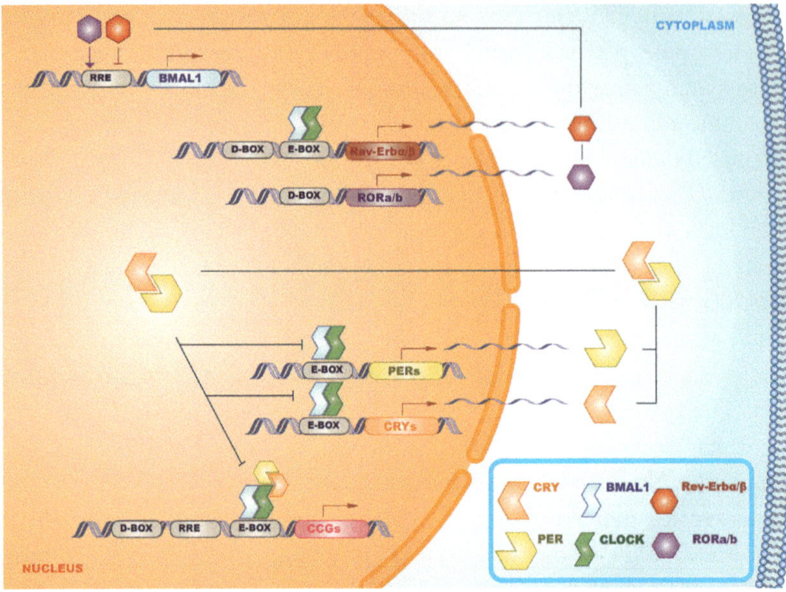

Fig. 1 Schematic representation of molecular clock mechanism in mammals. The mechanism is controlled by an intertwined transcription-translation feedback loops (TTFL).

interact with CLOCK/BMAL1 dimer and repress their activity and, in turn, suppress transcription of clock-controlled genes including their own (Fig. 1) (Kavakli, Gul, & Turkay, 2022; Mieda, 2019; Takahashi, 2016). Degradation of the PERs and CRYs by ubiquitin-dependent pathways results in restarting of the TTFL by relieving the repression on CLOCK/BMAL1 and restart the cycle again (Busino et al., 2007; Reischl et al., 2007; Siepka et al., 2007). There is a secondary loop, which regulates the transcription of the *Bmal1* gene, and, in turn, core the clock mechanism. Nuclear receptors retinoid acid receptor-related orphan receptors α, β, and γ (RORα, β, or γ) activate the transcription of the *Bmal1* gene whereas nuclear receptor subfamily 1, group D, member 1 (NR1D1, also known as REV-ERBα) and 2 (NR1D2, also known as REV-ERBβ) inhibit the transcription of the *Bmal1* gene (Preitner et al., 2002; Sato et al., 2004). Other auxiliary feedback loops also affect the circadian clock including the following proteins DBP, HLF, TEF, CBS, CHRONO, DEC1 (Bhlhb40), and DEC2 (Bhlhb41) (Anafi et al., 2014; Cal-Kayitmazbatir et al., 2021; Honma et al., 2002; Mohawk et al., 2012). Additionally, the combination of the gene targeting technologies and high-throughput screening of transcriptional targets reveal the novel interactions between the circadian clock and other cellular pathways (Emisoglu-Kulahli et al., 2021) and help to explain cellular dependent phenotypes such modulation of apoptosis and cellular migration by the circadian clock depending on the genetic background (Korkmaz et al., 2018).

Post-translational modifications (PTMs) of the core clock proteins control various features of clock components, including their stability, subcellular localization, transcriptional activity, and interactions with other proteins (Okamoto-Uchida et al., 2019). CKIε is encoded by the *tau* gene and is shown to be responsible for the phosphorylation of PER proteins (Lowrey et al., 2000). Because of a low rate of CKIε-dependent phosphorylation of PER2, the mutant hamster with *tau* mutations developed a short-period phenotype. Furthermore, the SNPs of *PER2* genes result in defected phosphorylation and associated with familial advanced sleep phase syndrome (FASPS) (Fu, Jones, Toh, Virshup, & Ptacek, 2001). Phosphorylation of the CRYs, BMAL1, and CLOCK are also shown to be important for the regulation and their degradation. For example, phosphorylation of CRY1 by adenosine monophosphate-activated protein kinase (AMPK) regulates its degradation (Lamia et al., 2009). Phosphorylation of BMAL1 and CLOCK are critical for their translocation into the nucleus and degradations (Brenna & Albrecht, 2020). Furthermore, other PTMs, SUMOylation and acetylation of BMAL1 and CLOCK, are important for the stability and activity (Okamoto-Uchida et al., 2019). All these findings clearly indicate that PTMs are critical in the regulation of the circadian clock, in turn, generation of the 24-h circadian rhythmicity at the molecular level.

3. Protein–protein interaction network

The first clock gene, named *Period* (*Per*) was discovered by Konopka and Benzer using mutagenesis and followed by screening Drosophila offspring for defects in circadian rhythm (Konopka & Benzer, 1971). Other clock components from different organisms are being discovered by mutagenesis coupled with forward genetics, biochemical methods, and yeast two-hybrid. Although the core clock proteins (CLOCK, BMAL1, CRYs, and PERs) are conserved in all cells, the differential outcome of peripheral tissues implies the existence of additional tissue-specific accessory proteins. For example, the hepatocyte nuclear factor 4A (HNF4A) is a nuclear receptor specifically expressed in the liver, where it regulates genes mainly involved in the hepatic gluconeogenic program and lipid metabolism (Chandra et al., 2013). HNF4A is involved in chromatin remodeling and facilitates the binding of CLOCK/BMAL1 dimer during the daytime to regulate clock-controlled gene expression in hepatocytes (Qu, Qu, Jia, & Kay, 2021). The deubiquitinating enzyme Ubiquitin specific protease 2 (USP2) interacts with CRY/PER and affects CLOCK/BMAL1 dimer transactivation in the brain, liver, and male genital tract (Kitamura & Hashimoto, 2021; Yang et al., 2012).

As explained below, core clock proteins interact with several different proteins including transcription factors, enzymes, receptors or structural proteins. Biochemical and genetics studies have shown that genetic variations affect the interaction of core clock proteins with each other and other proteins. For example, one of the *CRY1* gene variations (rs772585429; c.878G>A (p.Arg293His)) observed in the 1000 Genome Project inhibits the interaction between CRY1 and CLOCK proteins (Gul et al., 2020). One of the most common pathogenic variation in the cystathionine β-synthase (CBS) gene prevents interaction with CRY1 (Cal-Kayitmazbatir et al., 2021). A sleep disorder associated variation (rs121908635; c.1984A>G (p.Ser662Gly)) within *PER2* gene was shown to affect its phosphorylation via inhibition of binding with casein kinase Iε/Δ (CKI) (Fu et al., 2001). Another sleep disorder associated missense variation (rs201220841; c.778G>A (p.Ala260Thr)) enhances the affinity between CRY2 and the E3 ubiquitin ligase FBXL3, thus promoting its degradation (Hirano et al., 2016). Other numerous studies suggest the importance of interaction between core clock proteins and other partners in terms of cellular activities and possible associated diseases (Rijo-Ferreira & Takahashi, 2019). These interactions can be modulated by the small molecules as well. Feasibility of this approach is shown by a recent study, where they find a small molecule regulating the interaction between CLOCK and BMAL1 (Doruk et al., 2020). In this approach, they targeted PAS domains, responsible for the interaction between CLOCK and BMAL1, by virtual screening and identified a small molecule named CLK8. Their findings revealed that small molecules can be used to modulate protein-protein interaction in clock mechanism.

In this review, we will focus core clock mammalian proteins in terms of their function and interacting partners.

3.1 CLOCK

The *Clock* gene, the first identified member of the mammalian core clock gene, is discovered with a forward mutagenesis screen in the mouse and then followed by positional cloning (King et al., 1997; Vitaterna et al., 1994). The CLOCK belongs to the bHLH-PAS (basic Helix Loop Helix-Period Arnt Single-minded) family of transcription factors. PAS and bHLH domains are important for protein-protein interaction and DNA-protein interaction, respectively. CLOCK consists of PAS A and PAS B domains in addition to the bHLH domain. These bHLH-PAS transcription factors play roles in a wide array of cellular responses to environmental cues (such as hypoxia (Hypoxia inducible factor, HIF), neurogenesis (SIM1), synaptic plasticity (NPAS4) and circadian clock (BMAL1, CLOCK, and NPAS2)) (McIntosh, Hogenesch, & Bradfield, 2010). CLOCK dimerizes with BMAL1 via their PAS domains. This heterodimer translocates to the nucleus, where it binds to an *E*-box sequence (CACGTG) in the promoter regions of the other components of the core clock and clock-controlled genes (CCG) (Fig. 1).

Fig. 2A shows proteins network of CLOCK. The Mixed lineage leukemia 1 (MLL1), a histone methyltransferase (HMT), allows the recruitment of the CLOCK/BMAL1 dimer to chromatin (Katada & Sassone-Corsi, 2010). Other factors like transcription coactivators CREB binding protein (CBP) and p300 are required to stimulate the activity of the CLOCK/BMAL1 dimer on DNA (Etchegaray, Lee, Wade, & Reppert, 2003; Takahata et al., 2000). Further study showed such stimulation of the activity of the CLOCK/BMAL1 might be tissues specific (Hosoda et al., 2009). JumonjiC (JmjC) and ARID domain-containing histone lysine demethylase 1a (JARID1a) increases histone acetylation and recruits CLOCK/BMAL1 complex to the *Per2* promoter. Finally, thyroid hormone receptor-associated protein-150 (TRAP150) is a coactivator and promotes CLOCK/BMAL1 dimer to the promoter of the target genes (Lande-Diner, Boyault, Kim, & Weitz, 2013). *BHLHE40* and *BHLHE41* are important for the regulation of circadian clock. Products of both gene, DEC1 and DEC2, inhibit that activity of the CLOCK/BMAL1 (Honma et al., 2002). In fact, SNPs at *BHLHE40* locus are associated with short sleep period in human (He et al., 2009; Pellegrino et al., 2014). Additionally, CLOCK-Interacting Protein, Circadian (CIPC) is shown to interact with CLOCK in a yeast two-hybrid system and, in turn, to inhibit the CLOCK/BMAL1 driven transcription (Zhao et al., 2007).

CLOCK is engaged in several different cellular activities as well. CLOCK has an intrinsic acetylase activity, which can act on histones as well as its partner BMAL1 and is required for its activating role (Doi, Hirayama, & Sassone-Corsi, 2006; Hirayama et al., 2007). For example,

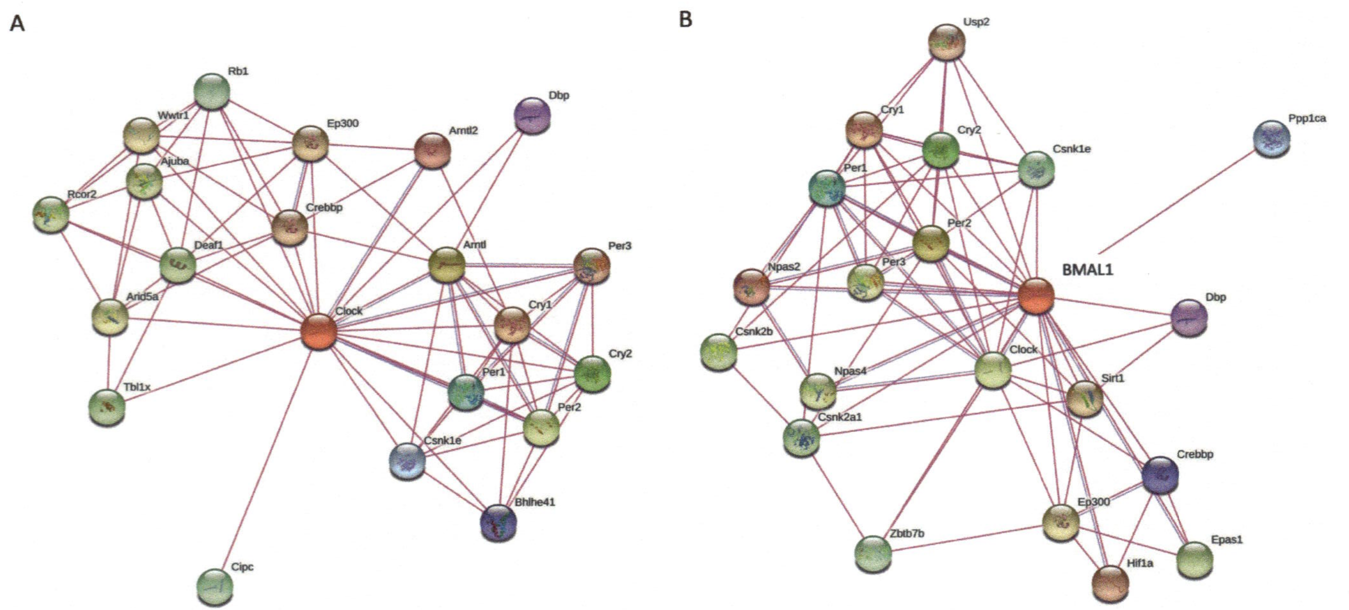

Fig. 2 Interaction networks of (A) CLOCK and (B) BMAL1 proteins generated by String. This network is generated with Interaction information obtained from experiments and with a confidence score of 0.4 or above are visualized. The red colored circle represents query proteins and the first shell of interactors.

CLOCK is important in the regulation of the arginine biogenesis and ureagenesis by acetylating the argininosuccinate synthase 1 (ASS1) (Lin et al., 2017). Also, a recent study revealed that CLOCK can bind to the mitochondrial carrier SLC25A10 and modulate mitochondrial activity. SLC25A10 is a dicarboxylate carrier and transports malate and succinate for the citric acid cycle, urea synthesis, and sulfur metabolism (Cai et al., 2019). Another study showed that CLOCK can interact with RANBP9 and its partner SF3B3 proteins and bind with mRNAs. RANBP9, a member of the Ran family, is important for male germ cell development through involving in alternative splicing in spermatogenesis. Thus, CLOCK involves in alternative splicing in spermatogenesis in testis (Yang et al., 2018).

3.2 BMAL1

Like CLOCK, BMAL1 is also a member of a family of transcriptional regulators that contain bHLH and PAS domains. As the member of the positive arm of the TTLF, BMAL1 makes heterodimers with CLOCK or NPAS4 (Ooe, Saito, Mikami, Nakatuka, & Kaneko, 2004).

Fig. 2B, generated by String from the experimental studies, shows proteins that interact with BMAL1. USP2b (45-kDa isoform of USP2) interacts with BMAL1 and regulates the stability and turnover of BMAL1. Additionally, the presence of the PER1 enhances the affinity between BMAL1 and USP2b (Scoma et al., 2011). Unlike BMAL1 turnover, USP2 regulates nucleocytoplasmic shuttling and nuclear retention of PER1 and, therefore, affects PER1 repressor activity (Yang, Duguay, Fahrenkrug, Cermakian, & Wing, 2014). Phosphorylation of BMAL1 by kinases is an important regulator for both stability and activity. Eide et al. showed that BMAL1 is a substrate for CKIε *in vitro*, and CKIε activity positively regulates BMAL1-dependent transcription from circadian promoters (Eide, Vielhaber, Hinz, & Virshup, 2002). On the other hand, co-immunoprecipitation experiments and real-time monitoring of protein–protein interactions revealed that CRY-mediated periodic binding of CK2β to BMAL1 inhibits BMAL1-Ser90 phosphorylation by CK2α (Tamaru et al., 2015). Wallach et al. performed yeast two hybrid assay to identify novel interacting proteins for BMAL1 (Wallach et al., 2013). In their screening, they identified the Protein Phosphatase 1 Catalytic Subunit Alpha (PPP1CA) which interacts with BMAL1. Further characterization of this interaction indicates that PPP1CA regulates BMAL1 abundancy in circadian dependent manner. Recently, the subunit of the

NF-κB RELA was shown to interact with BMAL1 and to inhibit the activity of the CLOCK/BMAL1 transactivation (Shen et al., 2021). Studies with knockout *Ikk*^{CA} male animals indicated that they have long period compared to control animals. CHRONO interacts with BMAL1 and represses CLOCK/BMAL1 driven transcription by reducing the stability of BMAL1 (Anafi et al., 2014). The CHRONO knockout mice exhibit a prolonged free-running circadian period.

ZBTB7B (Zinc finger and BTB domain-containing protein 7B) is a transcription regulator that acts as a key regulator of lineage commitment of immature T-cell precursors. The interaction between BMAL1 and ZBTB7B (Fig. 2B) is detected using affinity capture mass spectrometry analysis (Srivastava & Mishra, 2020). The functional significant of such interaction is yet be shown.

Interactions between BMAL1 and HIF1α and HIF2α (encoded by EPAS1) were shown by yeast two-hybrid system (Hogenesch, Gu, Jain, & Bradfield, 1998). Further study showed that interaction between them mediated the response to the cellular hypoxia that influence metabolic adaptation to hypoxia condition (Peek et al., 2017).

SIRT1, the NAD^+-dependent enzyme, carries out deacetylase function on core clock proteins of BMAL1 and PER2 (Asher et al., 2008; Nakahata et al., 2008). SIRT1 is proposed to function as a molecular rheostat of CLOCK-mediated acetylation of BMAL1, which causes the increased activity of the CRY repression on CLOCK/BMAL1 transactivation and, in turn enhancement of circadian amplitude (Nakahata et al., 2008). On the contrary, Asher et al. reported that SIRT1 exhibits deacetylase activity on PER2 rather than BMAL1 and, in turn, results in reduction of the circadian amplitude (Asher et al., 2008). A recent study supports that SIRT1 regulates circadian clock through PER2 (Foteinou et al., 2018). SIRT1 is a member of sirtuin family and serves important function for the regulation of cellular metabolism. It is, therefore, believed that SIRT1 provides link between cell metabolism and circadian clock.

There are also other protein in the networks (Fig. 2B) but further studies are required to enlighten their roles in circadian rhythm perspective.

3.3 Cryptochrome 1 and 2

Cryptochromes (CRYs) belong to a large family of cryptochrome/photolyase family (CPF) (Asimgil & Kavakli, 2012; Kavakli et al., 2017). CPF protein family performs wide range of functions depending on the type of organism. For example, photolyases repair UV-induced DNA damages while cryptochromes act as a transcriptional repressors of the circadian clock in mammals (Sancar, 2000). CRYs were initially discovered as blue light photoreceptors in *Arabidopsis thaliana* where they regulate hypocotyl elongation and various aspects of plant growth (Ahmad & Cashmore, 1993; Lin & Shalitin, 2003). Subsequent studies revealed that CRYs function as photoreceptor and transcriptional repressor in *Drosophila* circadian clock mechanism (Emery, So, Kaneko, Hall, & Rosbash, 1998; Stanewsky et al., 1998).

In mammals, CRYs were shown to be a repressor of CLOCK/BMAL1-driven transcription (Kume et al., 1999; Thresher et al., 1998; van der Horst et al., 1999; Vitaterna et al., 1999). The comparison of the primary structure of CRYs with photolyase from different organisms indicated that CRYs have an extended C-terminal region and the photolyase domain called PHR (Kavakli et al., 2022; Parico & Partch, 2020) (Fig. 3A). The C-terminal domain is intrinsically disordered, and its length varies from 10 to 300 amino acids depending on the type of organism. The C-terminal end of the CRYs enables them to interact with various type of proteins, such as FBXL3, PER2, and CBS in mammals (Cal-Kayitmazbatir et al., 2021; Chaves et al., 2006; Ozber et al., 2010; Xing et al., 2013). For example, residues between 493 and 512 of the mouse CRY2 were shown to be responsible for PER2 interaction, and the mutation of Arg-501 and Lys-503 residues of mCRY2 within this C-terminal region totally abolishes interaction with PER2 (Ozber et al., 2010).

The PHR domain has two distinct domains (α/β and helical domains) linked each other with a coil. The helical domain is required for the binding of FAD, and this region is also called a primary pocket in mammalian CRYs while α/β domain is required for the binding of the second chromophore of 5,10-methenyl-tetrahydrofolate (MTHF), called secondary pocket in mammalian CRYs (Parico & Partch, 2020; Park, Kim, Sancar, & Deisenhofer, 1995; Sancar, 2000). The PHR domains are important for the interaction with CLOCK/BMAL1 dimer (Xu et al., 2015). The secondary pocket is shown to be important for the binding of CLOCK protein (Michael et al., 2017; Rosensweig et al., 2018). Further studies indicate that there is allosteric regulation in mammalian CRY1 but not in CRY2 to regulate the binding of CLOCK to CRY1 (Gul et al., 2021; Ozcan, Gul, & Kavakli, 2021) (Fig. 3B). The Arg-293 residue in CRY1 regulates the dynamicity of the serine loop next to the secondary pocket which mediates allosteric modulation of CLOCK binding (Gul et al., 2021).

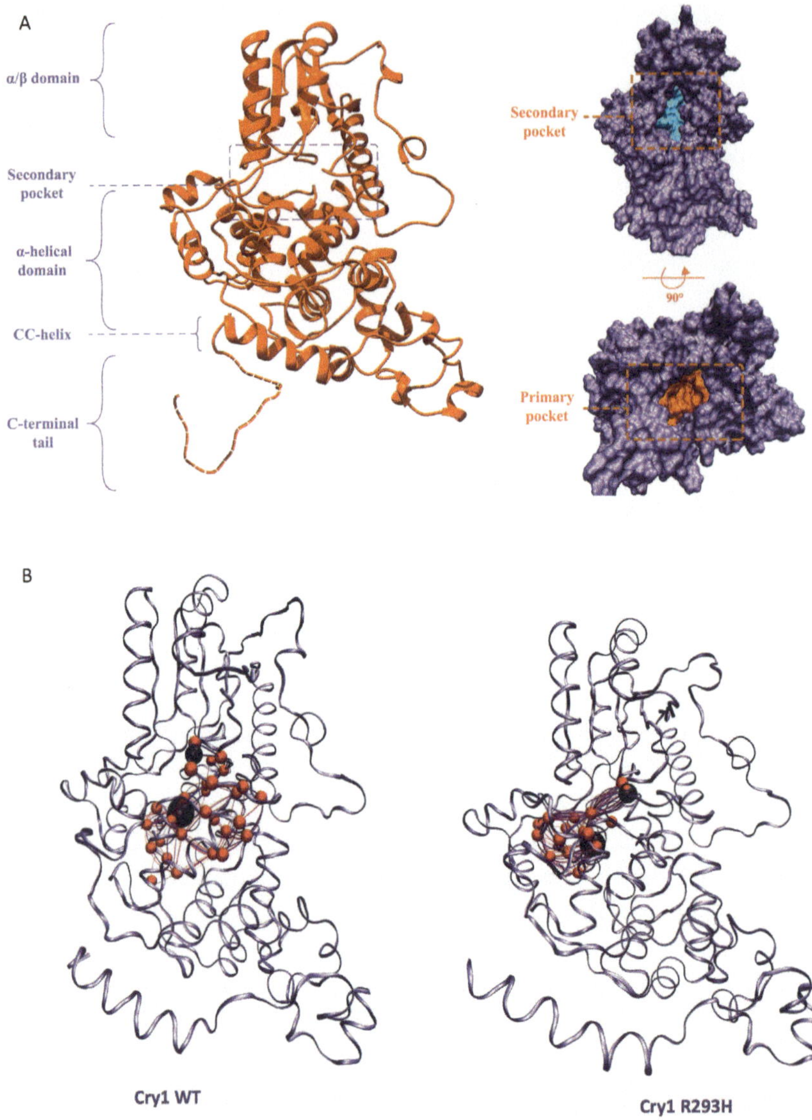

Fig. 3 Crystal structure of mammalian Cry1 (PHR domain) retrieved from Protein Data Bank (ID:5T5X), missing residues in PHR domain completed via homology modeling (RaptorX) and model simulated for 300 ns. Trajectory subjected to geometry-based clustering. (A) Cluster representative of most crowded cluster used for visualizations of PHR domain. The C-terminal tail is added artificially for representation. (B) Allosteric path in CRY1 regulates the dynamicity of the Ser-loop.

Fig. 4 shows protein–protein interaction network of mammalian CRY 1 and 2. As explained in above of this chapter, both CRY1 and CRY2 behave as a transcriptional repressor and interact with core clock proteins of CLOCK, BMAL1, and PERs to generate the TTFL mechanism of the circadian rhythm. The cellular CRYs stability and turn-over rate are important for the periodicity and are regulated via interaction with FBXL3 and FBXL21. FBXL3, is a part of the Skp1-Cul1-FBXL3 ubiquitin ligase complex, interacts with both CRYs and mediates their degradation by the proteasome system (Busino et al., 2007). Interaction between them is promoted by AMP-activated protein kinase (Lamia et al., 2009). In fact, mutations on the *Fbxl3* gene in mouse leads to circadian defected phenotypes, where animals have longer period of circadian rhythm (Godinho et al., 2007; Siepka et al., 2007). Another protein, FBXL21, also ubiquitinates both CRYs but with the opposite effect of the FBXL3. The FBXL21 stabilizes CRYs and retains them in the cytosol (Hirano et al., 2013).

Several studies have reported that CRYs interact with other proteins as well (Fig. 4). There are strong interactions between CRYs and cellular metabolism. For example, CRY1 but not CRY2 interacts with Cystathionine-β-synthase (CBS), a central enzyme in one-carbon metabolism. The CBS-CRY1 interaction enables to modulate cellular circadian physiology and metabolic control. Further metabolomic studies revealed the temporal change in one-carbon and transsulfuration pathways (Cal-Kayitmazbatir et al., 2021). CRYs affect blood glucose levels and gluconeogenesis in hepatocytes through glucagon (GC) signaling. CRY1 and CRY2 have been reported to interact with the alpha subunit of G-protein and to block the GC-mediated release of cAMP, which results in the repression of gluconeogenic gene expression (Zhang et al., 2010).

There are other proteins shown in Fig. 4 interact with CRYs from other organisms and are computationally found to be the interacting partners of the CRYs.

3.4 Period 1

The circadian clock mechanism consists of core clock proteins and their interacting proteins in mammals. This interaction can vary depending on the circadian time. Although there is one PER protein in the Drosophila circadian clock, three PER proteins operate in a similar TTLF mechanism to that of mammals (Ozturk et al., 2007). PERs are core clock proteins and

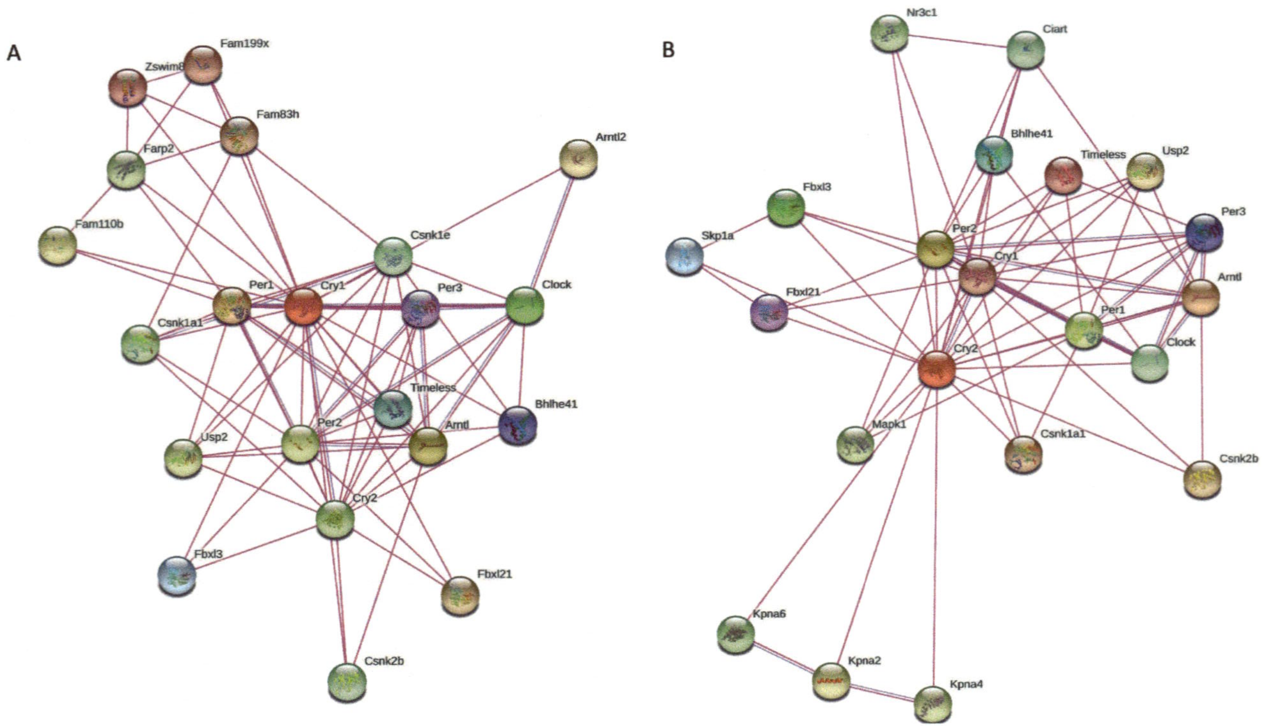

Fig. 4 Interaction networks of (A) Cryptochrome 1 (CRY1) and (B) Cryptochrome 2 (CRY2) proteins generated by String. This network is generated Interaction information obtained from experiments and with a confidence score of 0.4 or above are visualized. The red colored circle represents query proteins and the first shell of interactors.

function as a repressor in the negative arm of the TTFL. PER1, PER2, and PER3 interact with the other core clock components; CRY1, CRY2, BMAL1, and CLOCK/NPAS2 even though these interactions are regulated in a temporal manner (Fig. 5). Other proteins that interact with PER1 are given in Fig. 5A. Yagita et al. showed that PER proteins form homo- and hetero-dimers with each other (Yagita et al., 2000). Although PER2 and PER3 can form homodimers, PER1 is unable to form homodimers.

The amount of PER1 and its stability is under the control of post-translational modification. Phosphorylation of PERs by kinases is a signal for subsequent ubiquitination and proteolysis. For example, Casein kinase I ε/Δ (CKIε/Δ) phosphorylates PER1 (Camacho et al., 2001; Keesler et al., 2000). Subsequently, the ubiquitin-proteasome system recognizes phosphorylated PER1 and regulates the half-life of PER1 which is essential for maintanence of cellular oscillations (Eide et al., 2005). Phosphorylation of PER1 and PER2 by CKIε/Δ generates a recognition site for β-TRCP binding (Eide et al., 2005; Shirogane, Jin, Ang, & Harper, 2005). β-TRCP is an F-box protein and is the adaptor subunit of the SCF (Skp1-Cullin-F-box) E3 ubiquitin ligase complex (Pickart, 2001). Phosphorylated PER is ubiquitinated and degraded via the proteasomal pathway after ubiquitination by the b-TRCP-mediated ubiquitin ligase. Therefore, PER1 interacts with both β-TRCP1 and β-TRCP2 in a manner that depends on casein kinase 1 activity, and they control the PER1 turnover in mammalian cells (Eide et al., 2005; Shirogane et al., 2005).

PER1 is not ubiquitinated only for degradation purposes, but it is also ubiquitinated for other functional modifications. For example, PER1 is highly ubiquitinated in cells and it serves as a substrate of USP2 (deubiquitinating enzyme ubiquitin-specific peptidase 2). USP2 binds it and regulates nucleocytoplasmic shuttling and nuclear retention of PER1 and its repressive role on the clock transcription factors CLOCK and BMAL1 (Yang et al., 2012, 2014).

Type II protein arginine methyltransferase 5 (PRMT5) is known to be a transcriptional co-repressor due to its activity on the dimethylation of histone H3R8 and H4R3 and interacts with CRY1. Rhythmic recruitment of CRY1 and PRMT5 to the PER1 gene promoter coincides with the rhythmic dimethylation of histone H4R3 and PER1 gene repression (Na et al., 2012).

A recent study suggests that PER1 protein plays a role in cellular metabolism (Sun et al., 2020). PER1 interacts with GPX1 (Glutathione peroxidase 1) which regulates metabolic homeostasis under oxidative stress in liver.

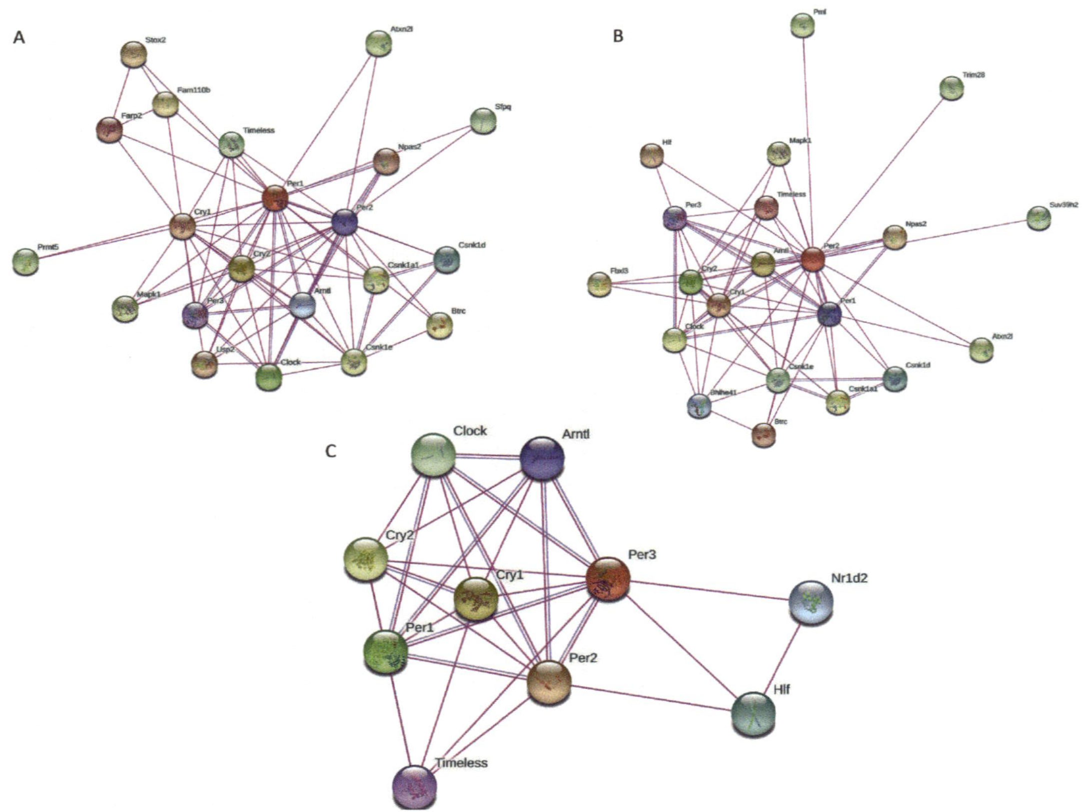

Fig. 5 Interaction networks of (A) Period 1 (PER1) and (B) Period 2 (PER2) and (C) Period 3 (PER3) proteins generated by String. This network is generated Interaction information obtained from experiments and with a confidence score of 0.4 or above are visualized. The red colored circle represents query proteins and the first shell of interactors.

In this interaction, PER1 regulates the enzymatic activity of the GPX1 by increasing clearance of the ROS (reactive oxygen species) in a circadian-dependent manner (Sun et al., 2020).

Splicing factor, proline- and glutamine-rich (SFPQ), which is a protein highly homologous to NONO, has been shown to bind PER proteins (Duong & Weitz, 2014). SFPQ recruits the mSIN3A histone deacetylase to the promoter site and assists PER1 repressive activity. BraInMap, represents the largest experiment-based protein interaction network for the central nervous system (CNS), identified the interactions between PER1 and STOX2, FARP2, and FAM110B as a high confidence Protein-Protein Interaction (Pourhaghighi et al., 2020).

3.5 Period 2 and 3

The PER2 interacts with the other core clock proteins in the negative arm of TTFL (Ye, Selby, Ozturk, Annayev, & Sancar, 2011) (Fig. 5B). PER2 binds to CRY1 via the α-helical domain of CRY1. The crystal structure of the CRY1/PER2 complex revealed that this complex is stabilized by four interfaces and a zinc ion. The binding of FBXL3 or PER2 to CRY is mutually exclusive and suggests that PER2 stabilizes CRY1 by shielding it from FBXL3 (Schmalen et al., 2014). Even though TIM is not the part of the core TTFL mechanism, TIM is shown to bind PER2 in the SCN (Barnes et al., 2003). Kurien et al. suggested that the TIM–PER2–CRY2 exists in the same complex in which TIM plays the role in modulating the levels of CRY2/PER2 (Kurien et al., 2019).

Hypoxia-inducible factor 1α (HIF-1α) is a master transcriptional regulator of extra-cellular oxygen levels (Hirota & Semenza, 2005) and oxygen homeostasis (Semenza, 1998). Even though PER2 has the potential to increase the levels of both transcripts and protein of HIF-1α (Eckle et al., 2012), Kobayashi et al. found a different mechanism where PER2 functions as an effector molecule for the recruitment of HIF-1α to its enhancer region (hypoxia-response element) (Kobayashi, Koyasu, Goto, Hiraoko, & Harada, 2018).

Like PER1, PER2 is also phosphorylated by CKIε/Δ, which generates a recognition site for β-TRCP binding (Eide et al., 2005; Shirogane et al., 2005). However, the affinity between the CKIε and PER2 is higher compared to affinity between CKIε and PER1 (Eide et al., 2002). Mammalian PER proteins act as a scaffold with distinct domains that simultaneously bind CKIε and CRYs. CRYs are phosphorylated by CKIε only when both proteins are bound to PER proteins. Lastly, PER2 phosphorylation is also essential for its proteasomal degradation (Eide et al., 2005; Meng et al., 2008). Immunoprecipitation assay also revealed that PER2 and PER1 is also phosphorylated by other kinases such as CKIα and ERK2 (Hirota et al., 2010). Phosphorylation of PERs is also important for nuclear entry with the help of other proteins (Vanselow et al., 2006). For example, PML physically interacts with PER2 and promotes its transport to the nucleus (Miki et al., 2012).

HP1γ-SUV39H histone methyltransferase interacts with PER2 in a protein complex. This complex recruits an HP1γ-SUV39H1/H2 histone methyltransferase to DNA-bound CLOCK/BMAL1 complex to di- and tri-methylate H3K9 histones to establish a local repressive chromatin state (Duong & Weitz, 2014). PER2 is also found to interact with the basic helix-loop-helix family, member e41 (BHLHE41, hDEC2) (Wallach et al., 2013) but the functional significance of such interaction needs to be investigated.

PER3 interacts with all core clock proteins, CRY1, CRY2, PER1, PER2, CLOCK, BMAL1 (Yagita et al., 2000) (Fig. 5C). It also interacts with most of the PER2 interactors such as TIMELESS (Engelen et al., 2013). PER3 interacts with Nuclear Receptor Superfamily 1 Group D (NR1D1/NR1D2), where they could negatively regulate the activity of CLOCK/BMAL1 (Wallach et al., 2013; Ye et al., 2011, 2014).

4. Conclusion

Methods that have been used to identify several core clock and clock modifiers are fruitful. However, these methods such as biochemical methods are far to capture proteins that participate in clock work in transient manner. It is clear that there are other candidates that are not identified yet. While core clock proteins move into the nucleus, they might interact with many proteins such as the ones that phosphorylate or ubiquitylate them in transient manner. Therefore, alternative methods can be used to identify additional proteins that might interact with core clock proteins. One such method, which is the BioID method, is developed to overcome technical problems to detect weak or transient interactions using conventional screening methods. This method is based on proximity-dependent cellular biotinylation by a promiscuous bacterial biotin ligase (*E. coli* BirA R118G, shortly called BioID) (Choi-Rhee, Schulman, & Cronan, 2004; Cronan, 2005) which is fused to a bait protein to biotinylate the proteins in close proximity in living cells (Roux, Kim, Raida, & Burke, 2012). Therefore, in contrast to conventional approaches, BioID method is one of the most suitable way to label interactions in the most native environment. Such an approach is used to identify PER2 interactome in human osteosarcoma cells which has revealed novel interactions in addition to the known ones (Gul et al., 2022). This study suggested that using different BioID enzymes such as BioID2, miniTurbo, or TurboID with different labeling periods can be tuned to increase the resolution of temporal changes of the circadian interactome. On the other hand, the BioID technique record/labels history of the PPIs. For example, the fusion of TurboID to CRYs, BMAL1, and CLOCK would allow labels proteins in close proximity from the resting state to their degradations by changing the labeling periods. Such an approach will enable us to identify transient interactors at the different circadian times and let us get complete mechanism of the circadian clock at molecular level. Even the TurboID- or APEX2-based BioID techniques were suggested to catch light-dependent or magnetic-field dependent interactions in Drosophila cells, it can be applied to investigate for a possible magnetic field and the circadian clock interaction in mammalian cells (Ozturk, 2021).

In conclusion, generation of circadian clock is very complex and requires involvement of the different proteins. Having a complete mechanism with new targets will enable us to understand molecular basis of clock-related diseases. These new targets can be utilized for the discovery of the small molecules that modulate protein-protein interactions.

Acknowledgment

We would like to thank Dr. Seref Gul and Onur Ozcan for helping us to prepare the figures. I.H.K. would like to thank TUBITAK 112T444 for the financial support.

References

Ahmad, M., & Cashmore, A. R. (1993). HY4 gene of *A. thaliana* encodes a protein with characteristics of a blue-light photoreceptor. *Nature, 366*(6451), 162–166. https://doi.org/10.1038/366162a0.

Allada, R., & Bass, J. (2021). Circadian mechanisms in medicine. *The New England Journal of Medicine, 384*(6), 550–561. https://doi.org/10.1056/NEJMra1802337.

Anafi, R. C., Lee, Y., Sato, T. K., Venkataraman, A., Ramanathan, C., Kavakli, I. H., et al. (2014). Machine learning helps identify CHRONO as a circadian clock component. *PLoS Biology, 12*(4), e1001840. https://doi.org/10.1371/journal.pbio.1001840.

Asher, G., Gatfield, D., Stratmann, M., Reinke, H., Dibner, C., Kreppel, F., et al. (2008). SIRT1 regulates circadian clock gene expression through PER2 deacetylation. *Cell, 134*(2), 317–328. https://doi.org/10.1016/j.cell.2008.06.050.

Asimgil, H., & Kavakli, I. H. (2012). Purification and characterization of five members of photolyase/cryptochrome family from *Cyanidioschyzon merolae*. *Plant Science, 185-186*, 190–198. https://doi.org/10.1016/j.plantsci.2011.10.005.

Barnes, J. W., Tischkau, S. A., Barnes, J. A., Mitchell, J. W., Burgoon, P. W., Hickok, J. R., et al. (2003). Requirement of mammalian Timeless for circadian rhythmicity. *Science, 302*(5644), 439–442. https://doi.org/10.1126/science.1086593.

Brenna, A., & Albrecht, U. (2020). Phosphorylation and circadian molecular timing. *Frontiers in Physiology, 11*, 612510. https://doi.org/10.3389/fphys.2020.612510.

Busino, L., Bassermann, F., Maiolica, A., Lee, C., Nolan, P. M., Godinho, S. I., et al. (2007). SCFFbxl3 controls the oscillation of the circadian clock by directing the degradation of cryptochrome proteins. *Science, 316*(5826), 900–904. https://doi.org/10.1126/science.1141194.

Cai, T., Hua, B., Luo, D., Xu, L., Cheng, Q., Yuan, G., et al. (2019). The circadian protein CLOCK regulates cell metabolism via the mitochondrial carrier SLC25A10. *Biochimica et Biophysica Acta (BBA)—Molecular Cell Research, 1866*(8), 1310–1321. https://doi.org/10.1016/j.bbamcr.2019.03.016.

Cal-Kayitmazbatir, S., Kulkoyluoglu-Cotul, E., Growe, J., Selby, C. P., Rhoades, S. D., Malik, D., et al. (2021). CRY1-CBS binding regulates circadian clock function and metabolism. *The FEBS Journal, 288*(2), 614–639. https://doi.org/10.1111/febs.15360.

Camacho, F., Cilio, M., Guo, Y., Virshup, D. M., Patel, K., Khorkova, O., et al. (2001). Human casein kinase Idelta phosphorylation of human circadian clock proteins period 1 and 2. *FEBS Letters, 489*(2–3), 159–165. https://doi.org/10.1016/s0014-5793(00)02434-0.

Chandra, V., Huang, P., Potluri, N., Wu, D., Kim, Y., & Rastinejad, F. (2013). Multidomain integration in the structure of the HNF-4alpha nuclear receptor complex. *Nature, 495*(7441), 394–398. https://doi.org/10.1038/nature11966.

Chaves, I., Yagita, K., Barnhoorn, S., Okamura, H., van der Horst, G. T., & Tamanini, F. (2006). Functional evolution of the photolyase/cryptochrome protein family: Importance of the C terminus of mammalian CRY1 for circadian core oscillator performance. *Molecular and Cellular Biology, 26*(5), 1743–1753. https://doi.org/10.1128/MCB.26.5.1743-1753.2006.

Choi-Rhee, E., Schulman, H., & Cronan, J. E. (2004). Promiscuous protein biotinylation by *Escherichia coli* biotin protein ligase. *Protein Science, 13*(11), 3043–3050. https://doi.org/10.1110/ps.04911804.

Cronan, J. E. (2005). Targeted and proximity-dependent promiscuous protein biotinylation by a mutant *Escherichia coli* biotin protein ligase. *Journal of Nutritional Biochemistry, 16*(7), 416–418. https://doi.org/10.1016/j.jnutbio.2005.03.017.

Doi, M., Hirayama, J., & Sassone-Corsi, P. (2006). Circadian regulator CLOCK is a histone acetyltransferase. *Cell, 125*(3), 497–508. https://doi.org/10.1016/j.cell.2006.03.033.

Doruk, Y. U., Yarparvar, D., Akyel, Y. K., Gul, S., Taskin, A. C., Yilmaz, F., et al. (2020). A CLOCK-binding small molecule disrupts the interaction between CLOCK and BMAL1 and enhances circadian rhythm amplitude. *The Journal of Biological Chemistry, 295*(11), 3518–3531. https://doi.org/10.1074/jbc.RA119.011332.

Duong, H. A., & Weitz, C. J. (2014). Temporal orchestration of repressive chromatin modifiers by circadian clock Period complexes. *Nature Structural & Molecular Biology, 21*(2), 126–132. https://doi.org/10.1038/nsmb.2746.

Eckel-Mahan, K. L., & Storm, D. R. (2009). Circadian rhythms and memory: Not so simple as cogs and gears. *EMBO Reports, 10*(6), 584–591. https://doi.org/10.1038/embor.2009.123.

Eckle, T., Hartmann, K., Bonney, S., Reithel, S., Mittelbronn, M., Walker, L. A., et al. (2012). Adora2b-elicited Per2 stabilization promotes a HIF-dependent metabolic switch crucial for myocardial adaptation to ischemia. *Nature Medicine, 18*(5), 774–U173. https://doi.org/10.1038/nm.2728.

Eide, E. J., Vielhaber, E. L., Hinz, W. A., & Virshup, D. M. (2002). The circadian regulatory proteins BMAL1 and cryptochromes are substrates of casein kinase Iepsilon. *The Journal of Biological Chemistry*, *277*(19), 17248–17254. https://doi.org/10.1074/jbc.M111466200.

Eide, E. J., Woolf, M. F., Kang, H., Woolf, P., Hurst, W., Camacho, F., et al. (2005). Control of mammalian circadian rhythm by CKIepsilon-regulated proteasome-mediated PER2 degradation. *Molecular and Cellular Biology*, *25*(7), 2795–2807. https://doi.org/10.1128/MCB.25.7.2795-2807.2005.

Emery, P., So, W. V., Kaneko, M., Hall, J. C., & Rosbash, M. (1998). CRY, a Drosophila clock and light-regulated cryptochrome, is a major contributor to circadian rhythm resetting and photosensitivity. *Cell*, *95*(5), 669–679. https://doi.org/10.1016/S0092-8674(00)81637-2.

Emisoglu-Kulahli, H., Gul, S., Morgil, H., Ozcan, O., Aygenli, F., Selvi, S., et al. (2021). Transcriptome analysis of the circadian clock gene BMAL1 deletion with opposite carcinogenic effects. *Functional & Integrative Genomics*, *21*(1), 1–16. https://doi.org/10.1007/s10142-020-00757-6.

Engelen, E., Janssens, R. C., Yagita, K., Smits, V. A. J., van der Horst, G. T. J., & Tamanini, F. (2013). Mammalian TIMELESS is involved in period determination and DNA damage-dependent phase advancing of the circadian clock. *PLoS One*, *8*(2), e56623. https://doi.org/10.1371/journal.pone.0056623.

Etchegaray, J. P., Lee, C., Wade, P. A., & Reppert, S. M. (2003). Rhythmic histone acetylation underlies transcription in the mammalian circadian clock. *Nature*, *421*(6919), 177–182. https://doi.org/10.1038/nature01314.

Foteinou, P. T., Venkataraman, A., Francey, L. J., Anafi, R. C., Hogenesch, J. B., & Doyle, F. J. (2018). Computational and experimental insights into the circadian effects of SIRT1. *Proceedings of the National Academy of Sciences of the United States of America*, *115*(45), 11643–11648. https://doi.org/10.1073/pnas.1803410115.

Franken, P., & Dijk, D. J. (2009). Circadian clock genes and sleep homeostasis. *European Journal of Neuroscience*, *29*(9), 1820–1829. https://doi.org/10.1111/j.1460-9568.2009.06723.x.

Fu, Y., Jones, C. R., Toh, K., Virshup, D., & Ptacek, L. J. (2001). An hPer2 phosphorylation site mutation in familial advanced sleep-phase syndrome. *American Journal of Human Genetics*, *69*(4), 597.

Godinho, S. I., Maywood, E. S., Shaw, L., Tucci, V., Barnard, A. R., Busino, L., et al. (2007). The after-hours mutant reveals a role for Fbxl3 in determining mammalian circadian period. *Science*, *316*(5826), 897–900. https://doi.org/10.1126/science.1141138.

Gul, S., Aydin, C., Ozcan, O., Gurkan, B., Surme, S., Baris, I., et al. (2020). The Arg-293 of Cryptochrome1 is responsible for the allosteric regulation of CLOCK-CRY1 binding in circadian rhythm. *The Journal of Biological Chemistry*, *295*(50), 17187–17199. https://doi.org/10.1074/jbc.RA120.014333.

Gul, S., Rahim, F., Isin, S., Yilmaz, F., Ozturk, N., Turkay, M., et al. (2021). Structure-based design and classifications of small molecules regulating the circadian rhythm period. *Scientific Reports*, *11*(1), 18510. https://doi.org/10.1038/s41598-021-97962-5.

Gul, H., Selvi, S., Yilmaz, F., Ozcelik, F., Olfaz-Aslan, S., Yazan, S., et al. (2022). Proteome analysis of the circadian clock protein PERIOD2. *Proteins*, *90*(6), 1315–1330. https://doi.org/10.1002/prot.26314.

He, Y., Jones, C. R., Fujiki, N., Xu, Y., Guo, B., Holder, J. L., Jr., et al. (2009). The transcriptional repressor DEC2 regulates sleep length in mammals. *Science*, *325*(5942), 866–870. https://doi.org/10.1126/science.1174443.

Hirano, A., Shi, G., Jones, C. R., Lipzen, A., Pennacchio, L. A., Xu, Y., et al. (2016). A cryptochrome 2 mutation yields advanced sleep phase in humans. *eLife*, *5*. https://doi.org/10.7554/eLife.16695.

Hirano, A., Yumimoto, K., Tsunematsu, R., Matsumoto, M., Oyama, M., Kozuka-Hata, H., et al. (2013). FBXL21 regulates oscillation of the circadian clock through ubiquitination and stabilization of cryptochromes. *Cell, 152*(5), 1106–1118. https://doi.org/10.1016/j.cell.2013.01.054.

Hirayama, J., Sahar, S., Grimaldi, B., Tamaru, T., Takamatsu, K., Nakahata, Y., et al. (2007). CLOCK-mediated acetylation of BMAL1 controls circadian function. *Nature, 450*(7172), 1086–U1013. https://doi.org/10.1038/nature06394.

Hirota, T., Lee, J. W., Lewis, W. G., Zhang, E. E., Breton, G., Liu, X. Z., et al. (2010). High-throughput chemical screen identifies a novel potent modulator of cellular circadian rhythms and reveals CKI alpha as a clock regulatory kinase. *PLos Biology, 8*(12), e1000559. https://doi.org/10.1371/journal.pbio.1000559.

Hirota, K., & Semenza, G. L. (2005). Regulation of hypoxia-inducible factor 1 by prolyl and asparaginyl hydroxylases. *Biochemical and Biophysical Research Communications, 338*(1), 610–616. https://doi.org/10.1016/j.bbrc.2005.08.193.

Hogenesch, J. B., Gu, Y. Z., Jain, S., & Bradfield, C. A. (1998). The basic-helix-loop-helix-PAS orphan MOP3 forms transcriptionally active complexes with circadian and hypoxia factors. *Proceedings of the National Academy of Sciences of the United States of America, 95*(10), 5474–5479.

Honma, S., Kawamoto, T., Takagi, Y., Fujimoto, K., Sato, F., Noshiro, M., et al. (2002). Dec1 and Dec2 are regulators of the mammalian molecular clock. *Nature, 419*(6909), 841–844. https://doi.org/10.1038/nature01123.

Hosoda, H., Kato, K., Asano, H., Ito, M., Kato, H., Iwamoto, T., et al. (2009). CBP/p300 is a cell type-specific modulator of CLOCK/BMAL1-mediated transcription. *Molecular Brain, 2*. https://doi.org/10.1186/1756-6606-2-34.

Johnson, C. H., & Golden, S. S. (1999). Circadian programs in cyanobacteria: Adaptiveness and mechanism. *Annual Review of Microbiology, 53*, 389–409. https://doi.org/10.1146/annurev.micro.53.1.389.

Katada, S., & Sassone-Corsi, P. (2010). The histone methyltransferase MLL1 permits the oscillation of circadian gene expression. *Nature Structural & Molecular Biology, 17*(12), 1414–1421. https://doi.org/10.1038/nsmb.1961.

Kavakli, I. H., Baris, I., Tardu, M., Gul, S., Oner, H., Cal, S., et al. (2017). The photolyase/cryptochrome family of proteins as DNA repair enzymes and transcriptional repressors. *Photochemistry and Photobiology, 93*(1), 93–103. https://doi.org/10.1111/php.12669.

Kavakli, I. H., Gul, S., & Turkay, M. (2022). Identification of novel small molecules targeting core clock proteins to regulate circadian rhythm. *Current Opinion in Chemical Engineering, 35*, 100730. https://doi.org/10.1016/j.coche.2021.100730.

Kavakli, I. H., & Sancar, A. (2002). Circadian photoreception in humans and mice. *Molecular Interventions, 2*(8), 484–492. https://doi.org/10.1124/mi.2.8.484.

Keesler, G. A., Camacho, F., Guo, Y., Virshup, D., Mondadori, C., & Yao, Z. (2000). Phosphorylation and destabilization of human period I clock protein by human casein kinase I epsilon. *Neuroreport, 11*(5), 951–955. https://doi.org/10.1097/00001756-200004070-00011.

King, D. P., Zhao, Y., Sangoram, A. M., Wilsbacher, L. D., Tanaka, M., Antoch, M. P., et al. (1997). Positional cloning of the mouse circadian clock gene. *Cell, 89*(4), 641–653. https://doi.org/10.1016/s0092-8674(00)80245-7.

Kitamura, H., & Hashimoto, M. (2021). USP2-related cellular signaling and consequent pathophysiological outcomes. *International Journal of Molecular Sciences, 22*(3). https://doi.org/10.3390/ijms22031209.

Kobayashi, M., Koyasu, S., Goto, Y., Hiraoko, M., & Harada, H. (2018). A circadian clock gene, PER2, activates HIF-1 as an effector molecule for recruitment of HIF-1 alpha to its enhancer regions. *Cancer Science, 109*, 978.

Konopka, R. J., & Benzer, S. (1971). Clock mutants of Drosophila melanogaster. *Proceedings of the National Academy of Sciences of the United States of America, 68*(9), 2112–2116. https://doi.org/10.1073/pnas.68.9.2112.

Korkmaz, T., Aygenli, F., Emisoglu, H., Ozcelik, G., Canturk, A., Yilmaz, S., et al. (2018). Opposite carcinogenic effects of circadian clock gene BMAL1. *Scientific Reports, 8*(1), 16023. https://doi.org/10.1038/s41598-018-34433-4.

Kramer, A., Yang, F. C., Snodgrass, P., Li, X. D., Scammell, T. E., Davis, F. C., et al. (2001). Regulation of daily locomotor activity and sleep by hypothalamic EGF receptor signaling. *Science, 294*(5551), 2511–2515. https://doi.org/10.1126/science.1067716.

Krishnaiah, S. Y., Wu, G., Altman, B. J., Growe, J., Rhoades, S. D., Coldren, F., et al. (2017). Clock regulation of metabolites reveals coupling between transcription and metabolism. *Cell Metabolism, 25*(4), 961–974 e964. https://doi.org/10.1016/j.cmet.2017.03.019.

Kume, K., Zylka, M. J., Sriram, S., Shearman, L. P., Weaver, D. R., Jin, X., et al. (1999). mCRY1 and mCRY2 are essential components of the negative limb of the circadian clock feedback loop. *Cell, 98*(2), 193–205. https://doi.org/10.1016/s0092-8674(00)81014-4.

Kurien, P., Hsu, P. K., Leon, J., Wu, D., McMahon, T., Shi, G. S., et al. (2019). TIMELESS mutation alters phase responsiveness and causes advanced sleep phase. *Proceedings of the National Academy of Sciences of the United States of America, 116*(24), 12045–12053. https://doi.org/10.1073/pnas.1819110116.

Lamia, K. A., Sachdeva, U. M., DiTacchio, L., Williams, E. C., Alvarez, J. G., Egan, D. F., et al. (2009). AMPK regulates the circadian clock by cryptochrome phosphorylation and degradation. *Science, 326*(5951), 437–440. https://doi.org/10.1126/science.1172156.

Lande-Diner, L., Boyault, C., Kim, J. Y., & Weitz, C. J. (2013). A positive feedback loop links circadian clock factor CLOCK-BMAL1 to the basic transcriptional machinery. *Proceedings of the National Academy of Sciences of the United States of America, 110*(40), 16021–16026. https://doi.org/10.1073/pnas.1305980110.

Lin, R., Mo, Y., Zha, H., Qu, Z., Xie, P., Zhu, Z. J., et al. (2017). CLOCK acetylates ASS1 to drive circadian rhythm of ureagenesis. *Molecular Cell, 68*(1), 198–209 e196. https://doi.org/10.1016/j.molcel.2017.09.008.

Lin, C., & Shalitin, D. (2003). Cryptochrome structure and signal transduction. *Annual Review of Plant Biology, 54*, 469–496. https://doi.org/10.1146/annurev.arplant.54.110901.160901.

Lowrey, P. L., Shimomura, K., Antoch, M. P., Yamazaki, S., Zemenides, P. D., Ralph, M. R., et al. (2000). Positional syntenic cloning and functional characterization of the mammalian circadian mutation tau. *Science, 288*(5465), 483–491. https://doi.org/10.1126/science.288.5465.483.

McIntosh, B. E., Hogenesch, J. B., & Bradfield, C. A. (2010). Mammalian Per-Arnt-Sim proteins in environmental adaptation. *Annual Review of Physiology, 72*, 625–645. https://doi.org/10.1146/annurev-physiol-021909-135922.

Meng, Q. J., Logunova, L., Maywood, E. S., Gallego, M., Lebiecki, J., Brown, T. M., et al. (2008). Setting clock speed in mammals: The CK1 epsilon tau mutation in mice accelerates circadian pacemakers by selectively destabilizing PERIOD proteins. *Neuron, 58*(1), 78–88. https://doi.org/10.1016/j.neuron.2008.01.019.

Michael, A. K., Fribourgh, J. L., Chelliah, Y., Sandate, C. R., Hura, G. L., Schneidman-Duhovny, D., et al. (2017). Formation of a repressive complex in the mammalian circadian clock is mediated by the secondary pocket of CRY1. *Proceedings of the National Academy of Sciences of the United States of America, 114*(7), 1560–1565. https://doi.org/10.1073/pnas.1615310114.

Mieda, M. (2019). The network mechanism of the central circadian pacemaker of the SCN: Do AVP neurons play a more critical role than expected? *Frontiers in Neuroscience, 13*, 139. https://doi.org/10.3389/fnins.2019.00139.

Miki, T., Xu, Z., Chen-Goodspeed, M., Liu, M., Van Oort-Jansen, A., Rea, M. A., et al. (2012). PML regulates PER2 nuclear localization and circadian function. *The EMBO Journal, 31*(6), 1427–1439. https://doi.org/10.1038/emboj.2012.1.

Mohawk, J. A., Green, C. B., & Takahashi, J. S. (2012). Central and peripheral circadian clocks in mammals. *Annual Review of Neuroscience, 35*, 445–462. https://doi.org/10.1146/annurev-neuro-060909-153128.

Na, J., Lee, K., Kim, H. G., Shin, J. Y., Na, W., Jeong, H., et al. (2012). Role of type II protein arginine methyltransferase 5 in the regulation of Circadian Per1 gene. *PLoS One*, 7(10), e48152. https://doi.org/10.1371/journal.pone.0048152.

Nakahata, Y., Kaluzova, M., Grimaldi, B., Sahar, S., Hirayama, J., Chen, D., et al. (2008). The NAD+-dependent deacetylase SIRT1 modulates CLOCK-mediated chromatin remodeling and circadian control. *Cell, 134*(2), 329–340. https://doi.org/10.1016/j.cell.2008.07.002.

Okamoto-Uchida, Y., Izawa, J., Nishimura, A., Hattori, A., Suzuki, N., & Hirayama, J. (2019). Post-translational modifications are required for circadian clock regulation in vertebrates. *Current Genomics, 20*(5), 332–339. https://doi.org/10.2174/1389202919666191014094349.

Ooe, N., Saito, K., Mikami, N., Nakatuka, I., & Kaneko, H. (2004). Identification of a novel basic helix-loop-helix-PAS factor, NXF, reveals a Sim2 competitive, positive regulatory role in dendritic-cytoskeleton modulator drebrin gene expression. *Molecular and Cellular Biology, 24*(2), 608–616. https://doi.org/10.1128/MCB.24.2.608-616.2004.

Ozber, N., Baris, I., Tatlici, G., Gur, I., Kilinc, S., Unal, E. B., et al. (2010). Identification of two amino acids in the C-terminal domain of mouse CRY2 essential for PER2 interaction. *BMC Molecular Biology, 11*, 69. https://doi.org/10.1186/1471-2199-11-69.

Ozcan, O., Gul, S., & Kavakli, I. H. (2021). Allosteric regulation of CRYs in mammalian circadian clock. In *Vol. 50. Computer aided chemical engineering* (pp. 2025–2031). Elsevier.

Ozturk, N. (2021). Light-dependent reactions of animal circadian photoreceptor cryptochrome. *FEBS Journal*. https://doi.org/10.1111/febs.16273.

Ozturk, N., Song, S. H., Ozgur, S., Selby, C. P., Morrison, L., Partch, C., et al. (2007). Structure and function of animal cryptochromes. *Clocks and Rhythms, 72*, 119–131.

Panda, S., Antoch, M. P., Miller, B. H., Su, A. I., Schook, A. B., Straume, M., et al. (2002). Coordinated transcription of key pathways in the mouse by the circadian clock. *Cell, 109*(3), 307–320. https://doi.org/10.1016/s0092-8674(02)00722-5.

Parico, G. C. G., & Partch, C. L. (2020). The tail of cryptochromes: An intrinsically disordered cog within the mammalian circadian clock. *Cell Communication and Signaling, 18*(1), 182. https://doi.org/10.1186/s12964-020-00665-z.

Park, H. W., Kim, S. T., Sancar, A., & Deisenhofer, J. (1995). Crystal structure of DNA photolyase from Escherichia coli. *Science, 268*(5219), 1866–1872. https://doi.org/10.1126/science.7604260.

Peek, C. B., Levine, D. C., Cedernaes, J., Taguchi, A., Kobayashi, Y., Tsai, S. J., et al. (2017). Circadian clock interaction with HIF1alpha mediates oxygenic metabolism and anaerobic glycolysis in skeletal muscle. *Cell Metabolism, 25*(1), 86–92. https://doi.org/10.1016/j.cmet.2016.09.010.

Pellegrino, R., Kavakli, I. H., Goel, N., Cardinale, C. J., Dinges, D. F., Kuna, S. T., et al. (2014). A novel BHLHE41 variant is associated with short sleep and resistance to sleep deprivation in humans. *Sleep, 37*(8), 1327–1336. https://doi.org/10.5665/sleep.3924.

Pickart, C. M. (2001). Mechanisms underlying ubiquitination. *Annual Review of Biochemistry, 70*, 503–533. https://doi.org/10.1146/annurev.biochem.70.1.503.

Pittendrigh, C. S. (1993). Temporal organization—reflections of a Darwinian clock-watcher. *Annual Review of Physiology*, *55*, 16–54.
Pourhaghighi, R., Ash, P. E. A., Phanse, S., Goebels, F., Hu, L. Z. M., Chen, S., et al. (2020). BraInMap elucidates the macromolecular connectivity landscape of mammalian brain. *Cell Systems*, *11*(2), 208. https://doi.org/10.1016/j.cels.2020.08.006.
Preitner, N., Damiola, F., Molina, L. L., Zakany, J., Duboule, D., Albrecht, U., et al. (2002). The orphan nuclear receptor REV-ERB alpha controls circadian transcription within the positive limb of the mammalian circadian oscillator. *Cell*, *110*(2), 251–260. https://doi.org/10.1016/S0092-8674(02)00825-5.
Qu, M., Qu, H., Jia, Z., & Kay, S. A. (2021). HNF4A defines tissue-specific circadian rhythms by beaconing BMAL1::CLOCK chromatin binding and shaping the rhythmic chromatin landscape. *Nature Communications*, *12*(1), 6350. https://doi.org/10.1038/s41467-021-26567-3.
Reischl, S., Vanselow, K., Westermark, P. O., Thierfelder, N., Maier, B., Herzel, H., et al. (2007). Beta-TrCP1-mediated degradation of PERIOD2 is essential for circadian dynamics. *Journal of Biological Rhythms*, *22*(5), 375–386. https://doi.org/10.1177/0748730407303926.
Rijo-Ferreira, F., & Takahashi, J. S. (2019). Genomics of circadian rhythms in health and disease. *Genome Medicine*, *11*(1), 82. https://doi.org/10.1186/s13073-019-0704-0.
Rosensweig, C., Reynolds, K. A., Gao, P., Laothamatas, I., Shan, Y., Ranganathan, R., et al. (2018). An evolutionary hotspot defines functional differences between CRYPTOCHROMES. *Nature Communications*, *9*(1), 1138. https://doi.org/10.1038/s41467-018-03503-6.
Roux, K. J., Kim, D. I., Raida, M., & Burke, B. (2012). A promiscuous biotin ligase fusion protein identifies proximal and interacting proteins in mammalian cells. *The Journal of Cell Biology*, *196*(6), 801–810. https://doi.org/10.1083/jcb.201112098.
Sancar, A. (2000). Cryptochrome: The second photoactive pigment in the eye and its role in circadian photoreception. *Annual Review of Biochemistry*, *69*, 31–67. https://doi.org/10.1146/annurev.biochem.69.1.31.
Sato, T. K., Panda, S., Miraglia, L. J., Reyes, T. M., Rudic, R. D., McNamara, P., et al. (2004). A functional genomics strategy reveals Rora as a component of the mammalian circadian clock. *Neuron*, *43*(4), 527–537. https://doi.org/10.1016/j.neuron.2004.07.018.
Schmalen, I., Reischl, S., Wallach, T., Klemz, R., Grudziecki, A., Prabu, J. R., et al. (2014). Interaction of circadian clock proteins CRY1 and PER2 is modulated by zinc binding and disulfide bond formation. *Cell*, *157*(5), 1203–1215. https://doi.org/10.1016/j.cell.2014.03.057.
Scoma, H. D., Humby, M., Yadav, G., Zhang, Q. J., Fogerty, J., & Besharse, J. C. (2011). The De-ubiquitinylating enzyme, USP2, is associated with the circadian clockwork and regulates its sensitivity to light. *PLoS One*, *6*(9), e25382. https://doi.org/10.1371/journal.pone.0025382.
Semenza, G. L. (1998). Hypoxia-inducible factor 1 and the molecular physiology of oxygen homeostasis. *Journal of Laboratory and Clinical Medicine*, *131*(3), 207–214. https://doi.org/10.1016/S0022-2143(98)90091-9.
Shen, Y., Endale, M., Wang, W., Morris, A. R., Francey, L. J., Harold, R. L., et al. (2021). NF-kappaB modifies the mammalian circadian clock through interaction with the core clock protein BMAL1. *PLoS Genetics*, *17*(11), e1009933. https://doi.org/10.1371/journal.pgen.1009933.
Shirogane, T., Jin, J., Ang, X. L., & Harper, J. W. (2005). SCFbeta-TRCP controls clock-dependent transcription via casein kinase 1-dependent degradation of the mammalian period-1 (Per1) protein. *The Journal of Biological Chemistry*, *280*(29), 26863–26872. https://doi.org/10.1074/jbc.M502862200.

Siepka, S. M., Yoo, S. H., Park, J., Song, W., Kumar, V., Hu, Y., et al. (2007). Circadian mutant overtime reveals F-box protein FBXL3 regulation of cryptochrome and period gene expression. *Cell*, *129*(5), 1011–1023. https://doi.org/10.1016/j.cell.2007.04.030.

Srivastava, A., & Mishra, R. K. (2020). Interactome of vertebrate GAF/ThPOK reveals its diverse functions in gene regulation and DNA repair. *Journal of Biosciences*, *45*.

Stanewsky, R., Kaneko, M., Emery, P., Beretta, B., Wager-Smith, K., Kay, S. A., et al. (1998). The cry(b) mutation identifies cryptochrome as a circadian photoreceptor in drosophila. *Cell*, *95*(5), 681–692. https://doi.org/10.1016/S0092-8674(00)81638-4.

Stephan, F. K., & Zucker, I. (1972). Circadian rhythms in drinking behavior and locomotor activity of rats are eliminated by hypothalamic lesions. *Proceedings of the National Academy of Sciences of the United States of America*, *69*(6), 1583–1586. https://doi.org/10.1073/pnas.69.6.1583.

Storch, K. F., Lipan, O., Leykin, I., Viswanathan, N., Davis, F. C., Wong, W. H., et al. (2002). Extensive and divergent circadian gene expression in liver and heart. *Nature*, *417*(6884), 78–83. https://doi.org/10.1038/nature744.

Sun, Q., Yang, Y., Wang, Z., Yang, X., Gao, Y., Zhao, Y., et al. (2020). PER1 interaction with GPX1 regulates metabolic homeostasis under oxidative stress. *Redox Biology*, *37*, 101694. https://doi.org/10.1016/j.redox.2020.101694.

Takahashi, J. S. (2016). Molecular architecture of the circadian clock in mammals. In P. Sassone-Corsi, & Y. Christen (Eds.), *A time for metabolism and hormones* (pp. 13–24). Cham (CH).

Takahashi, J. S. (2017). Transcriptional architecture of the mammalian circadian clock. *Nature Reviews. Genetics*, *18*(3), 164–179. https://doi.org/10.1038/nrg.2016.150.

Takahata, S., Ozaki, T., Mimura, J., Kikuchi, Y., Sogawa, K., & Fujii-Kuriyama, Y. (2000). Transactivation mechanisms of mouse clock transcription factors, mClock and mArnt3. *Genes to Cells*, *5*(9), 739–747. https://doi.org/10.1046/j.1365-2443.2000.00363.x.

Tamaru, T., Hattori, M., Honda, K., Nakahata, Y., Sassone-Corsi, P., van der Horst, G. T., et al. (2015). CRY drives cyclic CK2-mediated BMAL1 phosphorylation to control the mammalian circadian clock. *PLoS Biology*, *13*(11), e1002293. https://doi.org/10.1371/journal.pbio.1002293.

Thresher, R. J., Vitaterna, M. H., Miyamoto, Y., Kazantsev, A., Hsu, D. S., Petit, C., et al. (1998). Role of mouse cryptochrome blue-light photoreceptor in circadian photoresponses. *Science*, *282*(5393), 1490–1494. https://doi.org/10.1126/science.282.5393.1490.

van der Horst, G. T., Muijtjens, M., Kobayashi, K., Takano, R., Kanno, S., Takao, M., et al. (1999). Mammalian Cry1 and Cry2 are essential for maintenance of circadian rhythms. *Nature*, *398*(6728), 627–630. https://doi.org/10.1038/19323.

Vanselow, K., Vanselow, J. T., Westermark, P. O., Reischl, S., Maier, B., Korte, T., et al. (2006). Differential effects of PER2 phosphorylation: Molecular basis for the human familial advanced sleep phase syndrome (FASPS). *Genes & Development*, *20*(19), 2660–2672. https://doi.org/10.1101/gad.397006.

Vitaterna, M. H., King, D. P., Chang, A. M., Kornhauser, J. M., Lowrey, P. L., McDonald, J. D., et al. (1994). Mutagenesis and mapping of a mouse gene, clock, essential for circadian behavior. *Science*, *264*(5159), 719–725. https://doi.org/10.1126/science.8171325.

Vitaterna, M. H., Selby, C. P., Todo, T., Niwa, H., Thompson, C., Fruechte, E. M., et al. (1999). Differential regulation of mammalian period genes and circadian rhythmicity by cryptochromes 1 and 2. *Proceedings of the National Academy of Sciences of the United States of America*, *96*(21), 12114–12119. https://doi.org/10.1073/pnas.96.21.12114.

Wallach, T., Schellenberg, K., Maier, B., Kalathur, R. K., Porras, P., Wanker, E. E., et al. (2013). Dynamic circadian protein-protein interaction networks predict temporal organization of cellular functions. *PLoS Genetics*, *9*(3), e1003398. https://doi.org/10.1371/journal.pgen.1003398.

Washietl, S., Kellis, M., & Garber, M. (2014). Evolutionary dynamics and tissue specificity of human long noncoding RNAs in six mammals. *Genome Research*, *24*(4), 616–628. https://doi.org/10.1101/gr.165035.113.

Xing, W. M., Busino, L., Hinds, T. R., Marionni, S. T., Saifee, N. H., Bush, M. F., et al. (2013). SCFFBXL3 ubiquitin ligase targets cryptochromes at their cofactor pocket. *Nature*, *496*(7443), 64-+. https://doi.org/10.1038/nature11964.

Xu, H. Y., Gustafson, C. L., Sammons, P. J., Khan, S. K., Parsley, N. C., Ramanathan, C., et al. (2015). Cryptochrome 1 regulates the circadian clock through dynamic interactions with the BMAL1 C terminus. *Nature Structural & Molecular Biology*, *22*(6), 476–U470. https://doi.org/10.1038/nsmb.3018.

Yagita, K., Yamaguchi, S., Tamanini, F., van Der Horst, G. T., Hoeijmakers, J. H., Yasui, A., et al. (2000). Dimerization and nuclear entry of mPER proteins in mammalian cells. *Genes & Development*, *14*(11), 1353–1363.

Yamazaki, S., Numano, R., Abe, M., Hida, A., Takahashi, R., Ueda, M., et al. (2000). Resetting central and peripheral circadian oscillators in transgenic rats. *Science*, *288*(5466), 682–685. https://doi.org/10.1126/science.288.5466.682.

Yang, Y., Duguay, D., Bedard, N., Rachalski, A., Baquiran, G., Na, C. H., et al. (2012). Regulation of behavioral circadian rhythms and clock protein PER1 by the deubiquitinating enzyme USP2. *Biology Open*, *1*(8), 789–801. https://doi.org/10.1242/bio.20121990.

Yang, Y., Duguay, D., Fahrenkrug, J., Cermakian, N., & Wing, S. S. (2014). USP2 regulates the intracellular localization of PER1 and circadian gene expression. *Journal of Biological Rhythms*, *29*(4), 243–256. https://doi.org/10.1177/0748730414544741.

Yang, J. B., Zhang, Z. W., Zhang, Y. Y., Zheng, X. L., Lu, Y. L., Tao, D. C., et al. (2018). CLOCK interacts with RANBP9 and is involved in alternative splicing in spermatogenesis. *Gene*, *642*, 199–204. https://doi.org/10.1016/j.gene.2017.11.007.

Ye, R., Selby, C. P., Chiou, Y. Y., Ozkan-Dagliyan, I., Gaddameedhi, S., & Sancar, A. (2014). Dual modes of CLOCK:BMAL1 inhibition mediated by cryptochrome and period proteins in the mammalian circadian clock. *Genes & Development*, *28*(18), 1989–1998. https://doi.org/10.1101/gad.249417.114.

Ye, R., Selby, C. P., Ozturk, N., Annayev, Y., & Sancar, A. (2011). Biochemical analysis of the canonical model for the mammalian circadian clock. *The Journal of Biological Chemistry*, *286*(29), 25891–25902. https://doi.org/10.1074/jbc.M111.254680.

Zhang, R., Lahens, N. F., Ballance, H. I., Hughes, M. E., & Hogenesch, J. B. (2014). A circadian gene expression atlas in mammals: Implications for biology and medicine. *Proceedings of the National Academy of Sciences of the United States of America*, *111*(45), 16219–16224. https://doi.org/10.1073/pnas.1408886111.

Zhang, E. E., Liu, Y., Dentin, R., Pongsawakul, P. Y., Liu, A. C., Hirota, T., et al. (2010). Cryptochrome mediates circadian regulation of cAMP signaling and hepatic gluconeogenesis. *Nature Medicine*, *16*(10), 1152–1156. https://doi.org/10.1038/nm.2214.

Zhao, W. N., Malinin, N., Yang, F. C., Staknis, D., Gekakis, N., Maier, B., et al. (2007). CIPC is a mammalian circadian clock protein without invertebrate homologues. *Nature Cell Biology*, *9*(3), 268–U253. https://doi.org/10.1038/ncb1539.

CHAPTER SEVEN

Investigation of differentially expressed genes and dysregulated pathways involved in multiple sclerosis

S. Udhaya Kumar[a], Ankur Datta[a], Ramanathan Gnanasambandan[a], Salma Younes[b], Tamma Medha[a], Ramamoorthy Siva[a], C. George Priya Doss[a,*], and Hatem Zayed[b,*]

[a]Laboratory of Integrative Genomics, Department of Integrative Biology, School of Biosciences and Technology, Vellore Institute of Technology, Vellore, India

[b]Department of Biomedical Sciences, College of Health and Sciences, Qatar University, QU Health, Doha, Qatar

*Corresponding authors: e-mail address: georgepriyadoss@vit.ac.in; hatem.zayed@qu.edu.qa

Contents

Abstract

Multiple Sclerosis (MS) is a neurodegenerative autoimmune and organ-specific demyelinating disorder, known to affect the central nervous system (CNS). While genetic studies have revealed several critical genes and diagnostic biomarkers associated with MS,

Advances in Protein Chemistry and Structural Biology, Volume 131
ISSN 1876-1623
https://doi.org/10.1016/bs.apcsb.2022.05.003

the etiology of the disease remains poorly understood. This study is aimed at screening and identifying the key genes and canonical pathways associated with MS. Gene expression profiling of the microarray dataset GSE38010 was used to analyze two control brain samples (control 1; GSM931812, control 2; GSM931813), active inflammation stage samples (CAP1; GSM931815, CAP2; GSM931816) and late subsided stage samples (CP1; GSM931817, CP2; GSM931818) collected from patients ranging between 23 and 54 years and both genders. This analysis yielded a list of 58,866 DEGs (29,433 for active-inflammation stage and 29,433 for late-subsided Stage). The interactions between the DEGs were then studied using STRING, Cytoscape software, and MCODE was employed to find the genes that form clusters. Functional enrichment and integrative analysis were performed using ClueGO/CluePedia and MetaCore™. Our data revealed dysregulated key canonical pathways in MS patients. In addition, we identified three hub genes (*SCN2A, HTR2A,* and *HCN1*) that may serve as potential biomarkers for the prognosis of MS. Furthermore, the expression patterns of *HPCA* and *PLCB1* provide insights into the progressive stages of MS, indicating that these genes could be used in predicting MS progression. We were able to map potential biomarkers that could be used for the prognosis and diagnosis of MS.

Abbreviations

BP biological processes
CAP chronic active plaque
CNS central nervous system
CP chronic plaque
CSF Cerebro-Spinal Fluid
EA enrichment analysis
GEO gene expression omnibus
GO gene ontology
MF molecular function
MS multiple sclerosis

1. Introduction

Multiple Sclerosis (MS) is an inflammatory autoimmune disease affecting the central nervous system (CNS), characterized by the degradation of myelin sheaths, inflammation, and diffuse neurodegeneration. Inflammatory lesions appear in the white matter of the CNS, which negatively impacts the ability of the nerve cells to conduct electrical signals, thereby leading to a loss of sensation and other neural responses (O'Connor, Bar-Or, & Hafler, 2001). Reports claim that MS is known to affect approximately 2.5 million individuals globally and is considered the leading cause of acquired neurological impairment in adolescents (Koch-Henriksen & Sørensen, 2010).

The occurrence of MS happens to vary according to ethnic origin and geographic location, with northern Europeans and their descendants being more prone to developing the disorder (Koch-Henriksen & Sørensen, 2010).

Although the etiology of MS remains unclear, an increasing body of evidence suggests that genetic and environmental components collectively affect disease susceptibility (Liu et al., 2013; Pernice et al., 2019; Xu, Feng, Zhu, & Qiu, 2021). The genetic component of MS was revealed via twin and familial clustering studies (Sawcer, Franklin, & Ban, 2014). Siblings of MS affected individuals were reported to have a 10- to 20-fold greater risk of developing the disease (2–4%), in comparison to the general population (0.2%), with monozygotic or identical twins having an even greater risk (25–30%) compared to dizygotic or fraternal twins (3–7%) (Dyment, Sadnovich, & Ebers, 1997).

In recent years, the various developments and advancements in genome-wide association studies (GWAS) have fueled the characterization of several risk loci associated with MS susceptibility. Clinical analysis paired with computational techniques can enable scientists to better understand and comprehend the underlying mechanisms of MS and will help reveal robust diagnostic and prognostic biomarkers for MS. The most muscular susceptible locus of MS has been reported in chromosome 6p21.3 within the Major Histocompatibility Complex (MHC) (Ates, Kurt, Bozkurt, & Karaer, 2010).

Several studies have been published to validate a target gene to act as a potential biomarker for MS. *YKL-40* (also referred to as chitinase 3-like-1 or *CHI3L1*) is a glial activation marker that is expressed on reactive astrocytes, activated macrophages, chondrocytes, and many other cells (Bonneh-Barkay, Wang, Starkey, Hamilton, & Wiley, 2010; Cantó et al., 2015). Studies have reported increased expression of Cerebro-Spinal Fluid (CSF) *YKL-40* in Clinically Isolated Syndrome (CIS) patients converting to MS as compared to the non-converters (Comabella et al., 2010). In addition, *CXCL13*, a lymphoid chemokine family component responsible for regulating the migration and compartmentalization of B cells in secondary lymphoid organs under both homeostatic and inflammatory conditions (Cyster, 2005; Okada & Cyster, 2006). It was also highly expressed in samples collected from CSF of CIS, RRMS, SPMS, and PPMS patients compared to symptomatic controls (Stilund et al., 2015).

Incomparable advancements in bioinformatics and systems biology have provided scientists with more accurate high throughput tools to collate and better understand various diseases' prognostic and diagnostic aspects and identify molecular drug targets for the treatment of several disorders.

One such technique developed was microarray technology, which enabled researchers to conduct concurrent analysis of probes in a single batch process and save critical time (Dufva, 2009). Microarray technology has evolved into a strong and widely used technique for high-throughput comprehensive gene expression analysis, as well as for studying differentially expressed genes (DEGs) in patients suffering from various disorders (Balasundaram, Udhaya Kumar, Doss, & C., 2022; Udhaya Kumar et al., 2020; Udhaya Kumar et al., 2020; Udhaya Kumar et al., 2021). Investigators have since been successful in studying the expression profiles of multiple genes using microarray technology. One of the critical areas where microarray expression studies have gained momentum is identifying DEGs and exploring their viability to act as biomarkers for various diseases. In this study, we used a comprehensive bioinformatics pipeline to identify DEGs pertaining to the pathogenesis and progression of MS.

2. Methodology

2.1 Data retrieval

Gene Expression Omnibus (GEO) database was accessed to retrieve the dataset for this study. GEO is an open-access database that contains array- and sequence-based data on various disorders (Barrett et al., 2013). GEO also incorporates tools that help analyze curated gene expression data. The dataset we retrieved involves microarray expression profiles of MS studied in a total of five MS samples sub-divided into various progressive stages, namely, active plaque (AP; GSM931814), chronic active plaque 1 (CAP1; GSM931815), chronic active plaque 2 (CAP2; GSM931816), chronic plaque 1 (CP1; GSM931817), chronic plaque 2 (CP1; GSM931818) and two control brain samples (control 1; GSM931812, control 2; GSM931813) (Han et al., 2012). The samples for the clinical study involved subjects of both sexes, ages ranging from 23 to 54 years, and an autopsy interval varying from 4 to 24 h. The data used for this computational study was downloaded from the GEO database, and any of the authors facilitated no additional human experiments during this study.

2.2 Data processing and target gene prediction

The raw expression data of the GSE38010 acquired from the GEO repository was examined using the GEO2R tool, which is integrated along with the GEO database. GEO2R enables researchers to compare expression data of several sample groups expression data to identify the aberrantly expressed target genes for the corresponding sample groups. The GEO2R tools utilize the Bioconductor from R and Limma package (Barrett et al., 2013; Ritchie et al., 2015; Smyth, 2005). For primary data processing using GEO2R, we employed the Benjamini and Hochberg (False discovery rate) algorithm for adjusting the *P*-values of the DEGs for both MS affected samples and control samples (Aubert, Bar-Hen, Daudin, & Robin, 2004). The principal standards to acquire DEGs were set to $\text{logFC} > 1$ and $P < 0.05$ in order to identify the important DEGs that belong to the dataset, and are considered to be upregulated in the case of $\text{logFC} \geq 1$ and downregulated in the case of $\text{logFC} \leq -1$. The rest of the parameters for the analysis using GEO2R were set to their default settings. The concerned DEGs for respective sample groups were collected and showcased using a volcano plot. RStudio and the Calibrate function was used to produce the volcano plot. The web-based tool heatmapper was used to generate the heatmap, which indicates the expression profiles of the differentially expressed target genes belonging to the corresponding sample groups considered for analysis in this computational study. The average linkage algorithm was used to determine the clustering, and similarly, to compute the columns and rows distance, we utilized the Euclidean algorithm. Clustering and the dendrograms were applied to both; rows and columns (Babicki et al., 2016).

2.3 PPI network association and module analysis

A protein-protein interaction (PPI) network was generated via the web-based STRING tool to assess the synergistic interactions between the concerned target genes (Szklarczyk et al., 2017, 2019). The inconsistent associations of the target genes were eradicated by setting the cutoff limit to an interaction score of ≥ 0.4; hence, a network comprising strong and relevant interactions was established. Cytoscape (v3.9.1) enabled us to create a gene map of all the relevant DEGs for their respective sample groups (Shannon et al., 2003). To further enrich our search and determine the more closely associated target genes, the Molecular Complex Detection (MCODE) plugin was used, integrated into the Cytoscape software. The clusters were searched for in the whole network, and the degree cutoff of the networking score was set to 2. The other parameters that were considered while searching for the clusters were Kappa score (K-core) set to two, Max. Depth was set to 100. Finally, the node cutoff was set as 0.2 (Bader & Hogue, 2003).

2.4 ClueGO functional enrichment of DEGs

The DEGs acquired in this dataset were subjected to analysis using ClueGO v2.5.5/CluePedia v1.5.5 to attain GO BP, GO MF, and disease-associated pathways for the respective sample groups. ClueGO incorporates KEGG and GO databases to curate the pathway networks and other functionalities (Bindea et al., 2009). For the ClueGO analysis, terms and networks with *P*-values < 0.05 were significant. Furthermore, the inbuilt files' organic layout was used to represent the results better. The network specificity was set to medium for this enrichment experiment, with other parameters set as default.

2.5 MetaCore™ gene ontological analysis

GeneGo enables researchers to assess cellular pathways, protein interaction biological networks rapidly, and physiological pathway maps using data obtained from clinical studies (MetaCoreLogin | Clarivate Analytics). MetaCore™ (https://clarivate.com/products/metacore/) developed and maintained by Clarivate Analytics (based in London, United Kingdom), was used to conduct further enrichment of Gene Ontological terms, namely BP, MF, and KEGG pathway. The enriched pathways were identified using the pathway maps tool based on their hypergeometric distribution. We considered a probable threshold ($P < 0.05$) value, and a diagrammatic portrayal of the connections among the DEGs from the considered samples was produced. Calculating the hypergeometric *P*-value enabled us to predict the correlation between the genes and ontological phases. An FDR value of < 0.05 was set as the criteria to determine if important DEGs contributed to a regulatory pathway.

3. Results

3.1 Dataset retrieval and identification of DEGs

The GSE38010 dataset (Janus-like Opposing Roles of CD47 in autoimmune brain inflammation in human and mice) obtained from the GEO database comprises micro-array expression profiles of 2 control samples and 5 MS patient samples studied via Human Genome U133 Plus 2.0 arrays (Affymetrix). Table 1 presents the fundamental information of the subjects used for the clinical study by Han et al. (2012). The integrated GEO2R web

Table 1 Information on the patient sample data obtained from the GEO database.

Group	Accession	Title	Source name	Histology	Disease state	Tissue
Control	GSM931812	Control 1	White matter from age-matched healthy controls	Control	Healthy control	Brain white matter
	GSM931813	Control 2	White matter from age-matched healthy controls	Control	Healthy control	Brain white matter
Chronic active plaque (diseased)	GSM931815	Chronic active plaque 1 (CAP1)	MS brain lesion (following demyelination-active inflammation)	Chronic active plaque	Multiple sclerosis following demyelination-active Inflammation	Multiple sclerosis brain lesion
	GSM931816	Chronic active plaque 2 (CAP2)	MS brain lesion (following demyelination-active inflammation)	Chronic active plaque	Multiple Sclerosis following demyelination-active inflammation	Multiple sclerosis brain lesion
Chronic plaque (diseased)	GSM931817	Chronic plaque 1 (CP1)	MS brain lesion (after inflammation has subsided -late stage)	Chronic plaque	Multiple sclerosis after inflammation has subsided-late stage	Multiple sclerosis brain lesion
	GSM931818	Chronic plaque 2 (CP2)	MS brain lesion (after inflammation has subsided -late stage)	Chronic plaque	Multiple sclerosis after inflammation has subsided-late stage	Multiple sclerosis brain lesion

tool was employed for the primary processing of the microarray-based gene expression data, which provided us with the Top 250 DEGs for both the sample groups. The volcano plot in which the relevant genes which satisfy the values (P-value < 0.05, logFC ≥ 1, and logFC ≤ -1) are tagged and displayed as green color dots. Fig. 1A and B depict the relevant genes for sample group one (control vs. chronic active plaque) and sample group two (control vs. chronic plaque). The heatmap generated using the web-based heatmapper tool for sample group 1 (control vs. chronic active plaque), and sample group 2 (control vs. chronic plaque) is shown in Fig. 2A and B. Based on their hierarchy, the gene expression profiles depicted in those heat maps were subjected to clustering.

3.2 Network association and module analysis

The identified top 250 DEGs were then subjected to STRING web-based network generating tools to establish and visualize connections between DEGs on a functional and biological level. The interaction score limit was set to a confidence level of 0.4 for this study, and all the singleton genes were removed from consideration. We then exported the results obtained from STRING to Cytoscape. Fig. 3 represents the results obtained from the Cytoscape. Fig. 3A highlights the significant PPI network interactions, including 146 nodes (proteins) and 712 edges (association between the proteins), and Fig. 3B highlights the relevant interactions among 135 nodes and 676 edges. The MCODE plugin tool was used in the Cytoscape to enrich our understanding of the interactions further. We selected the top 2 clusters presented by the MCODE plugin for each sample group. Fig. 4A and B represent the top 2 clusters for the control vs. CAP (active inflammation stage) sample group with scores of 5.25, which contains nine nodes with 42 edges and five with five nodes with 20 edges, respectively. Fig. 4C and D feature the top 2 cluster networks for the control vs. CP (late-subsided stage) sample group with scores of 4.6, which contains 11 nodes with 46 edges, and 3.778, which contains ten nodes with 34 edges. The details of the clusters containing the name of the nodes and scores formed using MCODE are presented in Table 2. Among the top cluster for both sample groups, we observed common *SCN2A*, *SYN2*, and *SYT1* genes (Fig. 4A and C). Similarly, in the case of the 2nd most relevant cluster, *UNC13C, PCLO, and ERC2* were the DEGs that were common in the results obtained for both the sample groups (Fig. 4B and D).

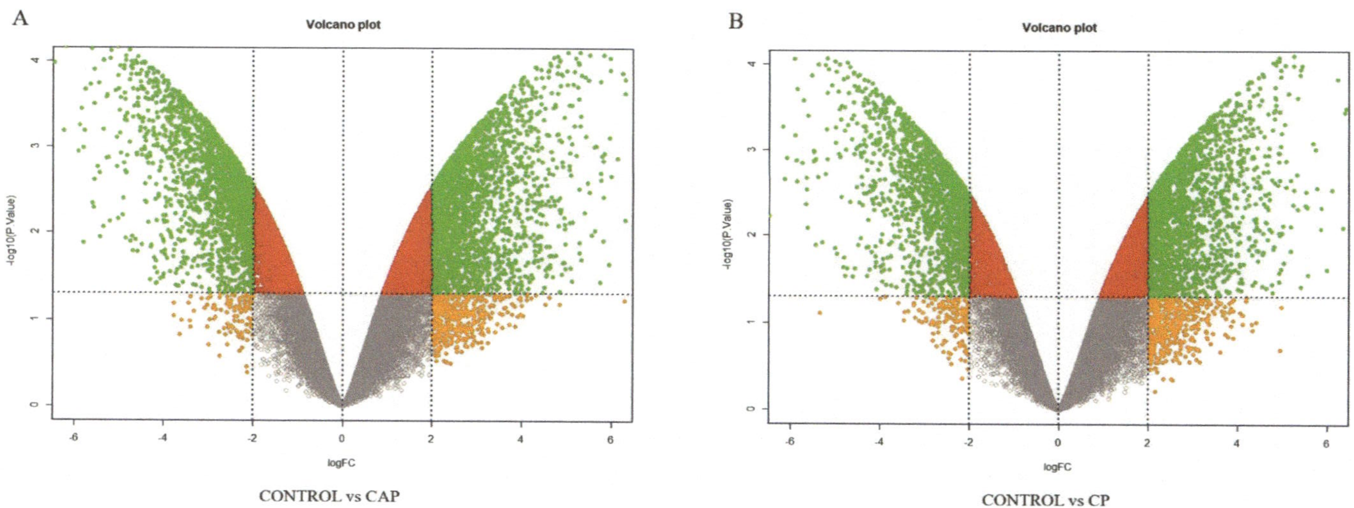

Fig. 1 Volcano plot representation of the DEGs for the considered sample groups, with x-axis highlighting the logFC values; y-axis highlighting the ($-\log_{10}$) of the *P*-value. The red dots depict genes with *P*-values < 0.05, and genes with a logFC value ≥ 2 and logFC value ≤ -2 have been depicted using yellow-colored dots; the important DEGs which satisfy both formerly mentioned values have been highlighted in green. The left panel represents the upregulated ones, whereas the right panel represents the downregulated ones. (A) Represents the target genes for the control vs. CAP (active inflammation stage) sample group that compares control (control 1; GSM931812, control 2; GSM931813) and chronic active plaque 1 (CAP1; GSM931815), chronic active plaque 2 (CAP2; GSM931816); (B) Represents the target genes for the control vs. CP (late-subsided stage) sample group that compares control (control 1; GSM931812, control 2; GSM931813) and chronic plaque 1 (CP1; GSM931817), chronic plaque 2 (CP1; GSM931818).

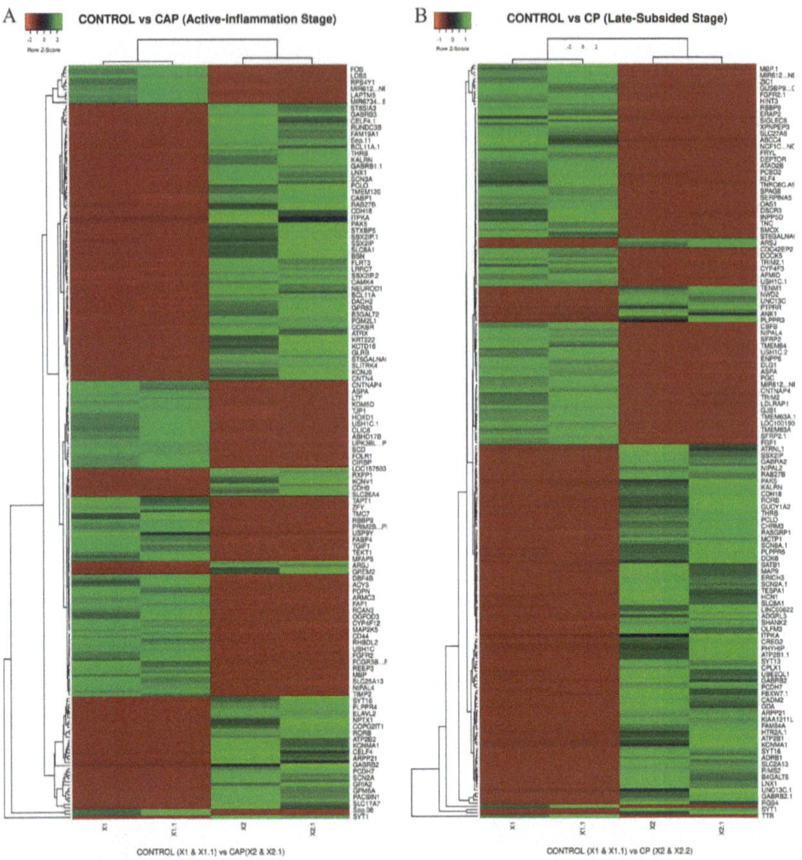

Fig. 2 Heat map for the present study, featuring the top 250 DEGs for the corresponding groups. The genes with elevated signals are depicted in green and the genes lowly express end in red. (A) Represents the expression patterns for the target genes for the control vs. CAP (active inflammation stage) group; (B) Represents the expression patterns for the target genes for the control vs. CP (late-subsided stage) group.

3.3 ClueGO analysis and representation

For a compendious representation of our findings, we used the ClueGo plugin of Cytoscape to map and visualize the DEGs that affect the key GO terms (BP, MF) and pathways (KEGG). The parameters set for our analysis using ClueGO were a two-sided hypergeometric-based test with a P-value threshold of 0.05, correction such as Benjamini-Hochberg, and the $\geq$0.4 of kappa-score. In carrying out the analysis, we obtain the enriched pathways and BP, MF (GO Terms) for both sample groups considered in this study.

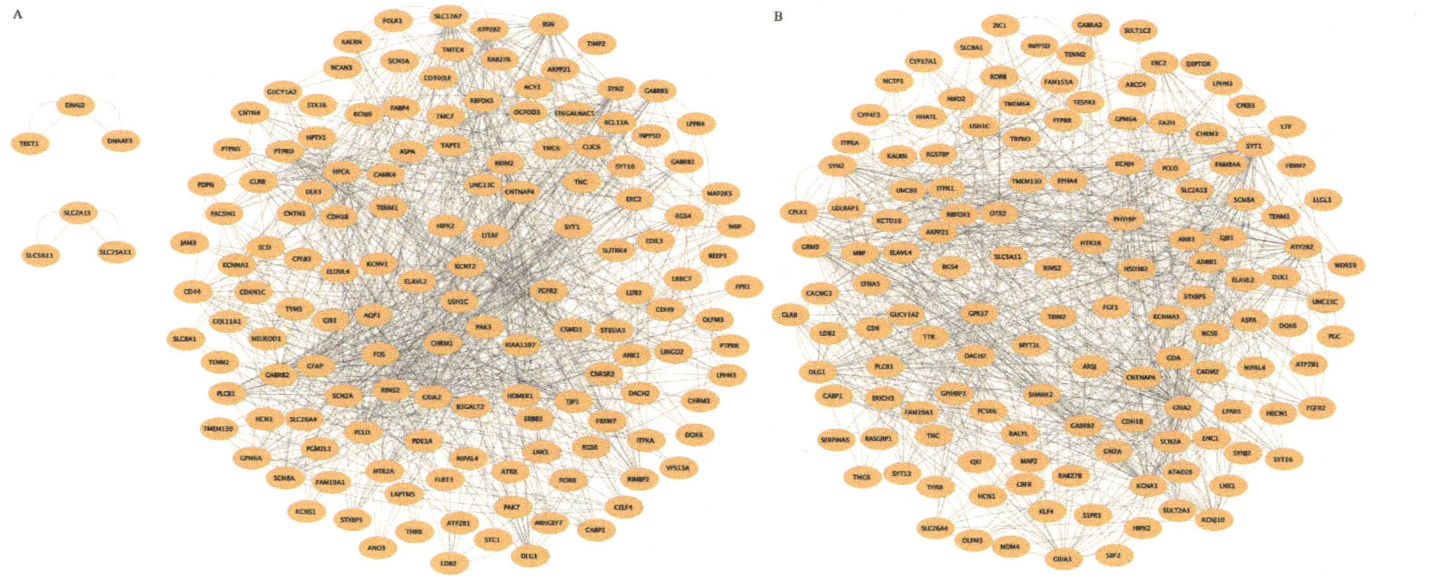

Fig. 3 The PPI network has been highlighted for the top 250 DEGs for the concerned sample groups. (A) Depicts the network interaction for the top 250 DEGs for control vs. CAP (active inflammation stage) group; (B) Depicts the network interaction for the top 250 DEGs for control vs. CP (late-subsided stage) group.

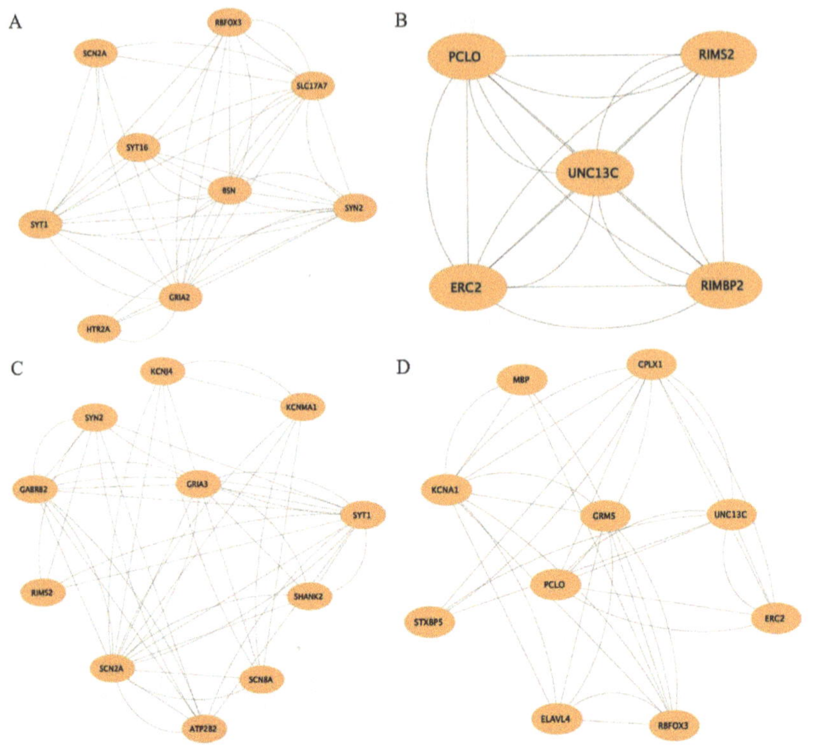

Fig. 4 (A) and (B) Present the top 2 enriched clusters for control vs. CAP (active inflammation stage) group, with a score of 5.25 and 5, respectively; (C) and (D) Display the top 2 enriched clusters for control vs. CP (late-subsided stage) group, with a score of 4.6 and 3.8 (3.788), respectively.

Table 2 Identified clusters from complex PPI networks using MCODE.

Group	Cluster	Score (density*#nodes)	Nodes	Edges	Node IDs
control vs CAP	1	5.25	9	42	*RBFOX3, BSN, HTR2A, SYN2, SCN2A, SYT1, SYT16, SLC17A7, GRIA2*
	2	5	5	20	*UNC13C, RIMS2, ERC2, PCLO, RIMBP2*
control vs CP	1	4.6	11	46	*SYN2, ATP2B2, KCNJ4, SCN2A, SCN8A, SHANK2, RIMS2, SYT1, KCNMA1, GABRB2, GRIA3*
	2	3.778	10	34	*RBFOX3, CPLX1, STXBP5, ELAVL4, MBP, KCNA1, UNC13C, ERC2, GRM5, PCLO*

The common GO terms observed for both the sample groups, active inflammation stage (control vs. CAP sample group) and late-subsided stage (control vs. CP sample group), were chemical synaptic transmission (GO:0007268), calmodulin-binding (GO:0005516) with genes overlapping in both sample groups for both processes. In the case of the active inflammation stage (control vs. CAP sample group), chemical synaptic transmission (GO:0007268), neurotransmitter receptor activity (GO:0030594), synapse organization (GO:0050808), gated-ion channel activity (GO:0022836), neural retina development (GO:0003407) were some of the other GO terms that were found to be enriched. The remaining terms are depicted in Fig. 5. Similarly, for the late-subsided stage (control vs. CP sample group), chemical synaptic transmission (GO:0007268), myelination (GO:0042552), calmodulin-binding (GO:0005516), central nervous system neuron differentiation (GO:0021953), calcium ion transmembrane transporter activity (GO:0015085), retinal cone cell differentiation (GO:0042670) were few of the essential terms that were enriched, with the other relevant processes shown in Fig. 6. By studying and analyzing the observed expression patterns of the DEGs, we are convinced that target genes regulate many cellular functions and pathways, and their aberrant expression can dysregulate the same. Thus, we aimed to expose and highlight those dysregulated functions and the responsible genes in this study.

3.4 MetaCore™ enrichment

To carry out a comprehensive enrichment analysis of the gene expression patterns of the involved genes, we used MetaCore™ software (Clarivate Analytics). Comparative enrichment analysis was carried out using the functional ontology tool. Thus, a lower *P*-value would suggest the particular term to be more significant and relevant. The results obtained from the functional enrichment analysis are depicted in the Fig. 7 series. Regulation of intrinsic membrane properties and excitability of cortical pyramidal neurons, muscle contraction - GPCRs in the regulation of smooth muscle tone, neurophysiological process - *HTR2A* signaling in the nervous system were the top 3 key canonical pathways observed, with the other pathways depicted in Fig. 7A. Fig. 7B shows the top 10 GO cellular processes, with nervous system development, chemical synaptic transmission modulation, and trans-synaptic signaling regulation being the three most relevant cellular processes. Neurophysiological process–transmission of nerve impulse, development-neurogenesis and

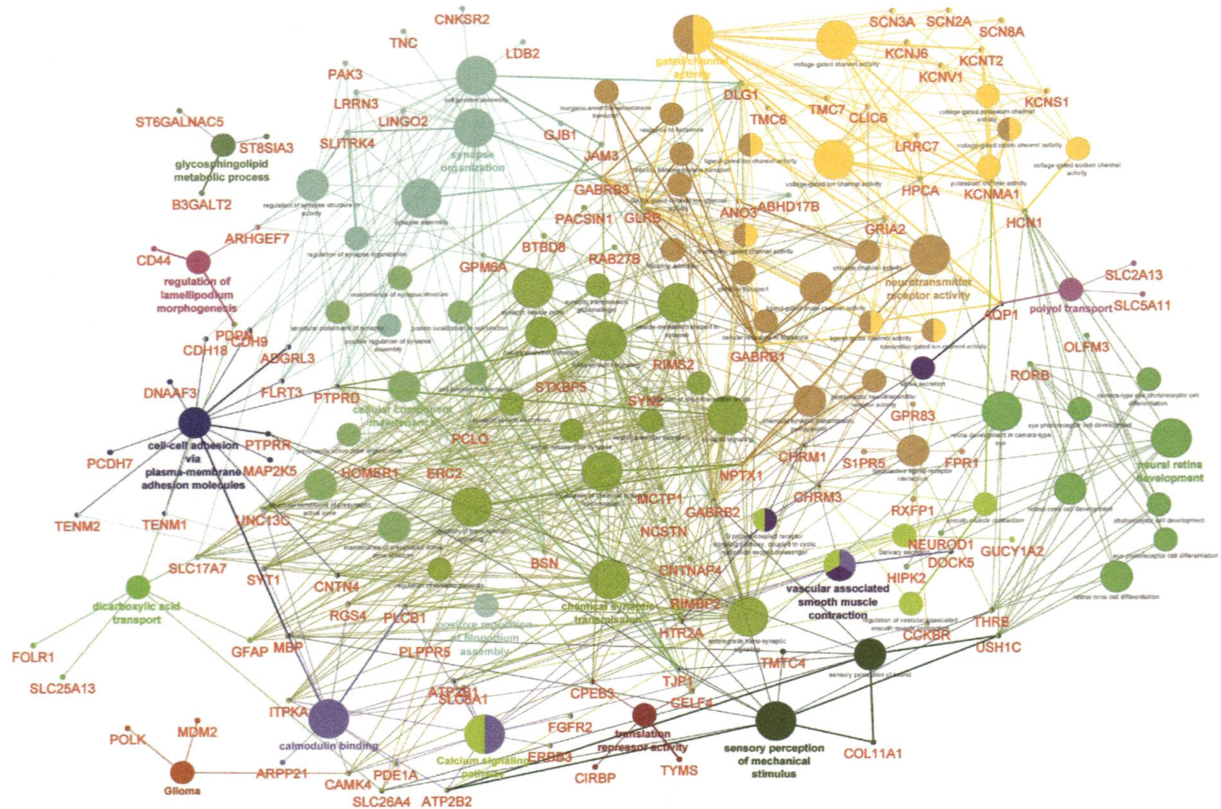

Fig. 5 Functional and pathway enrichment analysis of GO terms and also KEGG pathway. The key BP, MP, and pathways have been depicted along with the concerned DEGs for the control vs. CAP (active inflammation stage) group. Based on the value of *P* the node size varies. The node color code highlights the specific functions, the DEGs are known to affect. The color represents the various BP and MF and pathways. The bold fonts indicate the most vital functional GO term (MF and BP of each group) and KEGG pathway. The gene symbols are displayed in red font.

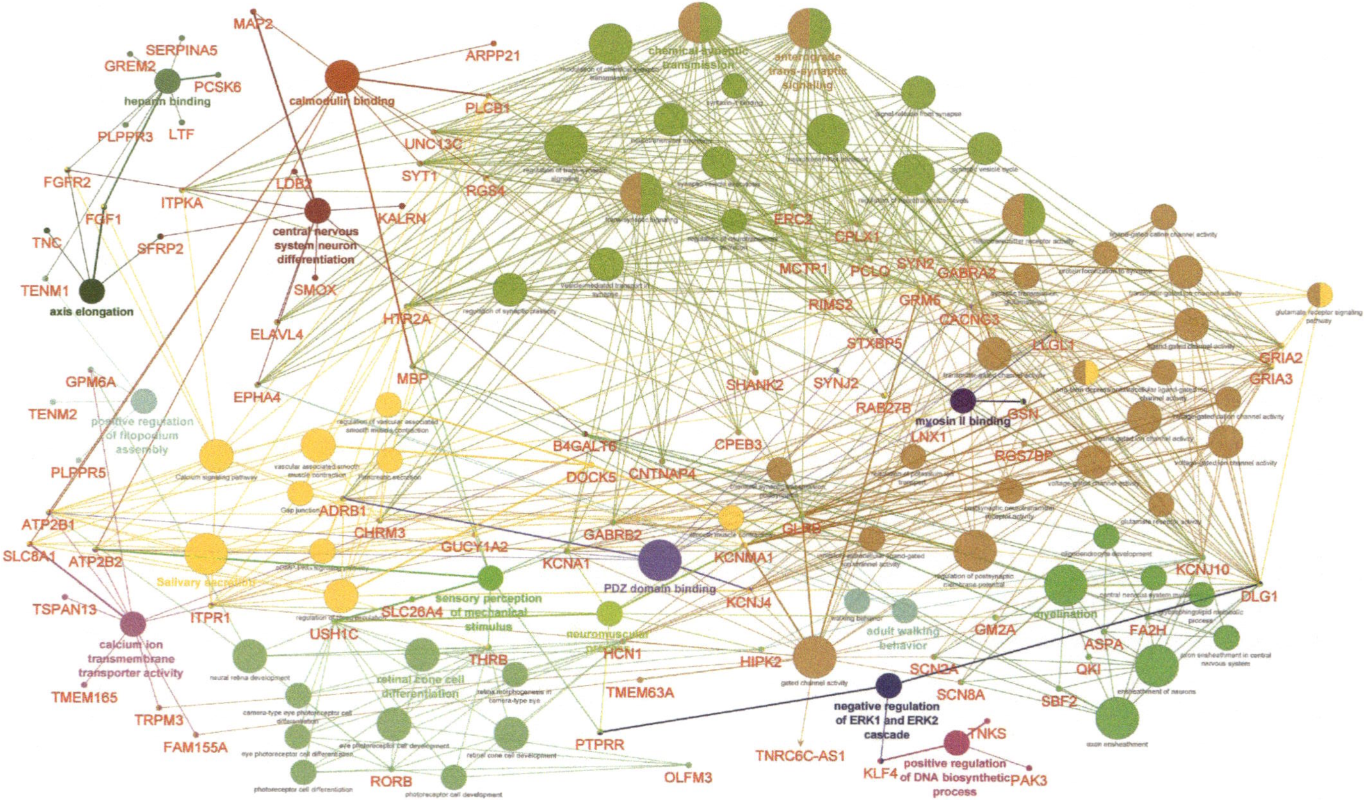

Fig. 6 Functional and pathway enrichment analysis of GO terms and also KEGG pathway. The key BP, MF, and pathways have been depicted along with the concerned DEGs for the control vs. CP (late-subsided stage) group. Based on the value of *P* the node size varies. The node color code highlights the specific functions, the DEGs are involved in. The color represents various BP and MF and pathways. The bold fonts indicate the most vital functional GO term (MF and BP of each group) and KEGG pathway. The gene symbols are displayed in red font.

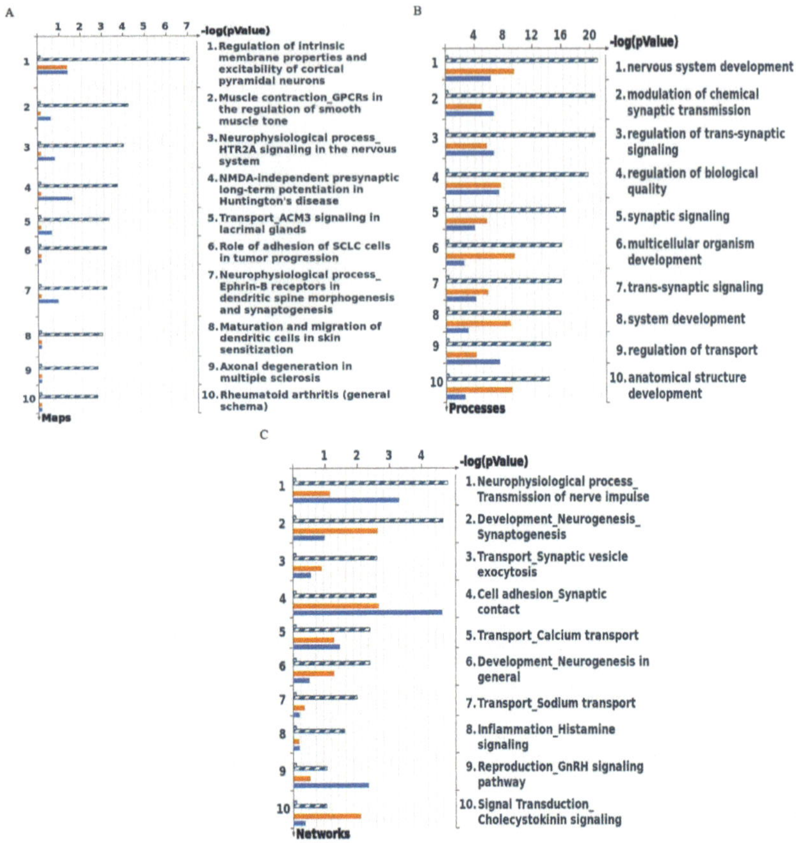

Fig. 7 The top 10 pathway maps (A), ontological processes (B), and metabolic networks (C) were selected and visualized via GeneGo software. The height of the bar for every chart depicted is directly correlated to the number of target genes that affect the corresponding parameter.

synaptogenesis, transport–synaptic vesicle exocytosis were the top 3 metabolic networks, with the other significant networks highlighted in Fig. 7C. GeneGo tool built into the MetaCore™ software was used to generate these pathway maps, and the pathway which was enriched was the regulation of intrinsic membrane properties and excitability. Fig. 8 represents the cascade of reactions involving the activation and inhibition of target genes involved in the pathway.

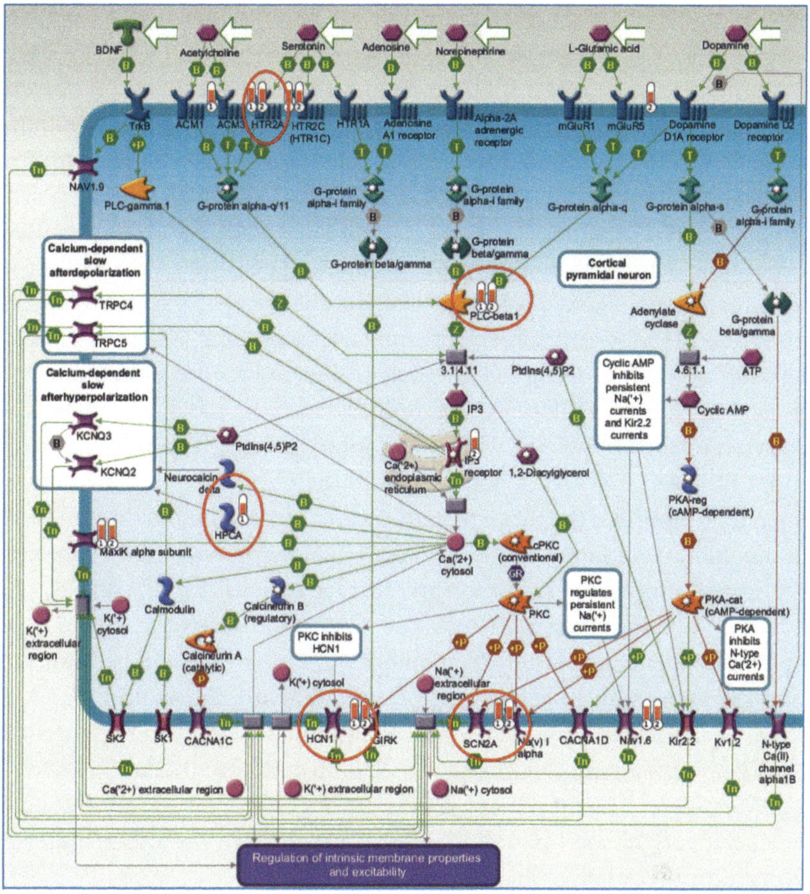

Fig. 8 The cascading mechanisms entangled in the regulation of intrinsic membrane properties and excitability. The key genes that are being considered have been circled in red. The association of proteins have been highlighted using arrows which depicts the stimulatory and inhibitory effect of the particular protein.

4. Discussion

Microarray technology has proven to be one of the pioneering technologies for studying and analyzing disease-affected patients' expression profiles. Several studies have used high-throughput computational tools for carrying out analytical studies on gene expression patterns and have been presented with positive results (Kumar, Kumar, Siva, Doss, & Zayed, 2019; Mishra et al., 2021; Udhaya Kumar et al., 2020, 2021). We set out to analyze the expression data obtained from samples of patients suffering from MS for this computational study. The GSE38010 dataset comprises two control samples and five MS patient samples, which are sub-categorized into active-inflammation stage and late-subsided stage based on the disease progression status. We investigated 58,866 genes (29,433 for active-inflammation stage and 29,433 for late-subsided stage), out of which only 470 DEGs (235 for both active-inflammation stage and late-subsided stage) were further studied using macro-analytical techniques to understand their interactions better. Following the primary analysis, the DEGs were then subjected to the web-based STRING tool to generate the interaction network of the top 250 DEGs obtained from the GEO2R tool. The network generated comprised the proteins as the nodes, and the interactions observed among them were in the form of edges. The network was then exported to Cytoscape for better visualization and further analysis (Fig. 3A and B). The MCODE plugin, integrated into the Cytoscape software (v3.9.1), was exploited to generate clusters to augment our knowledge and better understand the expression patterns of the clustered genes, shown in Fig. 4A–D. ClueGO, in integration with Cytoscape, was used to conduct the functional EA. The results are shown in Figs. 5 and 6. In addition to ClueGO, we also used MetaCore™ to enrich our results further. The GeneGo tool was used to generate a pathway map that gives us insight into the other molecules involved in the other cellular processes.

The functional analysis of the target genes using ClueGO provided us with meaningful observations regarding the key pathways, cellular processes, and functions for both the sample groups considered for this study. The common and overlapping cellular processes among the 2 sample groups were chemical synaptic transmission (GO:0007268) and calmodulin-binding (GO:0005516) (Figs. 5 and 6). The common genes affecting these two cellular processes in both the sample groups are Synaptotagmin 1 *(SYT1)*, Regulator of G Protein Signaling 4 (*RGS4*), Phospholipase C Beta 1 (*PLCB1*), Myelin Basic Protein (*MBP*), Hyperpolarization Activated Cyclic Nucleotide Gated Potassium Channel 1 *(HCN1)*. Mutations in the *SYT1* gene, previously known to be present at the neuronal synapses and responsible for the synchronous release of the neurotransmitters with Ca^{2+} ion influx can significantly hinder the synaptic transmission (Bradberry et al., 2020). *RGS4* is a member of the R4 family of proteins (Regulator of G-protein) and is responsible for modulating the physiological activity of hormones, neurotransmitters, and many other biomolecules (Squires, Montañez-Miranda, Pandya, Torres, & Hepler, 2018). *RGS4* is also known for the signaling activity that leads to the migration of lymphocytes and macrophages to and from sites of inflamed tissues and for the maturation of B-cells (Xie, Chan, & Druey, 2016). *PLCB1* is one of the two isozymes that belong to the larger phospholipase-C family and is responsible for hydrolyzing phosphatidylinositol 4,5-bisphosphate (PIP2), producing inositol 1,4,5-triphosphate (IP3) and diacylglycerol (DAG). These two molecules are essential in the efficient conduction of neuronal signals (Bush et al., 2011). *MBP* is known to be present in abundance in the CNS myelin, and its role in long-term role in the maintenance of myelin has also been well-established (Martinsen & Kursula, 2022). Espino-Paisán et al. studied the expression in a Spanish population and reported that relapse cases are higher in the male population (Espino-Paisán et al., 2020). *HCN1* gene was involved in the dysregulation of gated channel activity and neural retina Development. *HCN1* gene, belonging to the *HCN1–4* gene family and expressed in the CNS and also to an extent in the peripheral nervous system, is reported to modulate the activity of electrical signals that are received by the neurons present in the basal ganglia (Chang, Wang, Jiang, Shi, & Xie, 2019; Notomi & Shigemoto, 2004).

With respect to the active-inflammation stage, other key cytological processes that were dysregulated were synapse organization (GO:0050808), gated channel activity (GO:0022836), and neural retina development (GO:0003407) (Fig. 5). On analyzing the ClueGO charts, we noted that the Gamma-Aminobutyric Acid type A Receptor sub-unit Beta 3 (*GABRB3*) gene was involved in synapse organization and gated channel activity. *GABRB3* is a gated chloride ion channel responsible for inhibiting neurotransmitter release and the production of functional inhibitory GABAergic synapses (Wagstaff, Chaillet, & Lalande, 1991).

Among the recently stated processes, we observed that the Beta-1,4-Galactosyltransferase 6 (B4GALT6) gene was seen to regulate both, myelination and CNS neurons differentiation processes. An earlier study reported that B4GALT6 gene expression was upregulated in CNS MS lesions and was believed to be linked to neurodegeneration (Mayo et al., 2014). Likewise, for late-subsided stage MS, we observed myelination (GO:0042552), central nervous system neuron differentiation (GO:0021953), and calcium ion transmembrane transporter activity (GO:0015085) were some of the other key cellular functions that were affected (Fig. 6).

We further enriched our observations by correlating the results obtained from ClueGO with the results obtained from GeneGo. Bioinformatics software is developed and maintained by MetaCore™ software. Comparing the results obtained from both the functional enrichment tools, we observed that the key pathway map enriched was the regulation of intrinsic membrane properties and excitability of cortical pyramidal neurons (Fig. 8). The key target genes involved in this process were Sodium Voltage-Gated Channel Alpha Subunit 2 (*SCN2A*), 5-Hydroxytryptamine Receptor 2A (*HTR2A*), and *HCN1*. In addition to these three genes, another intriguing observation noted was the presence of genes named *HPCA* and *PLCB1*. Our results indicated that Hippocalcin (*HPCA*) was observed in the control vs. CAP sample group and seen to regulate voltage-gated channel activity, which leads us to believe that *HPCA* can be a diagnostic biomarker for detecting the active-inflammation stage in MS. On a similar note, the same was observed for Phospholipase C Beta 1 (*PLCB1*), wherein *PLCB1* was observed in the control vs. CP sample group only. It was seen to regulate calmodulin binding trans-synaptic signaling. Thus, the genes *HPCA* and *PLCB1* both possess the potential to be biomarkers for MS that could indicate the developmental stages of the patient and may also be used to study the progression of MS through various stages (Fig. S1 in the online version at https://doi.org/10.1016/bs.apcsb.2022.05.003).

Furthermore, we observed that *SCN2A* was the only DEG that overlapped with the results obtained from both the MCODE plugin and pathway maps obtained from MetaCore™. The cascade mechanisms involved in regulating intrinsic membrane properties and excitability of cortical pyramidal neurons are depicted in Fig. 8. *SCN2A, HTR2A,* and *HCN1* were the enriched target genes, with another keynote being that all the three genes are signal transduction molecules expressed on the membrane of the neuronal cells. Mutations in the *SCN2A* gene have been directly correlated to brain disorders, and the relation between the mutations and membrane excitability was studied via Voltage Clamp analysis (Berecki et al., 2018). In untreated patients with schizophrenia, *HTR2A* activity was upregulated, directly correlated to metabotropic glutamate receptors (mGluR2) (González-Maeso et al., 2008). Similarly, mutations in *HCN1* have been reported to cause increased excitability of the neurons (Bonzanni et al., 2018).

Based on the results obtained from our bioinformatics analysis, it is noteworthy to mention that the genes mentioned above could serve as potential biomarkers for MS. With substantial evidence reporting that the aberrant expression of *SCN2A*, *HTR2A*, and *HCN1* genes are directly affiliated with various brain disorders. The expression levels of these genes are provided in Figs. S2 and S3 in the online version at https://doi.org/10.1016/bs.apcsb.2022.05.003, respectively.

5. Conclusion

The findings of this report indicate that the identified DEGs might play a substantial role in the initiation, development, and progression of MS. The functional enrichment analysis identified three genes (*SCN2A*, *HTR2A*, and *HCN1*) that were significantly upregulated in diseased patient samples and could aid in the detection of MS. Additionally, two genes (*HPCA* and *PLCB1*) that could differentiate between the different stages of MS were identified. Indeed, the identified DEGs may help researchers understand how the molecular processes, including membrane excitability, signal transduction dysregulation, and inflammatory responses, are dysregulated in MS patients. A better understanding of MS pathophysiology will help scientists develop better treatment options based on personalized medicine.

Acknowledgments

The authors would like to take this opportunity to thank the management of Vellore Institute of Technology (VIT), Vellore, India, and Qatar University, Doha, Qatar, for providing the necessary facilities and encouragement to carry out this work.

Author contributions

UKS, HZ, and GPDC were involved in the study's design. UKS, AD, TM, SR, and GR were involved in the data collection and conducted the experiment. UKS, AD, SY, TM, SR, and GR were involved in the acquisition, analysis, and interpretation results. UKS and AD drafted the manuscript. GPDC and HZ supervised the entire study and were involved in the study design, the acquisition, analysis, and understanding of the data, and critically reviewed the manuscript. All authors edited and approved the submitted version of the article.

Conflict of interest

The authors have declared that no conflicts of interest exist.

Role of funding source

No funding agency was involved in the present study.

References

Ates, O., Kurt, S., Bozkurt, N., & Karaer, H. (2010). NRAMP1 (SLC11A1) Variants: Genetic Susceptibility to Multiple Sclerosis. *Journal of Clinical Immunology*, *30*(4), 583–586. https://doi.org/10.1007/s10875-010-9422-5.

Aubert, J., Bar-Hen, A., Daudin, J.-J., & Robin, S. (2004). Determination of the differentially expressed genes in microarray experiments using local FDR. *BMC Bioinformatics*, *5*(1), 125. https://doi.org/10.1186/1471-2105-5-125.

Babicki, S., Arndt, D., Marcu, A., Liang, Y., Grant, J. R., Maciejewski, A., et al. (2016). Heatmapper: Web-enabled heat mapping for all. *Nucleic Acids Research*, *44*(W1), W147–W153. https://doi.org/10.1093/nar/gkw419.

Bader, G. D., & Hogue, C. W. (2003). An automated method for finding molecular complexes in large protein interaction networks. *BMC Bioinformatics*, *4*(1), 2. https://doi.org/10.1186/1471-2105-4-2.

Balasundaram, A., Udhaya Kumar, S., Doss, G. P., & C. (2022). A computational model revealing the immune-related hub genes and key pathways involved in rheumatoid arthritis (RA). *Advances in Protein Chemistry and Structural Biology*, *129*, 247–273. https://doi.org/10.1016/bs.apcsb.2021.11.006.

Barrett, T., Wilhite, S. E., Ledoux, P., Evangelista, C., Kim, I. F., Tomashevsky, M., et al. (2013). NCBI GEO: Archive for functional genomics data sets—Update. *Nucleic Acids Research*, *41*(D1), D991–D995. https://doi.org/10.1093/nar/gks1193.

Berecki, G., Howell, K. B., Deerasooriya, Y. H., Cilio, M. R., Oliva, M. K., Kaplan, D., et al. (2018). Dynamic action potential clamp predicts functional separation in mild familial and severe de novo forms of SCN2A epilepsy. *Proceedings of the National Academy of Sciences of the United States of America*, *115*(24), E5516–E5525. https://doi.org/10.1073/pnas.1800077115.

Bindea, G., Mlecnik, B., Hackl, H., Charoentong, P., Tosolini, M., Kirilovsky, A., et al. (2009). ClueGO: A Cytoscape plug-in to decipher functionally grouped gene ontology and pathway annotation networks. *Bioinformatics*, *25*(8), 1091–1093. https://doi.org/10.1093/bioinformatics/btp101.

Bonneh-Barkay, D., Wang, G., Starkey, A., Hamilton, R. L., & Wiley, C. A. (2010). In vivo CHI3L1 (YKL-40) expression in astrocytes in acute and chronic neurological diseases. *Journal of Neuroinflammation*, 7, 34. https://doi.org/10.1186/1742-2094-7-34.

Bonzanni, M., DiFrancesco, J. C., Milanesi, R., Campostrini, G., Castellotti, B., Bucchi, A., et al. (2018). A novel de novo HCN1 loss-of-function mutation in genetic generalized epilepsy causing increased neuronal excitability. *Neurobiology of Disease*, *118*, 55–63. https://doi.org/10.1016/j.nbd.2018.06.012.

Bradberry, M. M., Courtney, N. A., Dominguez, M. J., Lofquist, S. M., Knox, A. T., Sutton, R. B., et al. (2020). Molecular basis for synaptotagmin-1-associated neurodevelopmental disorder. *Neuron*, *107*(1), 52–64.e7. https://doi.org/10.1016/j.neuron.2020.04.003.

Bush, W. S., McCauley, J. L., DeJager, P. L., Dudek, S. M., Hafler, D. A., Gibson, R. A., et al. (2011). A knowledge-driven interaction analysis reveals potential neurodegenerative mechanism of multiple sclerosis susceptibility. *Genes and Immunity*, *12*(5), 335–340. https://doi.org/10.1038/gene.2011.3.

Cantó, E., Tintoré, M., Villar, L. M., Costa, C., Nurtdinov, R., Álvarez-Cermeño, J. C., et al. (2015). Chitinase 3-like 1: Prognostic biomarker in clinically isolated syndromes. *Brain*, *138*(4), 918–931. https://doi.org/10.1093/brain/awv017.

Chang, X., Wang, J., Jiang, H., Shi, L., & Xie, J. (2019). Hyperpolarization-activated cyclic nucleotide-gated channels: An emerging role in neurodegenerative diseases. *Frontiers in Molecular Neuroscience*, *12*. https://www.frontiersin.org/article/10.3389/fnmol.2019.00141.

Comabella, M., Fernández, M., Martin, R., Rivera-Vallvé, S., Borrás, E., Chiva, C., et al. (2010). Cerebrospinal fluid chitinase 3-like 1 levels are associated with conversion to multiple sclerosis. *Brain, 133*(4), 1082–1093. https://doi.org/10.1093/brain/awq035.

Cyster, J. G. (2005). Chemokines, sphingosine-1-phosphate, and cell migration in secondary lymphoid organs. *Annual Review of Immunology, 23*, 127–159. https://doi.org/10.1146/annurev.immunol.23.021704.115628.

Dufva, M. (2009). Introduction to microarray technology. In M. Dufva (Ed.), *Vol. 529. DNA microarrays for biomedical research* (pp. 1–22). Humana Press. https://doi.org/10.1007/978-1-59745-538-1_1.

Dyment, D. A., Sadnovich, A. D., & Ebers, G. C. (1997). Genetics of multiple sclerosis. *Human Molecular Genetics, 6*(10), 1693–1698. https://doi.org/10.1093/hmg/6.10.1693.

Espino-Paisán, L., Agudo-Jiménez, T., Rosales-Martínez, I., López-Cotarelo, P., García-Martínez, M.Á., Domínguez-Mozo, M. I., et al. (2020). A polymorphism within the MBP gene is associated with a higher relapse number in male patients of multiple sclerosis. *Frontiers in Immunology, 11*. https://www.frontiersin.org/article/10.3389/fimmu.2020.00771.

González-Maeso, J., Ang, R. L., Yuen, T., Chan, P., Weisstaub, N. V., López-Giménez, J. F., et al. (2008). Identification of a serotonin/glutamate receptor complex implicated in psychosis. *Nature, 452*(7183), 93–97. https://doi.org/10.1038/nature06612.

Han, M. H., Lundgren, D. H., Jaiswal, S., Chao, M., Graham, K. L., Garris, C. S., et al. (2012). Janus-like opposing roles of CD47 in autoimmune brain inflammation in humans and mice. *Journal of Experimental Medicine, 209*(7), 1325–1334. https://doi.org/10.1084/jem.20101974.

Koch-Henriksen, N., & Sørensen, P. S. (2010). The changing demographic pattern of multiple sclerosis epidemiology. *The Lancet Neurology, 9*(5), 520–532. https://doi.org/10.1016/S1474-4422(10)70064-8.

Kumar, S. U., Kumar, D. T., Siva, R., Doss, C. G. P., & Zayed, H. (2019). Integrative bioinformatics approaches to map potential novel genes and pathways involved in ovarian cancer. *Frontiers in Bioengineering and Biotechnology*, 7, 391. https://doi.org/10.3389/fbioe.2019.00391.

Liu, M., Hou, X., Zhang, P., Hao, Y., Yang, Y., Wu, X., et al. (2013). Microarray gene expression profiling analysis combined with bioinformatics in multiple sclerosis. *Molecular Biology Reports, 40*(5), 3731–3737. https://doi.org/10.1007/s11033-012-2449-3.

Martinsen, V., & Kursula, P. (2022). Multiple sclerosis and myelin basic protein: Insights into protein disorder and disease. *Amino Acids, 54*(1), 99–109. https://doi.org/10.1007/s00726-021-03111-7.

Mayo, L., Trauger, S. A., Blain, M., Nadeau, M., Patel, B., Alvarez, J. I., et al. (2014). Regulation of astrocyte activation by glycolipids drives chronic CNS inflammation. *Nature Medicine, 20*(10), 1147–1156. https://doi.org/10.1038/nm.3681.

Mishra, S., Shah, M. I., Udhaya Kumar, S., Thirumal Kumar, D., Gopalakrishnan, C., Al-Subaie, A. M., et al. (2021). Network analysis of transcriptomics data for the prediction and prioritization of membrane-associated biomarkers for idiopathic pulmonary fibrosis (IPF) by bioinformatics approach. *Advances in Protein Chemistry and Structural Biology, 123*, 241–273. https://doi.org/10.1016/bs.apcsb.2020.10.003.

Notomi, T., & Shigemoto, R. (2004). Immunohistochemical localization of Ih channel subunits, HCN1-4, in the rat brain. *The Journal of Comparative Neurology, 471*(3), 241–276. https://doi.org/10.1002/cne.11039.

O'Connor, K. C., Bar-Or, A., & Hafler, D. A. (2001). The neuroimmunology of multiple sclerosis: Possible roles of T and B lymphocytes in immunopathogenesis. *Journal of Clinical Immunology, 21*(2), 81–92. https://doi.org/10.1023/a:1011064007686.

Okada, T., & Cyster, J. G. (2006). B cell migration and interactions in the early phase of antibody responses. *Current Opinion in Immunology*, *18*(3), 278–285. https://doi.org/10.1016/j.coi.2006.02.005.

Pernice, S., Pennisi, M., Romano, G., Maglione, A., Cutrupi, S., Pappalardo, F., et al. (2019). A computational approach based on the colored Petri net formalism for studying multiple sclerosis. *BMC Bioinformatics*, *20*(6), 623. https://doi.org/10.1186/s12859-019-3196-4.

Ritchie, M. E., Phipson, B., Wu, D., Hu, Y., Law, C. W., Shi, W., et al. (2015). Limma powers differential expression analyses for RNA-sequencing and microarray studies. *Nucleic Acids Research*, *43*(7), e47. https://doi.org/10.1093/nar/gkv007.

Sawcer, S., Franklin, R. J. M., & Ban, M. (2014). Multiple sclerosis genetics. *The Lancet Neurology*, *13*(7), 700–709. https://doi.org/10.1016/S1474-4422(14)70041-9.

Shannon, P., Markiel, A., Ozier, O., Baliga, N. S., Wang, J. T., Ramage, D., et al. (2003). Cytoscape: A software environment for integrated models of biomolecular interaction networks. *Genome Research*, *13*(11), 2498–2504. https://doi.org/10.1101/gr.1239303.

Smyth, G. K. (2005). limma: Linear models for microarray data. In R. Gentleman, V. J. Carey, W. Huber, R. A. Irizarry, & S. Dudoit (Eds.), *Bioinformatics and computational biology solutions using R and bioconductor* (pp. 397–420). Springer. https://doi.org/10.1007/0-387-29362-0_23.

Squires, K. E., Montañez-Miranda, C., Pandya, R. R., Torres, M. P., & Hepler, J. R. (2018). Genetic analysis of rare human variants of regulators of G protein signaling proteins and their role in human physiology and disease. *Pharmacological Reviews*, *70*(3), 446–474. https://doi.org/10.1124/pr.117.015354.

Stilund, M., Gjelstrup, M. C., Petersen, T., Møller, H. J., Rasmussen, P. V., & Christensen, T. (2015). Biomarkers of inflammation and axonal degeneration/damage in patients with newly diagnosed multiple sclerosis: Contributions of the soluble CD163 CSF/serum ratio to a biomarker panel. *PLoS One*, *10*(4), e0119681. https://doi.org/10.1371/journal.pone.0119681.

Szklarczyk, D., Gable, A. L., Lyon, D., Junge, A., Wyder, S., Huerta-Cepas, J., et al. (2019). STRING v11: Protein–protein association networks with increased coverage, supporting functional discovery in genome-wide experimental datasets. *Nucleic Acids Research*, *47*(D1), D607–D613. https://doi.org/10.1093/nar/gky1131.

Szklarczyk, D., Morris, J. H., Cook, H., Kuhn, M., Wyder, S., Simonovic, M., et al. (2017). The STRING database in 2017: Quality-controlled protein–protein association networks, made broadly accessible. *Nucleic Acids Research*, *45*(Database issue), D362–D368. https://doi.org/10.1093/nar/gkw937.

Udhaya Kumar, S., Madhana Priya, N., Thirumal Kumar, D., Anu Preethi, V., Kumar, V., Nagarajan, D., et al. (2021). An integrative analysis to distinguish between emphysema (EML) and alpha-1 antitrypsin deficiency-related emphysema (ADL)-A systems biology approach. *Advances in Protein Chemistry and Structural Biology*, *127*, 315–342. https://doi.org/10.1016/bs.apcsb.2021.02.004.

Udhaya Kumar, S., Rajan, B., Thirumal Kumar, D., Anu Preethi, V., Abunada, T., Younes, S., et al. (2020). Involvement of essential signaling cascades and analysis of gene networks in diabesity. *Genes*, *11*(11), E1256. https://doi.org/10.3390/genes11111256.

Udhaya Kumar, S., Saleem, A., Thirumal Kumar, D., Anu Preethi, V., Younes, S., Zayed, H., et al. (2021). A systemic approach to explore the mechanisms of drug resistance and altered signaling cascades in extensively drug-resistant tuberculosis. *Advances in Protein Chemistry and Structural Biology*, *127*, 343–364. https://doi.org/10.1016/bs.apcsb.2021.02.002.

Udhaya Kumar, S., Thirumal Kumar, D., Bithia, R., Sankar, S., Magesh, R., Sidenna, M., et al. (2020). Analysis of differentially expressed genes and molecular pathways in familial hypercholesterolemia involved in atherosclerosis: A systematic and bioinformatics approach. *Frontiers in Genetics*, *11*, 734. https://doi.org/10.3389/fgene.2020.00734.

Udhaya Kumar, S., Thirumal Kumar, D., Siva, R., George Priya Doss, C., Younes, S., Younes, N., et al. (2020). Dysregulation of signaling pathways due to differentially expressed genes from the B-cell transcriptomes of systemic lupus erythematosus patients—A bioinformatics approach. *Frontiers in Bioengineering and Biotechnology*, *8*, 276. https://doi.org/10.3389/fbioe.2020.00276.

Wagstaff, J., Chaillet, J. R., & Lalande, M. (1991). The GABAA receptor β3 subunit gene: Characterization of a human cDNA from chromosome 15q11q13 and mapping to a region of conserved synteny on mouse chromosome 7. *Genomics*, *11*(4), 1071–1078. https://doi.org/10.1016/0888-7543(91)90034-C.

Xie, Z., Chan, E. C., & Druey, K. M. (2016). R4 regulator of G protein signaling (RGS) proteins in inflammation and immunity. *The AAPS Journal*, *18*(2), 294–304. https://doi.org/10.1208/s12248-015-9847-0.

Xu, Z., Feng, X., Zhu, W., & Qiu, M. (2021). Identification of key genes and microRNAs for multiple sclerosis using bioinformatics analysis. *Medicine*, *100*(48), e27667. https://doi.org/10.1097/MD.0000000000027667.

CHAPTER EIGHT

Viral hijacking mechanism in humans through protein–protein interactions

Chandrabose Selvaraj[a,*], Gurunathan Rubha Shri[a], Rajendran Vijayakumar[b], Abdulaziz S. Alothaim[b], Saravanan Ramya[a], and Sanjeev Kumar Singh[a,*]

[a]Computer Aided Drug Design and Molecular Modeling Lab, Department of Bioinformatics, Science Block, Alagappa University, Karaikudi, Tamil Nadu, India
[b]Department of Biology, College of Science in Zulfi, Majmaah University, Majmaah, Saudi Arabia
*Corresponding authors: e-mail address: drbioselva@gmail.com; skysanjeev@gmail.com

Contents

Abstract

Numerous viruses have evolved mechanisms to inhibit or alter the host cell's apoptotic response as part of their coevolution with their hosts. The analysis of virus–host protein interactions require an in-depth understanding of both the viral and host protein structures and repertoires, as well as evolutionary mechanisms and pertinent biological facts. Throughout the course of a viral infection, there is constant battle for binding between virus and cellular proteins. Exogenous interfaces facilitating viral–host interactions are well known for constantly targeting and suppressing endogenous interfaces mediating intraspecific interactions, such as viral–viral and host–host connections. In these interactions, the protein–protein interactions (PPIs), are mostly shown as networks (protein interaction networks, PINs), with proteins represented as nodes and their interactions

Advances in Protein Chemistry and Structural Biology, Volume 131
ISSN 1876-1623
https://doi.org/10.1016/bs.apcsb.2022.05.001

represented as edges. Host proteins with a higher degree of connectivity are more likely to interact with viral proteins. Due to technical advancements, three-dimensional interactions may now be visualized computationally utilizing molecular modeling and cryo-EM approaches. The uniqueness of viral domain repertoires, their evolution, and their activities during viral infection make viruses fascinating models for research. This chapter aims to provide readers a complete picture of the viral hijacking mechanism in protein-protein interactions.

1. Introduction

Despite the fact that virus-induced metabolic reprogramming has been known for nearly half a century, the processes and effects have only just begun to be examined in depth. Viruses clearly need the help of host cells to spread (Thyrsted & Holm, 2021). They encourage anabolism in order to make macromolecules that are needed for virion replication and assembly. When viruses enter host cells, they activate metabolic reprogramming, which allows for maximum virus production (Eisenreich, Rudel, Heesemann, & Goebel, 2019). Tumorigenesis and fast cell proliferation are both connected with changes in the metabolism of nucleotides and lipids, which are also seen in a variety of virus infections. In the course of this chapter, we highlight key gaps in knowledge and unanswered questions in the virus metabolism field. Understanding the molecular process of viral infection can be achieved by analyzing the PPI (protein–protein interactions) data of human and virus proteins (Abu-Farha et al., 2020). Researchers have previously concentrated on determining the general characteristics of viral targets within the human PPI network. Especially, by using the system biological approaches and 3D interactions of PPI shows more clear evidence of protein interactions with proteins between the viral and host (Xing, Wallmeroth, Berendzen, & Grefen, 2016). A virus ability to infect and the body's response to it are both mediated by protein–protein interactions (PPI). Since human–virus PPIs can help us better understand human–virus connections and hence help us build extremely effective medications to stop the spread of infectious diseases, we should research these relationships. Human–virus PPI data has been rapidly accruing in recent years, providing an unprecedented opportunity for bioinformatics research on human–virus PPI interactions (Cook, Doncheva, Szklarczyk, von Mering, & Jensen, 2018).

2. Human–virus protein–protein interactions

Viruses that infect several hosts and are closely linked to their hosts are the most common. Few viruses, on the other hand, infect species that are as distantly related as humans and birds. Viruses can evolve to resist the immunological response of their hosts by undergoing adaptive mutations, which gives them an evolutionary advantage and the ability to infect additional hosts (Parrish et al., 2008). In general, only one host is needed for the vast majority of viruses and among those, most of the virus is studied for their disease-causing property. Each host's phenotype-specific virus is often produced in order to maximize the likelihood of transmission. In terms of pH, cell kinds, and temperature, hosts can vary widely. For example, arthropod-borne viruses are among the most multihost viruses in humans. It is well-known that many viruses are capable of significant recombination and can thus spread from animals to humans (Parrish et al., 2008). In comparison to arboviruses, viruses like influenza A do not have as wide a range of host habitats. Consequently, the host and viral proteins interact in a similar manner. The development of innovative antiviral medications will be aided by new insights into the mechanism of human–virus protein–protein interactions. With its small genome and ability to establish the function of human proteins, the virus has considerable advantages in order to complete its life cycle (Maginnis, 2018). Studying the interactions between viral glycoproteins and host receptors will help us better understand how viruses enter into cells. ACE2 (Angiotensin Converting Enzyme-2) has recently been discovered as a cellular receptor for the novel coronavirus SARS-CoV-2, which is entered into host cells via its spike glycoprotein as shown in the Fig. 1.

Several studies have demonstrated the existence of protein–protein interactions between humans and viruses. An infection by a virus is closely linked to a protein complex in the human body. Influenced host protein complexes regulate a wide range of biological activities by the majority of viral strains. As an example, the viral hijacking process in eukaryotic cells starts 4F factor to decrease the expression of important genes in host cells and ensures that viruses may build their proteins to elude the host immune response (Goodacre, Devkota, Bae, Wuchty, & Uetz, 2020). PPI databases have been employed in a substantially larger number of multihost interactions to date. 71 human viruses from 15 different viral families interacted with 43 nonhuman hosts, according to the findings. At least 100 of the

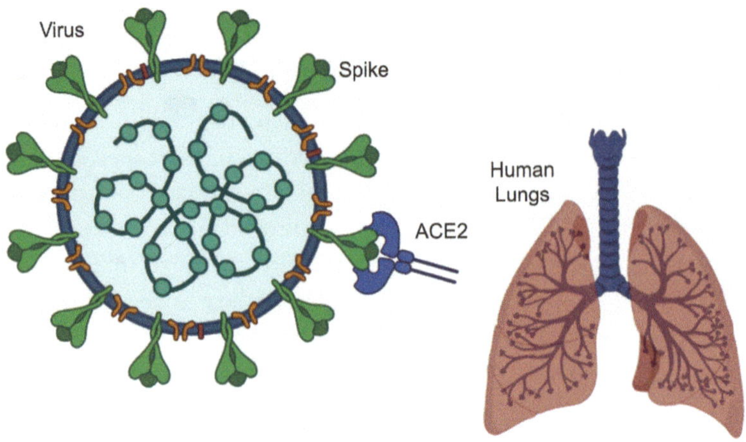

Fig. 1 Cartoonic representation of S-glycoprotein binds to the ACE2 receptor.

2006 h5n1 strains interacted with one another. Even so, it was discovered that nonhuman viruses were able to interact with human proteins with a minimal diversity of proteins (Valkov, Sharpe, Marsh, Greive, & Hyvonen, 2012). There are 525 protein–protein interactions between 39 nonhuman viruses belonging to various viral families. Humans have a greater proteome than most other hosts because of the vast number of protein–protein interactions. In contrast, viruses' ability to affect host networks (e.g., human protein complexes) has rarely been examined from a computational perspective (Mendez-Rios & Uetz, 2010). Despite the fact that human protein complexes have been linked to viral infection, previous research have not thoroughly explored human-virus PPIs from the perspective of human protein complexes. Antiviral drug development benefits from studies on human-virus PPIs in addition to a deeper understanding of viral infection. Viral proteases and polymerases can be inhibited by virus-targeting medications, while host-targeting therapies can affect the functioning of host proteins that are involved in viral life cycles (Yang, Fu, Lian, Dong, & Zhang, 2019).

3. Interfaces of protein interactions in virus-host systems

A protein's binding site plays an important role in its ability to interact with other proteins; hence, binding sites must be in direct physical contact and stable or transient, such binding sites are known as interfaces. An amino acid's three-dimensional structure has a significant impact on its ability to recognize binding partners (Wright & Dyson, 2015). As their electrostatic and structural qualities are impaired or improved, both harmful and beneficial mutations on binding sites greatly influence binding affinity. Protein-binding sites are defined as either endogenous or exogenous interfaces when they connect a virus to its host. In order to mediate the connection between viral and host proteomes, these endogenous interfaces are extremely important. Protein–protein interactions between viruses and their hosts are mediated through exogenous interfaces, which belong to different proteomes (Wu et al., 2021). Many new antiviral medicines are being developed based on this information, including the development of effective and potent small compounds that can inhibit the key interaction of the viral process, in order to gain a better understanding of virus-host protein–protein interaction. When HIV-1 enters a human host, the Maraviroc is a powerful viral-host interaction blocker because it is efficiently bound at the interface level. This medication binds to CCR5 receptors, which are critical for HIV-1 infection, and prevents GP120 from interacting with them (Zhu, Meng, Wang, & Wang, 2015). The interaction between humans and viruses involved both the characteristics of amino acids and the organization of amino acid residues in the context of the entire protein sequence. In other circumstances, the former semantic context of whole protein analysis is greatly concentrated for the prediction of effective protein–protein interactions (Cui, Fang, & Han, 2012). As more studies are recently out on SARS-CoV-2 spike interactions with ACE2 through the mode of protein–protein interactions, we have examined the surface interactions of RBD (Receptor-Binding Domain) interactions with ACE2, as shown in the Fig. 2. RBD is shown to serve as a binding interface between Spike-glycoprotein (S-glycoprotein) with receptors ACE2 and ACE3 and extracts its kinetic and thermodynamic properties.

4. Protein domains in the context of virus–host interactions

The domains of a protein are distinct structures that have specific biological roles, and the majority of proteins are made of numerous domains. For molecular interactions to occur, protein domains are important building blocks, and the domain repertoire of different viruses differs according to their genome's molecular structure, which is the basis for their classification. There are tiny genome viruses in about two-thirds of viral families (no more than 20,000 nt). Most viruses that encode protein domains can interact with

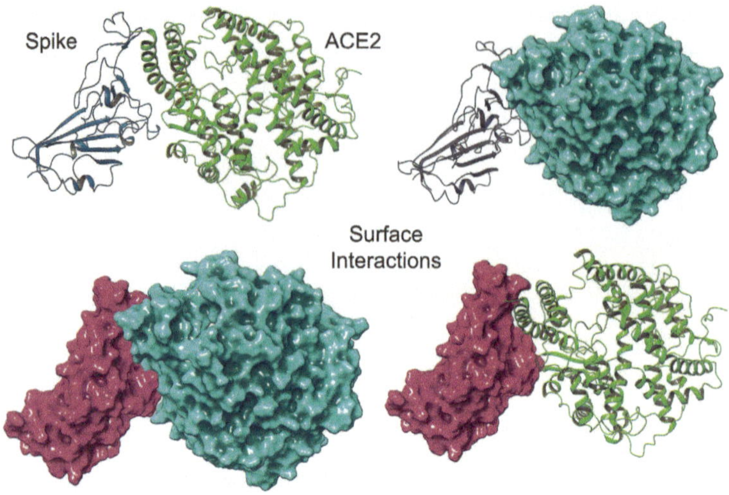

Fig. 2 Surface interactions of RBD in SARS-CoV-2 with ACE2 in human showing the surface level protein–protein interactions.

several host targets and disrupt a lot of biological functions as a direct result of the high level of genome compaction (Vogel, Bashton, Kerrison, Chothia, & Teichmann, 2004). Due to their ability to evolve through convergence and imitating the host-binding sites, viral domains are able to compete with host factors normally involved in other cellular processes by targeting their protein targets. Domain–domain interactions are better understood in RNA viruses because of the encoded domain and the functions that are carried out. "Evolutionary arms race" is a term used to describe how proteins constantly lose and reacquire binding sites in order to escape or maximize interspecific protein–protein interactions in the human–virus relationship (Brito & Pinney, 2017). Host proteins evolve offensively when they alter their ability to bind and neutralize viral components in an effort to maintain their recognition capabilities in that particular location. Virus proteins frequently interact with human immune system proteins in a way that favors viral evolution by attacking the immune system. Virological antagonists, which undergo mutations to prevent viral proteins from binding from their interfaces, target human proteins and study defensive mechanisms in human proteins (Alcami, Ghazal, & Yewdell, 2002). As a result of this continual interaction, fresh mutations can be fixed on viral surfaces, perhaps compensating for host evasion. Human-viral arms race, endogenous and exogenous interactions display distinct evolutionary patterns. The host interfaces facilitate the human-viral protein–protein interactions, which are less varied than those directly targeted by viral proteins. The function and structure of this protein can be dramatically altered by modest modifications of certain residues. This protein's intraspecific binding affinity can also be affected. Co-evolving host–host interactions are particularly susceptible to this type of evolution, as mutations can be detrimental and strong purifying selection occurs to keep the binding sites intact. Although some endogenous binding locations overlap with exogenous interfaces, shared amino acid residues of the endogenous interface can evolve quicker because of competition with viral antagonists (Franzosa & Xia, 2011).

5. Protein interaction networks underlying viral processes

Host cell viruses rely on molecular interactions between host proteins to replicate. These interactions are often represented as networks, in which nodes and edges reflect proteins and the physical interactions that exist between them. Multiple proteins encoded by the viral genomes, both in DNA and RNA, are critical for regulating gene expression, replicating the genome, and transmitting the virus to human cells (Liu et al., 2020). Viral transcription factors (TFS) and other regulators play a critical role in gene expression because of their capacity to influence the expression of both viral and host genes. Virus-encoded proteins (vTRs) can influence gene transcription in two ways: directly and indirectly, by interacting with other proteins and nucleic acids in the host cell itself. Virus-encoded proteins (vTRs) can be divided into two major categories: primary vTRs and secondary vTRs. In the primary vTRs, particular target genes are controlled while in the secondary vTRs, nucleic acid transport and transcriptional regulators play an important role (Xie et al., 2015). Examples of viral families that encode the number of proteins required for both transcriptional control and viral genome replication include adenovirus E2A expression, Herpes simplex virus type 1 UL29, and EB (Epstein–Barr) virus BMRF1. Secondary vTRs from a few viral families, such as herpesvirus and polyomavirus, perform other activities that are more specialized to the DNA repair mechanism (Conway, Zolotukhin, Muzyczka, Hayward, & Byrne, 1997). The secondary vTRs are more abundant in RNA viruses because of the compact architecture of the RNA genome. This results in gene products with a wide range of functions. The E6 and E7 of the vTRs of type 5 human papillomavirus (HPV) have pairwise protein amino acid identity as low as 30%, similar to other types of viral proteins (Graham, 2010).

6. Intraviral interactomes

In the case of phages, only a small percentage of viral proteins interact with each other, which suggests that the other genes are primarily interacting with the human protein. For example, of the 89 EBV proteins, only 27 interact with each other, while 40 interact with host proteins. While herpesvirus-encoded ribonucleotide reductases, for example, largely interact with substrates, other viral proteins interact with other proteins and also with the host proteins, thus it makes sense to begin interactions with among the viral genome proteins (Mendez-Rios & Uetz, 2010). A variety of virus–host interactions have been documented, and this could indicate that viral proteins are particularly rich in disordered regions, which are known to facilitate interactions with a wide range of other proteins.

7. Different patterns of virus interaction

Viruses tend to evolve quickly, and their diversity may alter how they interact with other organisms. Viral strain variety has been shown in numerous studies to influence the way viruses interact, such as in the case of human papillomaviruses. Using 94 and 88 human target proteins, Neveu et al. (2012) investigated the interactions of two key viral proteins, E6 and E7, from 11 distinct strains. Interestingly, depending on the strain from which they were derived, each viral protein interacts in a different fashion with different proteins (Neveu et al., 2012). For example, the different E6 proteins have varying affinity patterns for their biological targets such as Smad2 and Smad3 by more than 10 times. The tissue specificity of this virus can be explained, however, by the finding that the target proteins have a varied expression profile in different tissues. It was also studied in 2012 by Gulbahçe et al. that examined the influence of the E6 and E7 viral proteins on the expression levels of distinct target cells. Researchers found that the virus and human protein interact with each other, influencing the expression of many target proteins in human cells, which underscores the complexity of combining these two types of proteins in a study (Gulbahce et al., 2012). An evolutionary arms race is taking place in the human–virus interaction system as a result of the constant evolution of human–virus interactions in an effort to avoid or enhance interspecific protein interactions (Guo et al., 2020). Host proteins evolve offensively when they alter their ability to bind and neutralize viral factors in an effort to maintain their recognition capabilities in that particular region. Proteins from the human immune system interact with viral proteins frequently through an offensive strategy. Whereas viral antagonists that undergo mutations to prevent viral proteins from binding to their interface target human proteins, which exhibit defensive mechanisms (Thakur, Mikkelsen, & Jungersen, 2019). Because of this constant interaction, new mutations on viral interfaces may be more likely to be able to stick around, possibly compensating for host evasion. In this complex human-viral arms race, different evolutionary patterns emerge at endogenous and exogenous interfaces (Duggal & Emerman, 2012). There is less variation in human-viral protein interactions compared to interactions between viral proteins and interfaces that are specifically targeted by viruses. Small alterations to the protein's residues can have a large impact on its structure and function (Jubb et al., 2017). Affinity for intraspecific binding of this protein may also be altered as a result of these modifications. In coevolving host–host interfaces, mutations can be deleterious and strong purifying selection acts to maintain the integrity of the binding sites. Although some endogenous binding sites overlap with exogenous interfaces, sharing the amino acid residues of the endogenous interface can evolve faster because of competition with viral antagonists (Pandurangan & Blundell, 2020).

8. Viral protein interaction with human cellular proteins

Based on few literatures, the total of 11 proteins are encoded in the influenza virus in addition to structural proteins and surface glycoproteins. Protein's Hemagglutinin (HA) and Neuraminidase are required for viral survival and replication. It uses Hemagglutinin glycoproteins to adhere to the host cell throughout the infection process. The HA1 and HA2 subdomains are involved in receptor binding and cell fusion, respectively (Kosik & Yewdell, 2019). It is the monomer receptors of hemagglutinin-binding sites that impact the attachment of viral protein to the sialic acid residues of glycoconjugates on human cell surface proteins. One of the most critical components of this process is Neuraminidase, which helps Hemagglutinin recognize and bind to host cells. The viral infection process begins with the fusion of viral proteins with host cells, and the viral genome is then transported to the nucleus of the host cell (Stencel-Baerenwald, Reiss, Reiter, Stehle, & Dermody, 2014). The surface antigens HBsAg of the Hepatitis B virus are critical in initiating the viral infection and generating antibodies against HBsAg to prevent the viral infection. The viral infection and spread are aided by another protein, HBx, which also plays an important role. In hepatocellular carcinomas, HBx proteins are more prevalent than other viral proteins. Several mechanisms and pathways, including p53 suppression, are activated during infection by HBx proteins, which contribute to the development of hepatocarcinogenesis (Tarocchi, Polvani, Marroncini, & Galli, 2014).

9. HIV infection role through proteins

The human immunodeficiency virus (HIV) has a small genome and consequently relies significantly on the cellular machinery of the host to replicate. It is critical to determine which host proteins and complexes physically interact with viral proteins in order to fully comprehend how HIV rewires the host's cellular machinery while infection is taking place (Jager et al., 2011). Viral proteins interact heavily with host immune cells during HIV infection and attack the host immune system's saboteurs to overcome the immunological response. In particular, it attacks CD4+ T cells and other macrophages as well as dendritic cells. The HIV genome contains two copies of ssRNAs encased in a conical p24 protein capsid (Vidya Vijayan, Karthigeyan, Tripathi, & Hanna, 2017). The virus genome is encoded by nine genes that are part of the molecular machinery that allows the virus to spread within the host cell. Additionally, the capsid contains the viral replication and progression enzymes reverse transcriptase, integrase, and protease. The glycoproteins GP120 and GP41 in the virus envelope play an important role in virus structure construction. The attachment of virus particles to target cells, the first stage in HIV infection, is made possible by a variety of non-specific processes. Host proteins interacting with integrin $\alpha4\beta7$ bring the viral envelope protein into proximity to the host cell. A heterodimeric complex of glycoproteins GP41 and GP120 enhanced viral binding to CD4 protein through the interaction of cell surface proteins, where GP120 play an important role in host immune response evasion (Checkley, Luttge, & Freed, 2011).

10. SARS-CoV-2 infection role through proteins

The ongoing outbreak of coronavirus disease 2019 (COVID-19) pandemic, caused by severe acute respiratory syndrome coronavirus 2 (SARS-CoV-2), infecting the human lower respiratory tract and progress into an acute respiratory distress syndrome, which was reported first on

12 December 2019 in Wuhan, Hubei province, China. As of March 2021, there are > 114 million reported cases of SARS-CoV-2 infection and >2.53 million deaths worldwide by the World Health Organization (WHO). The novel coronavirus (2019-nCoV) or human coronavirus 2019 (hCoV-2019) was renamed as SARS-CoV-2 because of its similarity (about 79%) with previously identified SARS-CoV from the family Coronaviridae (Selvaraj et al., 2021). The SARS-CoV-2 is also a zoonotic virus, the whole genome sequence analysis has shown only 4% variation in comparison with SARS-related coronavirus (SARSr-CoV); RaTG13 from a bat (*Rhinolophus affinis*), explains its emergence. Pangolin-CoV from Malayan pangolins (*Manis javanica*), scaly anteater is the second closest CoV with only 9% variability, a potential intermediate host has also shown to be a natural host several other SARS-CoV-2-related viruses (Selvaraj et al., 2020). The virus–host interactome, a collection of physical interactions between viral and cellular proteins, plays a significant role in this process. To gain a better understanding of viral replication cycles, scientists have employed PPI mapping to identify the molecular modules that viruses use to replicate (Enjuanes, Almazan, Sola, & Zuniga, 2006). Understanding how viruses are identified by the immune system, but also escape immune defense through the evolution of countermeasures, is facilitated by virus–host interactomics. Genomic sequence and viral protein structures are currently regarded essential pieces of information for determining the molecular characteristics of a virus. According to Gorden et al. (2020), human cells were used to express 26 of the 29 SARS-CoV-2 proteins, and they used affinity-purification mass spectrometry to identify the human proteins that physically associated with each of the SARS-CoV-2 proteins, resulting in 332 high-confidence protein–protein interactions (Gordon et al., 2020). There are open reading frames coding for structural and non-structural proteins in the viral RNA genome of SARS-CoV-2. These include spike (S), nucleocapsid (N), membrane (M), and envelope (E) proteins. There are two subunits in the SARS-CoV-2 protein S1 and S2. With its N-terminal domain (NTD) and C-terminal receptor binding domain, the S1 subunit is able to recognize and bind to its target molecules (RBD). The RBD is responsible for SARS-CoV-2 infectivity because it binds to the angiotensin-converting enzyme 2 (ACE2) receptor and facilitates viral entry (Duan et al., 2020). Membrane fusion is promoted by the S2 subunit of S protein, while ACE2-mediated viral attachment is enhanced by the S1 subunit. According to the World Health Organization (WHO), SARS-CoV-2 VOCs (Variant of Concern) have a greater transmissibility and a wider range of symptoms (Harvey et al., 2021). Due to changes in

the S protein, these variants have a greater ability to survive and reproduce. SARS-CoV-2 infects a host cell, attaches through the mode by spike proteins, and begins replicating its own RNA once inside. SARS-CoV-2 mutates on a regular basis as a result of the virus's rapid replication over time in a large population, which allows for more mutations (Tang et al., 2022). It is more likely that the virus will develop a mutation that is beneficial to its survival as it spreads, transmits, and reproduces. On the viral surface, the SARS-CoV-2 spike protein performs a function and is the first point of contact between the virus and its host. SARS-CoV-2 Alpha, Beta, Gamma, Delta, and Omicron variants have arisen as a result of this mutation in the S protein (Cohen et al., 2022; Shah, Firmal, Alam, Ganguly, & Chattopadhyay, 2020).

SARS-CoV-2 proteins have recently been shown to target a wide range of human proteins, and researchers have identified the most important human proteins in the PPI network. Researchers working on SARS-CoV-2 have discovered a protein–protein interaction (PPI) between the virus and the host protein, and they believe that this information could be useful in developing antiviral drugs and predicting the virus behavior. Those interactions are provided as narrative in the Fig. 3 shows the

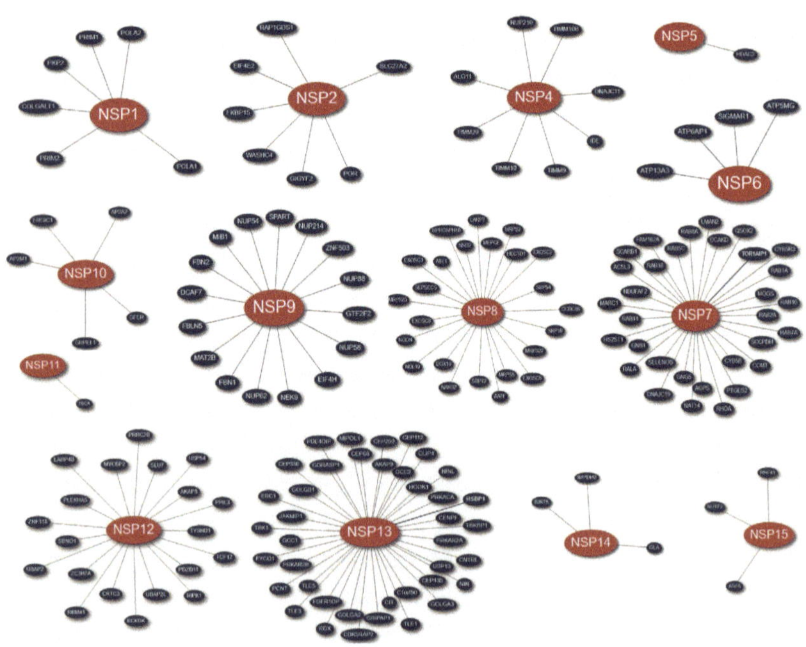

Fig. 3 Protein–protein interactions between the SARS-CoV-2 and Human proteins showing the viral hijacking mechanism.

interaction pattern between the viral protein's interactions with multiple host proteins. The interactions show the non-structural proteins tend to show the interactions with different proteins in human host for invading the host mechanism. Researchers have also found that, most of the receptor proteins from these interactions are belongs to cancer drug targets.

11. Conclusion

Protein–protein interactions (PPIs) are usually represented as networks (protein interaction networks, PINs), with proteins as nodes and their interactions as edges. Host proteins that are more connected to viral proteins are more likely to interact. As part of viral coevolution with their hosts, many viruses have evolved mechanisms to inhibit or alter apoptotic response. Virus and cellular proteins compete for binding throughout a viral infection. Intraspecific interactions such as viral–viral and host–host interactions are well-known for their ability to target and suppress endogenous interfaces. It is important to understand both the viral and host protein structures and repertoires, evolutionary mechanisms, and relevant biological facts when analyzing virus–host protein interactions. They are fascinating research models due to their uniqueness in viral domain repertoires, evolution, and infection behavior. An overview of viral-human protein–protein interactions, binding mode, hijacking mechanism, and recognition pattern can be found in this chapter.

Acknowledgment

The authors C.S. and S.K.S. thankfully acknowledge the Tamil Nadu State Council for Higher Education (TANSCHE) for the research grant (Au/S.o. (P&D): TANSCHE Projects: 117/2021). The author would like to thank Deanship of Scientific Research at Majmaah University for supporting this work.

Reference

Abu-Farha, M., Thanaraj, T. A., Qaddoumi, M. G., Hashem, A., Abubaker, J., & Al-Mulla, F. (2020). The role of lipid metabolism in COVID-19 virus infection and as a drug target. *International Journal of Molecular Sciences*, *21*(10). https://doi.org/10.3390/ijms21103544.

Alcami, A., Ghazal, P., & Yewdell, J. W. (2002). Viruses in control of the immune system. Workshop on molecular mechanisms of immune modulation: Lessons from viruses. *EMBO Reports*, *3*(10), 927–932. https://doi.org/10.1093/embo-reports/kvf200.

Brito, A. F., & Pinney, J. W. (2017). Protein-protein interactions in virus-host systems. *Frontiers in Microbiology*, *8*, 1557. https://doi.org/10.3389/fmicb.2017.01557.

Checkley, M. A., Luttge, B. G., & Freed, E. O. (2011). HIV-1 envelope glycoprotein biosynthesis, trafficking, and incorporation. *Journal of Molecular Biology*, *410*(4), 582–608. https://doi.org/10.1016/j.jmb.2011.04.042.

Cohen, A. A., van Doremalen, N., Greaney, A. J., Andersen, H., Sharma, A., Starr, T. N., et al. (2022). Mosaic RBD nanoparticles protect against multiple sarbecovirus challenges in animal models. *bioRxiv*. https://doi.org/10.1101/2022.03.25.485875.

Conway, J. E., Zolotukhin, S., Muzyczka, N., Hayward, G. S., & Byrne, B. J. (1997). Recombinant adeno-associated virus type 2 replication and packaging is entirely supported by a herpes simplex virus type 1 amplicon expressing rep and cap. *Journal of Virology*, *71*(11), 8780–8789. https://doi.org/10.1128/JVI.71.11.8780-8789.1997.

Cook, H. V., Doncheva, N. T., Szklarczyk, D., von Mering, C., & Jensen, L. J. (2018). Viruses.STRING: A virus-host protein-protein interaction database. *Viruses*, *10*(10). https://doi.org/10.3390/v10100519.

Cui, G., Fang, C., & Han, K. (2012). Prediction of protein-protein interactions between viruses and human by an SVM model. *BMC Bioinformatics*, *13*(Suppl 7), S5. https://doi.org/10.1186/1471-2105-13-S7-S5.

Duan, L., Zheng, Q., Zhang, H., Niu, Y., Lou, Y., & Wang, H. (2020). The SARS-CoV-2 spike glycoprotein biosynthesis, structure, function, and antigenicity: Implications for the Design of Spike-Based Vaccine Immunogens. *Frontiers in Immunology*, *11*, 576622. https://doi.org/10.3389/fimmu.2020.576622.

Duggal, N. K., & Emerman, M. (2012). Evolutionary conflicts between viruses and restriction factors shape immunity. *Nature Reviews. Immunology*, *12*(10), 687–695. https://doi.org/10.1038/nri3295.

Eisenreich, W., Rudel, T., Heesemann, J., & Goebel, W. (2019). How viral and intracellular bacterial pathogens reprogram the metabolism of host cells to allow their intracellular replication. *Frontiers in Cellular and Infection Microbiology*, *9*, 42. https://doi.org/10.3389/fcimb.2019.00042.

Enjuanes, L., Almazan, F., Sola, I., & Zuniga, S. (2006). Biochemical aspects of coronavirus replication and virus-host interaction. *Annual Review of Microbiology*, *60*, 211–230. https://doi.org/10.1146/annurev.micro.60.080805.142157.

Franzosa, E. A., & Xia, Y. (2011). Structural principles within the human-virus protein-protein interaction network. *Proceedings of the National Academy of Sciences of the United States of America*, *108*(26), 10538–10543. https://doi.org/10.1073/pnas.1101440108.

Goodacre, N., Devkota, P., Bae, E., Wuchty, S., & Uetz, P. (2020). Protein-protein interactions of human viruses. *Seminars in Cell & Developmental Biology*, *99*, 31–39. https://doi.org/10.1016/j.semcdb.2018.07.018.

Gordon, D. E., Jang, G. M., Bouhaddou, M., Xu, J., Obernier, K., White, K. M., et al. (2020). A SARS-CoV-2 protein interaction map reveals targets for drug repurposing. *Nature*, *583*(7816), 459–468. https://doi.org/10.1038/s41586-020-2286-9.

Graham, S. V. (2010). Human papillomavirus: Gene expression, regulation and prospects for novel diagnostic methods and antiviral therapies. *Future Microbiology*, *5*(10), 1493–1506. https://doi.org/10.2217/fmb.10.107.

Gulbahce, N., Yan, H., Dricot, A., Padi, M., Byrdsong, D., Franchi, R., et al. (2012). Viral perturbations of host networks reflect disease etiology. *PLoS Computational Biology*, *8*(6), e1002531. https://doi.org/10.1371/journal.pcbi.1002531.

Guo, H., Hu, B. J., Yang, X. L., Zeng, L. P., Li, B., Ouyang, S., et al. (2020). Evolutionary arms race between virus and host drives genetic diversity in bat severe acute respiratory syndrome-related coronavirus spike genes. *Journal of Virology*, *94*(20). https://doi.org/10.1128/JVI.00902-20.

Harvey, W. T., Carabelli, A. M., Jackson, B., Gupta, R. K., Thomson, E. C., Harrison, E. M., et al. (2021). SARS-CoV-2 variants, spike mutations and immune escape. *Nature Reviews. Microbiology*, *19*(7), 409–424. https://doi.org/10.1038/s41579-021-00573-0.

Jager, S., Cimermancic, P., Gulbahce, N., Johnson, J. R., McGovern, K. E., Clarke, S. C., et al. (2011). Global landscape of HIV-human protein complexes. *Nature*, *481*(7381), 365–370. https://doi.org/10.1038/nature10719.

Jubb, H. C., Pandurangan, A. P., Turner, M. A., Ochoa-Montano, B., Blundell, T. L., & Ascher, D. B. (2017). Mutations at protein-protein interfaces: Small changes over big surfaces have large impacts on human health. *Progress in Biophysics and Molecular Biology*, *128*, 3–13. https://doi.org/10.1016/j.pbiomolbio.2016.10.002.

Kosik, I., & Yewdell, J. W. (2019). Influenza hemagglutinin and neuraminidase: Yin(−) Yang proteins coevolving to thwart immunity. *Viruses*, *11*(4). https://doi.org/10.3390/v11040346.

Liu, X., Hong, T., Parameswaran, S., Ernst, K., Marazzi, I., Weirauch, M. T., et al. (2020). Human virus transcriptional regulators. *Cell*, *182*(1), 24–37. https://doi.org/10.1016/j.cell.2020.06.023.

Maginnis, M. S. (2018). Virus-receptor interactions: The key to cellular invasion. *Journal of Molecular Biology*, *430*(17), 2590–2611. https://doi.org/10.1016/j.jmb.2018.06.024.

Mendez-Rios, J., & Uetz, P. (2010). Global approaches to study protein-protein interactions among viruses and hosts. *Future Microbiology*, *5*(2), 289–301. https://doi.org/10.2217/fmb.10.7.

Neveu, G., Cassonnet, P., Vidalain, P. O., Rolloy, C., Mendoza, J., Jones, L., et al. (2012). Comparative analysis of virus-host interactomes with a mammalian high-throughput protein complementation assay based on Gaussia princeps luciferase. *Methods*, *58*(4), 349–359. https://doi.org/10.1016/j.ymeth.2012.07.029.

Pandurangan, A. P., & Blundell, T. L. (2020). Prediction of impacts of mutations on protein structure and interactions: SDM, a statistical approach, and mCSM, using machine learning. *Protein Science*, *29*(1), 247–257. https://doi.org/10.1002/pro.3774.

Parrish, C. R., Holmes, E. C., Morens, D. M., Park, E. C., Burke, D. S., Calisher, C. H., et al. (2008). Cross-species virus transmission and the emergence of new epidemic diseases. *Microbiology and Molecular Biology Reviews*, *72*(3), 457–470. https://doi.org/10.1128/MMBR.00004-08.

Selvaraj, C., Dinesh, D. C., Krafcikova, P., Boura, E., Aarthy, M., Pravin, M. A., et al. (2021). Structural understanding of SARS-CoV-2 drug targets, active site contour map analysis and COVID-19 therapeutics. *Current Molecular Pharmacology*. https://doi.org/10.2174/1874467214666210906125959.

Selvaraj, C., Panwar, U., Dinesh, D. C., Boura, E., Singh, P., Dubey, V. K., et al. (2020). Microsecond MD simulation and multiple-conformation virtual screening to identify potential anti-COVID-19 inhibitors against SARS-CoV-2 Main protease. *Frontiers in Chemistry*, *8*, 595273. https://doi.org/10.3389/fchem.2020.595273.

Shah, V. K., Firmal, P., Alam, A., Ganguly, D., & Chattopadhyay, S. (2020). Overview of immune response during SARS-CoV-2 infection: Lessons from the past. *Frontiers in Immunology*, *11*, 1949. https://doi.org/10.3389/fimmu.2020.01949.

Stencel-Baerenwald, J. E., Reiss, K., Reiter, D. M., Stehle, T., & Dermody, T. S. (2014). The sweet spot: Defining virus-sialic acid interactions. *Nature Reviews. Microbiology*, *12*(11), 739–749. https://doi.org/10.1038/nrmicro3346.

Tang, H., Gao, L., Wu, Z., Meng, F., Zhao, X., Shao, Y., et al. (2022). Multiple SARS-CoV-2 variants exhibit variable target cell infectivity and ability to evade antibody neutralization. *Frontiers in Immunology*, *13*, 836232. https://doi.org/10.3389/fimmu.2022.836232.

Tarocchi, M., Polvani, S., Marroncini, G., & Galli, A. (2014). Molecular mechanism of hepatitis B virus-induced hepatocarcinogenesis. *World Journal of Gastroenterology*, *20*(33), 11630–11640. https://doi.org/10.3748/wjg.v20.i33.11630.

Thakur, A., Mikkelsen, H., & Jungersen, G. (2019). Intracellular pathogens: Host immunity and microbial persistence strategies. *Journal of Immunology Research*, *2019*, 1356540. https://doi.org/10.1155/2019/1356540.

Thyrsted, J., & Holm, C. K. (2021). Virus-induced metabolic reprogramming and innate sensing hereof by the infected host. *Current Opinion in Biotechnology*, *68*, 44–50. https://doi.org/10.1016/j.copbio.2020.10.004.

Valkov, E., Sharpe, T., Marsh, M., Greive, S., & Hyvonen, M. (2012). Targeting protein-protein interactions and fragment-based drug discovery. *Topics in Current Chemistry*, *317*, 145–179. https://doi.org/10.1007/128_2011_265.

Vidya Vijayan, K. K., Karthigeyan, K. P., Tripathi, S. P., & Hanna, L. E. (2017). Pathophysiology of CD4+ T-cell depletion in HIV-1 and HIV-2 infections. *Frontiers in Immunology*, *8*, 580. https://doi.org/10.3389/fimmu.2017.00580.

Vogel, C., Bashton, M., Kerrison, N. D., Chothia, C., & Teichmann, S. A. (2004). Structure, function and evolution of multidomain proteins. *Current Opinion in Structural Biology*, *14*(2), 208–216. https://doi.org/10.1016/j.sbi.2004.03.011.

Wright, P. E., & Dyson, H. J. (2015). Intrinsically disordered proteins in cellular signalling and regulation. *Nature Reviews. Molecular Cell Biology*, *16*(1), 18–29. https://doi.org/10.1038/nrm3920.

Wu, S., Tian, C., Liu, P., Guo, D., Zheng, W., Huang, X., et al. (2021). Effects of SARS-CoV-2 mutations on protein structures and intraviral protein-protein interactions. *Journal of Medical Virology*, *93*(4), 2132–2140. https://doi.org/10.1002/jmv.26597.

Xie, D., Han, L., Luo, Y., Liu, Y., He, S., Bai, H., et al. (2015). Exploring the associations of host genes for viral infection revealed by genome-wide RNAi and virus-host protein interactions. *Molecular BioSystems*, *11*(9), 2511–2519. https://doi.org/10.1039/c5mb00309a.

Xing, S., Wallmeroth, N., Berendzen, K. W., & Grefen, C. (2016). Techniques for the analysis of protein-protein interactions in vivo. *Plant Physiology*, *171*(2), 727–758. https://doi.org/10.1104/pp.16.00470.

Yang, S., Fu, C., Lian, X., Dong, X., & Zhang, Z. (2019). Understanding human-virus protein-protein interactions using a human protein complex-based analysis framework. *mSystems*, *4*(2). https://doi.org/10.1128/mSystems.00303-18.

Zhu, J. D., Meng, W., Wang, X. J., & Wang, H. C. (2015). Broad-spectrum antiviral agents. *Frontiers in Microbiology*, *6*, 517. https://doi.org/10.3389/fmicb.2015.00517.

CHAPTER NINE

Glycoprotein molecular dynamics analysis: SARS-CoV-2 spike glycoprotein case study

João Victor Paccini Coutinho[a,†], Janaina Macedo-da-Silva[a,†], Simon Ngao Mule[a], Thales Kronenberger[b,c,d,e], Livia Rosa-Fernandes[a], Carsten Wrenger[a], and Giuseppe Palmisano[a,f,*]

[a]Department of Parasitology, Institute of Biomedical Sciences, University of São Paulo, São Paulo, Brazil
[b]Department of Internal Medicine VIII, University Hospital Tuebingen, Tuebingen, Germany
[c]Department of Pharmaceutical and Medicinal Chemistry, Institute of Pharmaceutical Sciences, Eberhard-Karls-Universität, Tuebingen, Germany
[d]Cluster of Excellence iFIT (EXC 2180) "Image-Guided and Functionally Instructed Tumor Therapies", University of Tuebingen, Tuebingen, Germany
[e]Tuebingen Center for Academic Drug Discovery & Development (TüCAD2), Tuebingen, Germany
[f]Faculty of Science and engineering, Macquarie University, Sydney, NSW, Australia
*Corresponding author: e-mail addresses: palmisano.gp@gmail.com; palmisano.gp@usp.br

Contents

Abstract

Molecular Dynamics (MD) is a method used to calculate the movement of atoms and molecules broadly applied to several aspects of science. It involves computational simulation, which makes it, at first glance, not easily accessible. The rise of several

† These authors contributed equally to the work.

ISSN 1876-1623
https://doi.org/10.1016/bs.apcsb.2022.05.004

automated tools to perform molecular simulations has allowed researchers to navigate through the various steps of MD. This enables to elucidate structural properties of proteins that could not be analyzed otherwise, such as the impact of glycosylation. Glycosylation dictates the physicochemical and biological properties of a protein modulating its solubility, stability, resistance to proteolysis, interaction partners, enzymatic activity, binding and recognition. Given the high conformational and compositional diversity of the glycan chains, assessing their influence on the protein structure is challenging using conventional analytical techniques. In this manuscript, we present a step-by-step workflow to build and perform MD analysis of glycoproteins focusing on the SPIKE glycoprotein of SARS-CoV-2 to appraise the impact of glycans in structure stabilization and antibody occlusion.

Abbreviations

Glycoproteins	Molecular dynamics
SARS-CoV-2	Protein structure
Glycans	Antigen-Antibody interaction

1. Introduction

Severe acute respiratory syndrome coronavirus 2 (SARS-CoV-2) is a highly pathogenic virus and the agent responsible for the pandemic that began in late 2019 (Baloch, Baloch, Zheng, & Pei, 2020), which has already led to the death of more than 6 million people in the world and 434 million cases (WHO Coronavirus COVID-19 Dashboard, n.d.). Like other viruses in the *Coronovidae* family, SARS-CoV-2 also encodes the SPIKE (S) glycoprotein, which has an important role in infection, including virus entry into the target cells (Ou et al., 2020). This glycoprotein is also essential in the host's immune evasion, representing the main target of therapeutic interventions (Duan et al., 2020). Multiple SARS-CoV-2 variants of concern arose since the beginning of the pandemic, some of which initiated a new wave of infections. These variants include delta (Planas et al., 2021), omicron (Wolter et al., 2022), P1 (Naveca et al., 2021), among others (Hacisuleyman et al., 2021) and have mutations mainly in the S protein, which facilitates successful infection with higher transmission rate. The most relevant mutations in protein S are D614G, present in all variants of Concern (VOC) and it is responsible for accumulating higher viral load and infectivity (Volz et al., 2021). On the other hand, the N501Y mutation promotes new stabilizing hydrogen bonds between the receptor binding domain (RBD) and ACE-2 receptor, with ~10 times higher binding affinity (Liu, Wei, et al., 2021; Liu, Zhang, et al., 2021). The presence of HV69-70del is required to induce this effect. The absence of this deletion does not allow immune evasion in B.1.1.7 strains (Meng et al., 2021). The L452R, T478K, and P681H/R mutations promote an increase in infectivity, possibly due to an increase in the positive electrostatic potential of the surface and greater steric impediments (Pascarella et al., 2021; Tian, Sun, Zhou, & Ye, 2022). In contrast, the E484K mutation promotes less antibody neutralization and greater escape from the immune system (Lippi, Mattiuzzi, & Henry, 2022; Liu, Wei, et al., 2021).

Another additional layer of structural complexity comes from the conformational changes induced by post-translational modifications (PTMs). Protein glycosylation is a PTM that can have profound effects on the structure and function of a glycoprotein (Macedo-da-Silva, Santiago, Rosa-Fernandes, Marinho, & Palmisano, 2021; Reily, Stewart, Renfrow, & Novak, 2019; Schjoldager, Narimatsu, Joshi, & Clausen, 2020). N-linked glycosylation (N-X-S/T motif, where X is any amino acid residue except proline) is characterized by the addition of oligosaccharides of variable size and complexity to the side chains of asparagine, whereas in O-linked glycosylation, sugar addition occurs at the serine and threonine residues (Aebi, 2013; Hanisch, 2001). The glycan chains can impact in different ways on the protein proprieties, such as stabilizing loops and conformations (Wormald & Dwek, 1999), providing occlusion against antibodies (Ab) of a host cell (Watanabe et al., 2020), promoting correct folding of the protein (Dalziel, Crispin, Scanlan, Zitzmann, & Dwek, 2014), or enhancing lectin-based cell-adhesion (Lasky, 1991). The glycan composition and occupancy of glycosites can vary significantly. This, combined with the high flexibility of glycans, makes a definition of a glycosylated structural model very challenging (Lis & Sharon, 1993; Nagae & Yamaguchi, 2012). Both N- and O-glycosylation represents a key process in viral proteins in multiple aspects of their pathobiology. Since viruses have no metabolic pathways to perform this modification by themselves, they hijack the host-cell glycosylation machinery to their benefit. Glycosylation has been shown to control the infection of several viruses (Sugrue, 2007; Vigerust & Shepherd, 2007). The role of O-linked glycosylation during viral infection is poorly understood. There are reports in the literature that indicate CD99 binding to herpes simplex virus type 1 B glycoprotein

(Wang et al., 2009) and the role of the glycan shield and occlusive neutralization during viral infection. Moreover, O-linked glycan epitopes have be shown to be highly immunogenic in Gammaherpesvirus (Machiels et al., 2011). On the other hand, N-glycan alterations during viral infections have been extensively described, from host cell entry, particle release, and immune evasion as seen in the SPIKE protein of Coronavirus, influenza HA, Ebola GP, and Lassa-virus GPC (Hastie et al., 2017; Kwon et al., 2015; Lee et al., 2014; Mohan, Li, Ye, Compans, & Yang, 2012; Walls et al., 2016; Watanabe, Bowden, Wilson, & Crispin, 2019; Zhao et al., 2016). Due to that, viral glycoproteins represent the major components of the viral envelope and have important roles associated to host cell recognition, attachment, infection, invasion, transmission and immune evasion (Banerjee & Mukhopadhyay, 2016; Cook & Lee, 2013; Watanabe et al., 2019).

Currently, mass-spectrometry-based approaches are the primary choice for the analytical characterization of both site and structure-specific glycosylation (Oliveira, Thaysen-Andersen, Packer, & Kolarich, 2021; Pasing, Sickmann, & Lewandrowski, 2012). This enables a more accurate protein modeling in its biological context. In general, the enrichment of glycopeptides from complex samples is performed by applying strategies based on TiO_2 (Larsen, Jensen, Jakobsen, & Heegaard, 2007; Palmisano et al., 2010), lectin affinity chromatography (Yang & Hancock, 2004), and hydrophilic interaction liquid chromatography (HILIC) (Ongay, Boichenko, Govorukhina, & Bischoff, 2012), followed by glycan release or analysis of intact glycopeptides (Parker, Gupta, Cordwell, Larsen, & Palmisano, 2011; Thaysen-Andersen & Packer, 2014; Thaysen-Andersen, Packer, & Schulz, 2016). For N-linked, the enzyme N-glycosidase F (PNGase F) is usually used, which cleaves between GlcNAc and asparagine residues of N-linked glycoproteins and glycopeptides (Kaji et al., 2003). The release of O-linked occurs mainly by reductive alkaline β-elimination (Wilkinson & Saldova, 2020). The isolated portion of N/O-glycans can be derivatized by permethylation (Kang, Mechref, & Novotny, 2008), and then analyzed by mass spectrometry techniques, as well the glycopeptides, which includes matrix-assisted laser desorption/ionization time-of-flight (MALDI-TOF-MS) (Reiding, Blank, Kuijper, Deelder, & Wuhrer, 2014) or by electrospray ionization (ESI-MS/MS) (Morelle & Michalski, 2007; Palmisano, Larsen, Packer, & Thaysen-Andersen, 2013; Palmisano, Melo-Braga, Engholm-Keller, Parker, & Larsen, 2012). Ion mobility mass spectrometry allows access to the native glycocode giving information on the monosaccharide composition and glycosidic linkages (Schindler et al., 2017). Moreover, novel software solutions have been applied for a comprehensive characterization of thousands of intact glycopeptides in different biological matrices (Alocci et al., 2019; Campbell, 2017; Chernykh, Kawahara, & Thaysen-Andersen, 2021; Kawahara et al., 2021). Native MS allows to transmit large protein ions such as intact glycoprotein complexes which allow the determination of subunits stoichiometry, ligand binding and infer glycoprotein function (Struwe & Robinson, 2019; Wu & Robinson, 2022). Recently, a new method, termed limited deglycosylation assay, was introduced to probe 3D conformational changes of glycoproteins on a proteome-wide scale using limited amount of sample (Mule et al., 2021). This method takes advantage of the accessibility-dependent PNGaseF cleavage to quantitative study changes in glycoprotein conformational changes. X-ray crystallography and cryoEM have limitation in solving the detailed structure of intact glycoproteins due to stoichiometry, heterogeneity and flexibility of the glycan chains. This is known as the glycosylation problem (Chang et al., 2007; Davis & Crispin, 2010). Another approach used in the structural study of glycoproteins is nuclear magnetic resonance (NMR) spectroscopy, a versatile, quantitative, and non-destructive analytical technique (Unione, Ardá, Jiménez-Barbero, & Millet, 2021). NMR provides absolute quantification information for both glycan and glycoprotein. However, the application of the technique has limitations that include the need for large amounts of samples, glycoprotein size and access to adequate labeling (Valverde, Quintana, Santos, Ardá, & Jiménez-Barbero, 2019). NMR overcomes one of the main challenges encountered in glycoproteomics, which consists of differentiating the stereochemistry of a glycan, with isobaric species such as glucose (Glc), galactose (Gal) and mannose (Man) in a single analysis (Hargett et al., 2021). Taken together, the analytical toolbox available to the scientific community allows structural characterization of glycoproteins and correlation with their functions.

The complexity of glycans has always been a challenge to be overcome in molecular modeling, mainly due to the great variety of monomer units that can be organized into stereo and regiospecific-linked oligosaccharides within each glycosylation site. Initial techniques applied to carbohydrate modeling, allowed the elucidation of glycans' conformational and spatial preferences (Lemieux & Koto, 1974; Peters, Meyer, Stuike-Prill, Somorjai, & Brisson, 1993). Subsequently, the development of force fields capable of refining carbohydrate modeling during MD analysis changed the scope of oligosaccharide chain analysis in these simulations, taking into account their fluctuating nature and their interactions with solvent molecules (Woods, 2018). Development of reliable force-field parameters from experimental data suffers from the lack of training data, given the shortage of 3D structures of branched carbohydrates. These groups are often not resolved or removed from the initial crystal structures, due to considerable thermal fluctuations and flexibility. Among the computational tools to perform MD analysis of glycans and evaluate the force fields, there are AMBER (Pearlman et al., 1995), GLYCAM (Kirschner et al., 2008) and CHARMM (Vanommeslaeghe et al., 2010). MD simulations can reveal several features about the mechanisms and biological functions of glycoproteins and the effects that glycans have on them, as well as evaluate optimal conformations for binding lectins (Cheong, Shim, Kang, & Kim, 1999) or elucidate protein stability in a given conformation (Casalino et al., 2020).

In view of the great importance of the SPIKE glycoprotein in the success of SARS-CoV-2 infection, understanding the structure and impact of adding glycans to this protein can provide important insights to elucidate the mechanisms of viral infection, and highlight possible therapeutic targets. Based on this, we present a step-by-step bioinformatics pipeline to perform the MD of the SARS-CoV-2 SPIKE protein, including carbohydrate structures, solvent, and antibody accessibility. The analysis of the energy states of the performed simulation indicated that it correctly followed the parameters entered and allowed to evaluate and quantify the differential protection of the glycan shield in residues exposed to the solvent. In addition, it was also possible to identify a wide variety of protein-glycan hydrogen bonds and evaluate their interactions within the structure. MD of glycoproteins presents some challenges due to the increase in atoms and in the waterbox size, together with the extension and mobility of the glycans. The MD simulation steps reported here can be applied to other glycoproteins to evaluate various effects that glycosylation can impose on structure and accessibility with potential implications on the biological activities.

2. Methods

2.1 MD simulation workflow applied to SARS-CoV-2 SPIKE glycoprotein

A step-by-step bioinformatic pipeline to perform MD analysis is shown in Fig. 1. This workflow was applied to a structural model of the SPIKE glycoprotein published by Woo et al., 2020. The models were based on

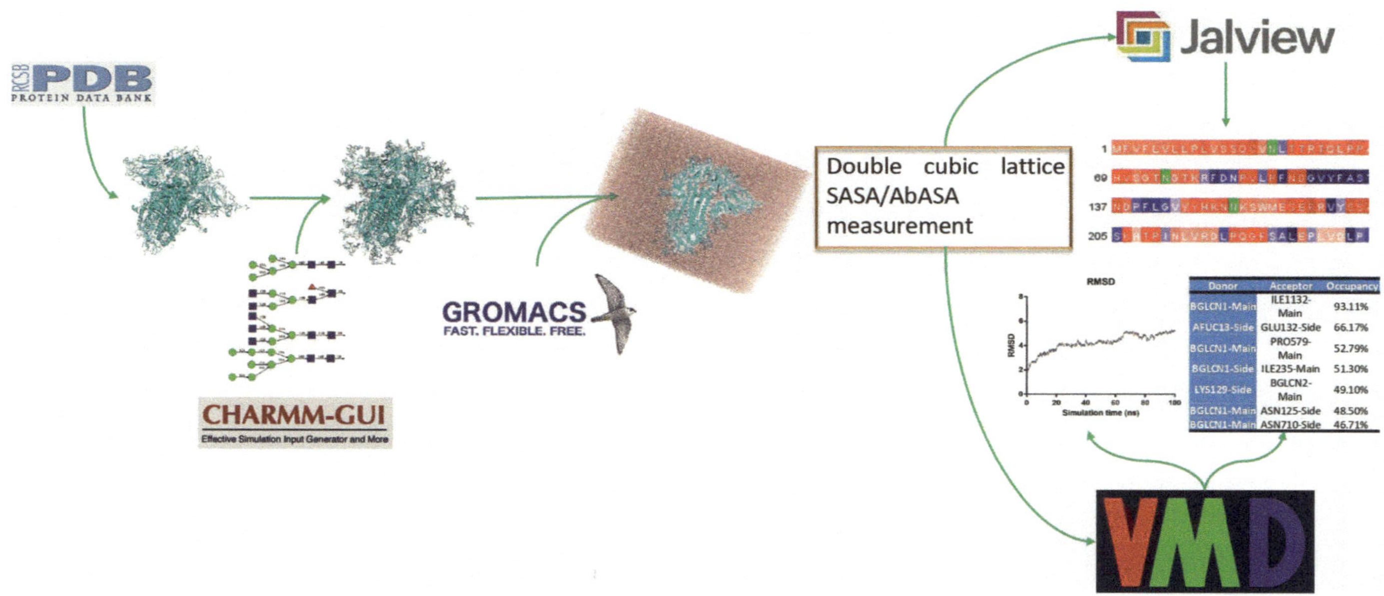

Fig. 1 Computational workflow and software tools used to perform molecular dynamics and evaluate residue specific exposure to solvent and antibody (Ab). The protein of interest structure is downloaded from PDB, in-silico glycosylated and processed to obtain MD simulation using CHARMM-GUI tool. Then, GROMACS is used to run both the simulation (*e.g.*, 100 ns as used in this study) and perform the solvent accessible surface area and average antibody accessible surface area (SASA/AbASA) calculations. VMD software is used to visualize trajectories and RMSD (root mean square deviation) and RMSF (root mean square fluctuation) calculations and Jalview to visualize the calculated proprieties of the protein.

the PDB-6vxx (https://doi.org/10.2210/pdb6vxx/pdb) crystallographic structure and presented a structural resolution of 2.80 Å (Walls et al., 2020). The chosen structure presents a range of modeled residues, which covers most of the head portion of SPIKE, disregarding the membrane and intracellular domains, not considered for this workflow. The detailed structural characterization of the glycosylation sites and glycan occupancy was based on Watanabe et al. (Watanabe et al., 2020; Woo et al., 2020). Mass spectrometric data were used to determine the most frequent glycan structures in each glycosite. A structural model of the protein of interest and site-specific glycan structures are needed to perform MD simulations as described below.

In general, it is advisable to use high-resolution structures whenever available. When selecting a structure for simulations, it is important to analyze the coverage of the structural motifs including the conserved ones, as well as to compare the conformation of individual amino-acids from the catalytic site. The information regarding catalytic/relevant residues could be found in the literature and obtained using site-directed mutagenesis, or from the conserved known motifs of the protein class of interest. The use of multiple structures as starting points as well as multiple simulation replicas as an ensemble is highly advised.

2.2 Computational tools to model glycoproteins

There are multiple computational platforms to model a glycoprotein, such as GLYCAM, Amber, Glycosilator, among others. For this work CHARMM-GUI was used (Jo et al., 2017; Jo, Kim, Iyer, & Im, 2008; Park et al., 2019), due to its web-based intuitive interface for the addition of glycans, an integrated tool for modeling missing residues belonging to unmodelled/low confidence regions, and MD model preparation. It is relevant to mention that full missing loops can be treated by *ab initio* or homology models, despite this being out of the scope of this manuscript (Rodriguez, Chinea, Lopez, Pons, & Vriend, 1998).

The CHARMM-GUI tool can be used to model a large array of biologically relevant structures, such as lipids, glycans, glycoproteins, lipopolysaccharide, solutions and others. For the modeling of glycoproteins, the "Glycan Reader and Modeler" was used, Fig. 2A (Jo, Song, Desaire, MacKerell, & Im, 2011; Park et al., 2019, 2017). PDB models without the oligosaccharide chains can be directly downloaded from the Research Collaboratory for Structural Bioinformatics (RCSB), but can also be uploaded by the user. A relevant note should be made regarding

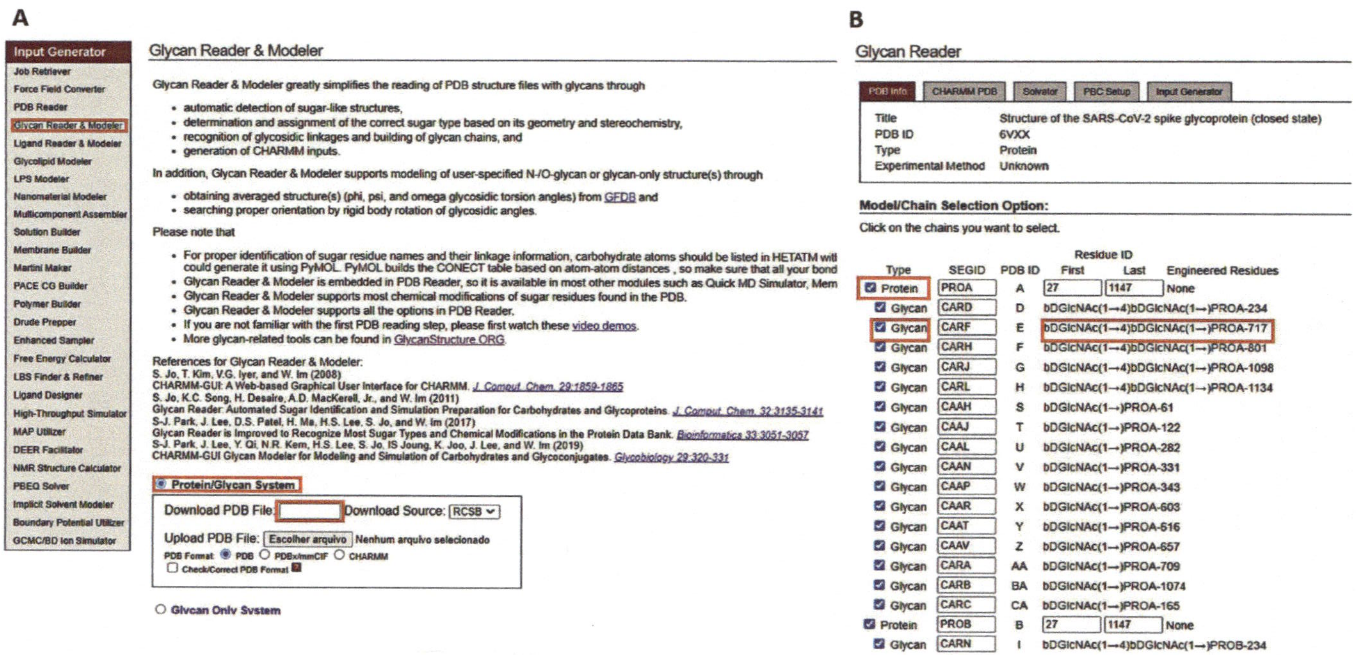

Fig. 2 Defining the model to build the glycoprotein of interest. (A) Depiction of the CHARMM-GUI "Glycan Reader and Modeler" tool used to build glycoproteins. The PDB protein model code is inputted in the "Download PDB File" field on the "Protein/Glycan System." (B) The tool will proceed to download and identify protein chains, connected glycans, ligands, experimental method for structure acquisition and name. The user is able to select all the elements of the model of interest before proceeding for glycoprotein modeling. The glycans code will be displayed by the glycosylated residue site and may be edited on further steps.

uploading glycosylated models. The glycans should be listed as hetero atoms (HETATM) with a CONECT table indicating the bonds among the glycans and the ASN side chain for N-linked glycoproteins. These changes can be added to the PDB by using PyMol software to build a new .pdb file with the coordinates (Fig. 2A).

Once the model is uploaded, the software will detect the protein sequence, chains, missing residues, disulfide bridges, connected glycans, and ligands, all of which might be edited/added/removed in the "Glycan reader" step by deselecting the marked boxes (Fig. 2B).

Upon advancing to the next section, the user will be prompted to edit/fix the protein. Any missing residues and disulfide bridges can be edited before MD simulations and the CHARMM-GUI can perform this modeling in an integrated manner (Fig. 3A and B). The glycans can also be edited in the glycosylation section, where a pop-up menu will bring an editable glycan builder, in which the user can input the appropriate glycans of interest by adding the sugars with the correct linkage (Fig. 3C). With the disulfide bridges added, missing residues modeled and glycans inserted, the glycosylated model is ready to be generated. The CHARMM-GUI will build the structure with the requested properties and prompt the user to the solvation tool, to build a waterbox for the solvent and ions of the protein.

The standard parameters found in the tool will generate a waterbox to match the protein size with an additional 10 Å to the edge and place a physiological concentration of KCl using the Monte-Carlo method, to correct for electrostatic interactions (Fig. 4A).

Then, after setting the periodic boundary conditions (Fig. 4B) and selecting the "force field" CHARMM36m (Huang et al., 2016) to be used for topology generation and inserting standard parameters for the equilibration of the system, the CHARMM-GUI will generate input files for several MD simulation programs as described below (Fig. 4C). The selection of the forcefield of choice can vary on the simulation specificities, but CHARMM36m was chosen due to its improved accuracy for backbone conformational ensembles, besides improved H-bond J coupling. Selecting the GROMACS box as "input generation option," Fig. 4C, will add a folder with all the different files needed for GROMACS MD simulations:

- MD ready model, with glycans, ions and periodic boundaries;
- Topology files;
- Molecular dynamics parameter files (.mdp) for minimization, equilibration and a 100 ns md run.

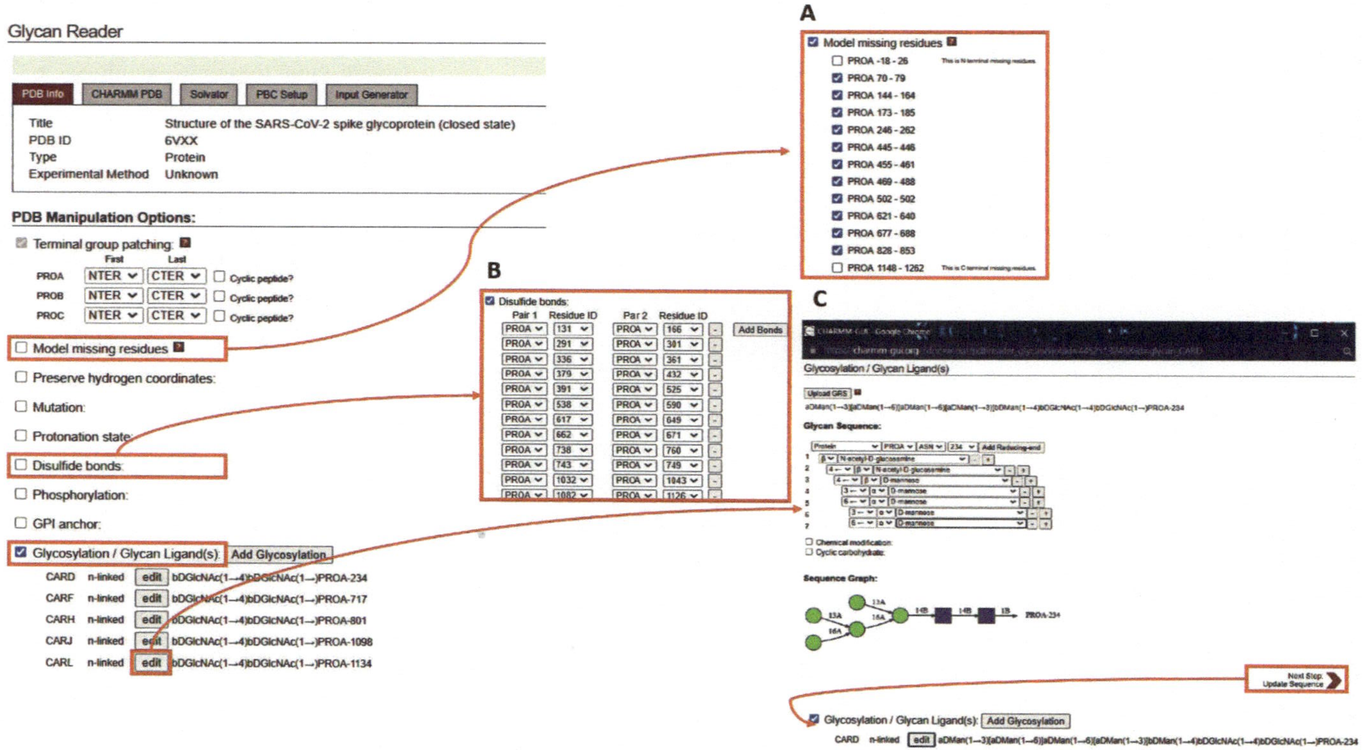

Fig. 3 Editing steps of CHARMM-GUI Glycan reader. The information and selected chains of the downloaded model are displayed and available to be edited at this moment. Preceding the simulation, multiple factors must be accounted for the glycoprotein during its modeling, such as modeling missing residues from low resolution regions of the model, for which the user can use the built-in tool to mark missing residues to model (A), checking for the presence of all structural disulfide bridges and adding missing ones (B), adding explicit hydrogens in the model and, finally, adding and editing glycans (C). Upon selecting a glycan to edit, a pop-up window emerges, allowing for precise glycan editing parameters, including sugar monomer selection and glycosidic bond position. Upon updating the sequence, the glycan code will be updated to reflect the new glycan incorporated in the model.

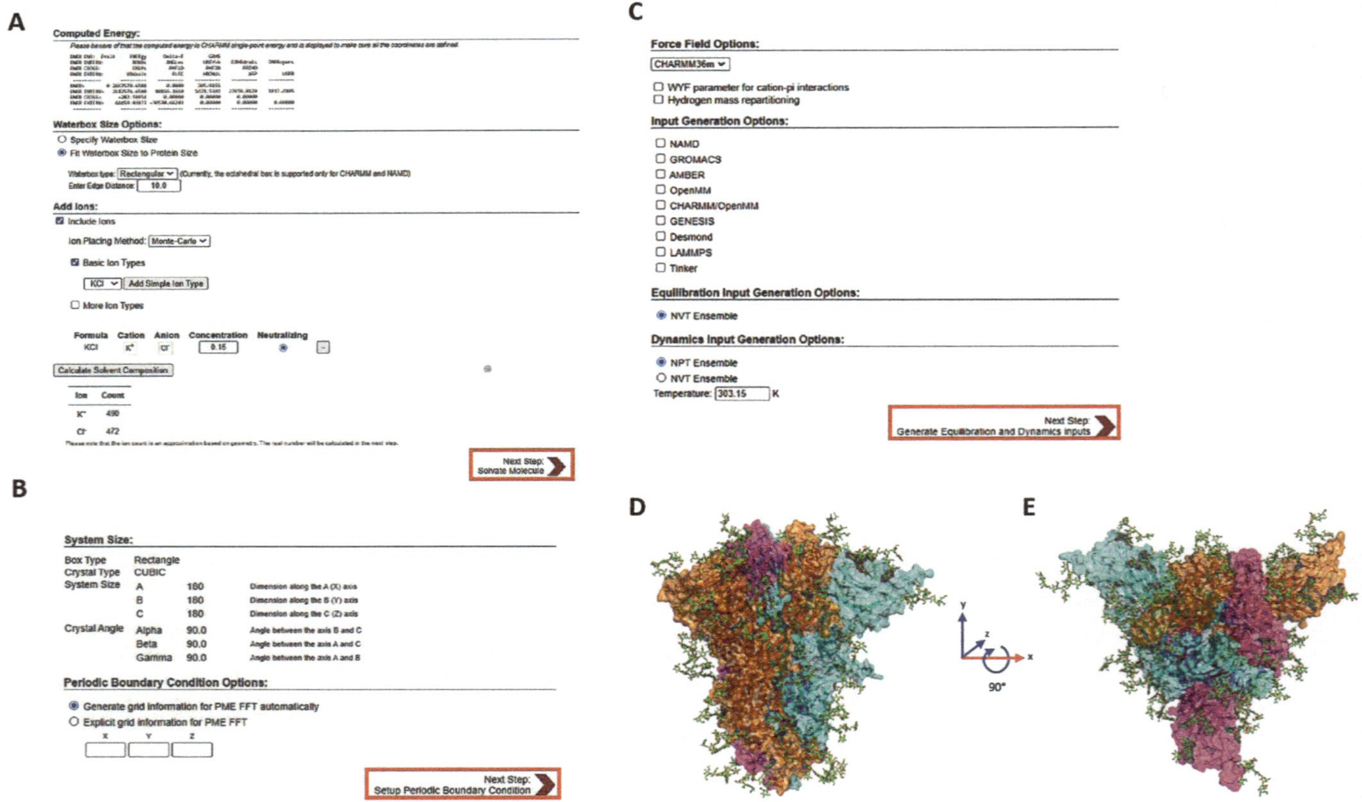

Fig. 4 See figure legend on opposite page.

The final model for MD simulation can be visualized using Pymol or other molecular visualization tools such as Chimera, Coot or VMD. The file is located inside the GROMACS folder named "step3_input.gro" (Fig. 4D and E).

2.3 MD simulation parameters

There are several MD programs available for use, such as AMBER, CHARMM and LAMMPS. In this guide, we will present the use of GROMACS software (Bauer, Hess, & Lindahl, 2022a, 2022b; Van Der Spoel et al., 2005) for our simulations. The .mdp output from CHARMM-GUI can be used as input for the simulation.

In ordsser to proceed with the MD simulation, the generated model needs to be compiled with all the topologies, restrains, run parameters, output options and pressure/temperature coupling algorithms parameters in a singular file, named the .tpr file. This compilation is done by the `gmx grommp` command as follow:

-f: Molecular dynamics parameters file (.mdp)
-o: name of the output portable binary run input file, necessary for the MD run (.tpr)
-c: structure file, the generated glycoprotein model (.gro; .pdb; .tpr; etc)
-p: topology file (.top)
-n: index file, used to set/create a new set of atoms or just using the GROMACS default groups.ndx

Fig. 4 Solvation and molecular dynamics input generation. After editing the structure for MD preparation, a solvent box with a physiological concentration of ions is added to the protein to both neutralize and add ionic strength to the system (A). The waterbox size, ion position method, salt type and concentration parameters can be selected in this section. In this step is critical to make sure that the waterbox encompass the whole protein and the glycans. The Periodic boundary conditions are set to define the "system size" during the simulation (B). Lastly, CHARMM-GUI generates the topology files using the selected "Force field" and generates inputs for an array of commonly used molecular dynamics programs, including parameter files for both equilibration and dynamics (C). The output structure of glycosylated protein with the solvent and ions occluded is presented as side view (D) and top view (E). Each chain of the protein is represented in either cyan, magenta or orange (water molecules were occluded to facilitate the visualization).

It is important to mention that the files: .mdp, .gro/.pdb, .top and others are obtained from the CHARMM-GUI output. The .mdp file contains detailed information on the parameters for MD that can be run in the default mode or adapted to best fit the simulation needs.

All of the .mdp options and their respective inputs can be checked in the GROMACS documentation (https://manual.gromacs.org/documentation/2018/user-guide/mdp-options.html). A full discussion is outside the scope of this article, so we will focus on selected parameters/options to be tailored to the simulation needs. The parameters can be divided in five categories: (a) preprocessing and run options; (b) output control; (c) neighbor search; (d) bonds and constrains and e) pressure and temperature coupling.

(a) Molecular dynamics preprocessing and run options

The preprocessing and run parameters define directories to include in the topology beyond the .itp files inside the .top file (using the -include input) and add position restrictions for user-defined atoms (using the -define input). The run options include the `integrator` algorithms for running the simulation. Among the options available for the `integrator` there are: the `md` algorithm, that uses the leap-frog method for integrating equations of motion, or the `steep` algorithm, for energy minimization using the steepest descent, which would also depend of a `emtol` input for tolerance of energy variation before declaring the system minimized. Lastly, the `nsteps` and the `dt` parameters can be set specifying the number of molecular dynamic steps to be taken in the simulation and the time step for integration in picoseconds.

(b) Molecular dynamics output files control options

The output parameters must be set in order to specify the frequency GROMACS will write information in the trajectory and log files. These parameters are set as an integral number for the number of steps that elapse in the simulation before writing: `nstlog`—energies to the log file; `nstvout`—velocities to trajectory file; `nstxcout`—coordinates to trajectory; `nstfout`—forces to trajectory; `nstcalcenergy` is used to set the number of steps between calculating energies. The `nstcalcenergy` is relevant only with `integrator` set to dynamics simulation algorithms and must be a multiple of `nstenergy`, a parameter for writing energies to the energy file.

(c) Neighbor searching, electrostatics and interatomic interactions

The interatomic interactions are accounted during the simulations by determining the interacting atomic vicinity using a list of user defined distances and other parameters. They account for both short and long distances interactions.

These parameters can be used as default or modified according to the user needs. For example, the `nstlist` is the frequency of update of the neighbor list; `rlist` sets the cutoff for short-range interactions (in nm); `rvdw` distance set to apply to van der waals interactions; `Cutoff-scheme` sets the algorithm to search for the short range interactions list, based on the `rvdw` and `rlist` values set; `vdwtype` set the algorithm for van der waals interactions search using both `rlist` and `rvdw`; `rvdw-switch` and `vdw-modifier` work in tandem to apply a smooth transition between short and vdw interations, based on the algorithm chosen for vdw modifier and the distance set in `rvdw-switch`, which must be smaller than the one set for `rvdw`. Lastly, `colombtype` sets the algorithm for electrostatics interactions and `rcoulomb` sets the distance for the Coulomb electrostatic interactions cut-off.

(d) Setting up bonds and constrains options

This class of parameters set constrains in otherwise variable components of the simulation to facilitate the simulation processing. The constrain parameters can encompass either only the bonds or bonds and angles, and can be set to affect all (`all-bonds;all-angles`), only bonds with H-atoms (`h-bonds;h-angles`) or none of them (`none`). The `constraint _algorithm` must also be selected between the `LINCS` (Hess, Bekker, Berendsen, & Fraaije, 1997) or `SHAKE` algorithm.

(e) Applying temperature and pressure coupling algorithms

These parameters define the algorithms, using both `tcoupl` and `pcoupl` to temperature and pressure respectively, and other variables for temperature and pressure equilibration during the simulation. It is important to define both `tau-t` and `tau-p` for both couplings to set the time constant for coupling (in picoseconds) and the target temperature (`ref_t` in Kelvin) and pressure (`ref_p` in bar) for the system. For this manuscript focusing on the SARS-CoV-2 SPIKE protein, the algorithms used were the `nose-hoover` (Evans & Holian, 1998) thermostat and the `Parrinello-Rahman` (Parrinello & Rahman, 1980) barostat for temperature and pressure, respectively. The selection and use of compatible ensembles for adequate monitoring of these variables is of high importance to take in account during the simulations (Hollingsworth & Dror, 2018; Hünenberger, 2005) but the usage of these ensembles is already validated and successfully used in past studies (Casarotto et al., 2021; Lupala, Ye, Chen, Su, & Liu, 2022; Mehdipour & Hummer, 2021).

Glycosides linked to proteins are often on the solvent exposed surface and are known to be involved in the formation of water mediated hydrogen bonds (Stanca-Kaposta et al., 2008; Tachibana et al., 2004). Since the water-box generation often creates a vacuum space between solute (protein) and solvent, it is relevant to allow for additional equilibration with restraints on the glycoprotein structure. This allows the water molecules in the vicinity of the carbohydrate-protein interface to relax and reconstitute the solvation shell, therefore reducing many artificial conformational changes in the glycosides.

2.4 Molecular dynamics simulation run

With the .tpr file generated by the `gmx grommp` command for the simulation, the command `gmx mdrun` can be used to generate the .xtc file.

This command requires:

-deffnm: set the default file name for all file output options
-nt: total number of threads to start (make sure to never start more threads them your hardware as of number of cores)
-s: portable xdr input file outputted by grompp (.tpr)
-pin: tries to generate thread affinities to optimize run time of the simulation
-dlb: optimizes the ratios of threads distributed between long and short distances interactions calculations
-cpi: checkpoint file for appending to an interrupted run (.cpt)
-append: append to a previously available output file when starting from a checkpoint file

The parameters in the .mpd files must be correctly set to perform 4 different steps:

- Minimization: a step performed to minimize the energy of the system, promoting solvent interactions and structure relaxation to represent better a biological context. The critical parameters to be set in this step is that the `integrator` for the simulation must be set to `steep`, the h-bonds can be used as constrains and no temperature and pressure coupling must be implemented in this step.
- Temperature equilibration: `integrator` set to `md` and apply the `tcoupl` using the `nose-hoover` ensemble to simulate the system in a specific temperature and modulate the forces in effect in the simulation taking in account the temperature effect on them. `Tau_t` was set to update every 1 ps.

- Pressure equilibration: same parameters set for the temperature equilibration, with addition of `pcoupl=Parrinello-Rahman` to encompass the pressure effect in the system. `tau_p` was set to update every 2 ps.
- Production: upon completing the equilibration step, the system may be released of all constrains and allowed to run for a defined simulation time (in this manuscript 100 ns were used for this step), using a longer time step for integration of 2 fs.

Using this segmented relaxation of the system before production is of high relevance, as adding multiple variables in the step could lead to artifactual trajectories of protein dynamics (Walton & VanVliet, 2006).

2.5 Processing of trajectories and molecular dynamics outputs

The production step generates the .xtc and .trr files. These files must be preprocessed before the analysis. Indeed, during the simulation, the protein could present atoms jumping across the periodic boundary conditions (PBC) and that could present itself as artifacts to calculate solvent accessibility and simulation RMSD/RMSF, among other analyses. To correct this artifact, the PBC must be recentered in the simulation cell.

To perform this correction, it is needed to generate an index using it to envelop all the sugars and protein chains in the same group, so the common center of mass can be centered in the PBC later on. That requires the extraction of the first frame of the trajectory, which is extracted by using the `gmx trjconv` command:

```
-f: trajectory file (.xtc)
-s: structure file (.gro)
-o: output file name
-dump 0 To extract the first frame of the simulation
-sep to isolate the initial frame of the production step
```

Then, to build the index for all the glycans in the protein, a group was created using `gmx make_ndx -f firstframe.gro` and selecting all the glycans added to the structure during the model building.

To recenter the glycoprotein in the trajectory file in the simulation box, the `gmx trjconv` command is used twice during this step.

1st: removing all jumps of bonded atoms across the PBC, making the glycoprotein as a whole but allowing atoms to diffuse out of the PBC.

```
-pdc nojump
-n: generated .ndx file with the protein and sugars grouped together
```

2nd: set the whole glycoprotein center of mass to the center of the PBC using the nojump.tpr file and the group of glycans and protein described in the step before.

```
-pbc mol
-center
-n: generated .ndx file with the protein and sugars grouped together
```

The trajectory file with the PBC recentered in the glycoprotein can be used to check if the simulation variables agree with the input parameters. To perform this, the command `gmx energy` was used together with the .edr file as input to plot:

- Potential energy: in all steps of the simulation to check if the energy minimization was successful and if the energy in the system kept itself regular through the equilibrations and productions.
- Temperature: in the equilibrations and production steps to check if the system was correctly kept at the inputted temperature of `ref_t`.
- Pressure and density: in the pressure equilibration and production steps to check it the system was accurately equilibrated at the inputted pressure of `ref_t` and if the density of the system followed that trend.

The output plots for the energy, temperature, pressure and density for the SPIKE glycoproteins analyzed in this manuscript are reported in Fig. 5.

3. Trajectory file analysis (RMSD/RMSF/H-bridges)

The recentered trajectory file (.xtc or .trr) was analyzed using the VMD v.1.9.4. (Humphrey, Dalke, & Schulten, 1996) to perform trajectory visualization and other analyses to evaluate the behavior and interactions of the different segments of the glycoproteins along the simulation.

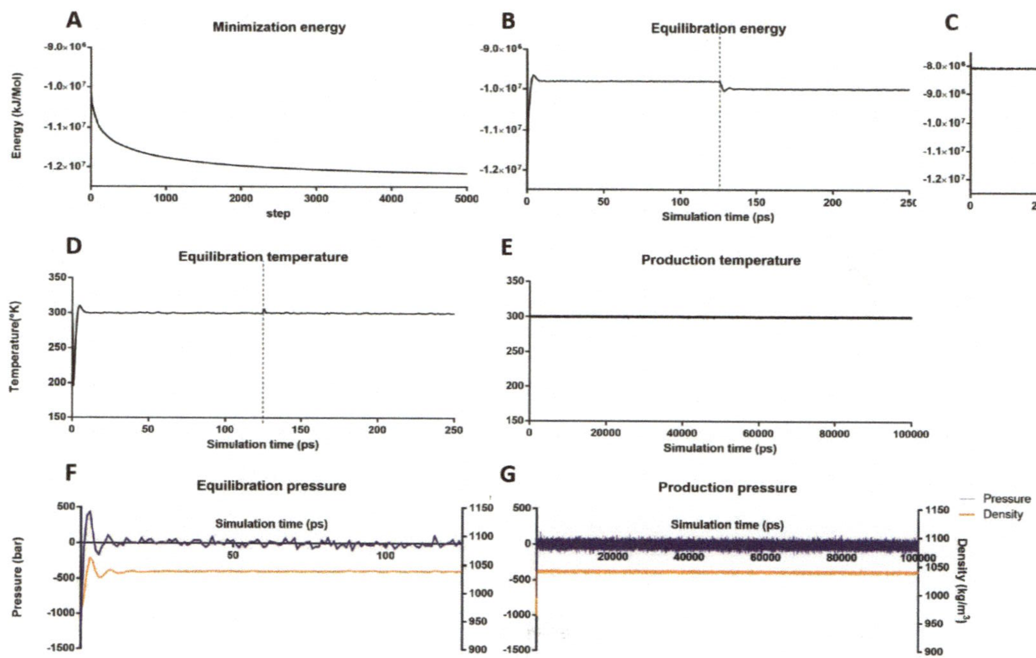

Fig. 5 output of the `gmx energy` for energy minimization (A), constant number volume and temperature (NVT) and constant number pressure and temperature (NPT) equilibration (first and second part of B, respectively), and production steps (C), for temperature during the NVT and NPT equilibration (D) and production steps (E), and pressure and density for the NPT (F) and production steps (G).

To load the trajectory file into the VMD software, a "new molecule" must be created and add the trajectory file. The molecule file browser (File > new molecule) is used to open the firstframe.gro and then loading the .xtc file into the molecule (Fig. 6A and B). The VMD will start loading frame by frame into the molecule. To improve the observation of the trajectory (representation > graphics…), we chose to hide the waters and ions of the representation, change the depiction of the protein to NewCartoon, and apply a 5 frames trajectory smoothing window to observe the trajectory (Fig. 6C–J).

To perform the calculations of RMSF/RMSD and search for H-bonds we used some command lines in the Tk console of VMD (Extensions">"Tk console) as described below.

```
##RMSD/RMSF setup
set start [atomselect top "protein and backbone" frame 0]
set current [atomselect top "protein and backbone"]
set num_steps [molinfo top get numframes]
set carbA [atomselect top "name CA"]
##RMSD
set outfile [open rmsd.dat w]
for {set frame 0} {$frame < $num_steps} {incr frame} {
    $current frame $frame
    set trans_mat [measure fit $current $start]
    $current move $trans_mat
    set rmsd [measure rmsd $current $start ]
  puts $outfile "$frame $rmsd"
}
close $outfile
##RMSF
set outfile [open rmsf.dat w]
set sel [atomselect top "name CA"]
set rmsf [measure rmsf $sel first 0 last [expr
{$num_steps - 1}] step 1]
for {set i 0} {$i < [$sel [expr {$num_steps - 1}]]} {incr i} {
  puts $outfile "[expr {$i+1}] [lindex $rmsf $i]"
}
close $outfile
## H-bonds
package require hbonds
hbonds -sel1 [atomselect top protein] -sel2 [atomselect top
"all not water and not ion and not protein"] -writefile
yes -plot yes
```

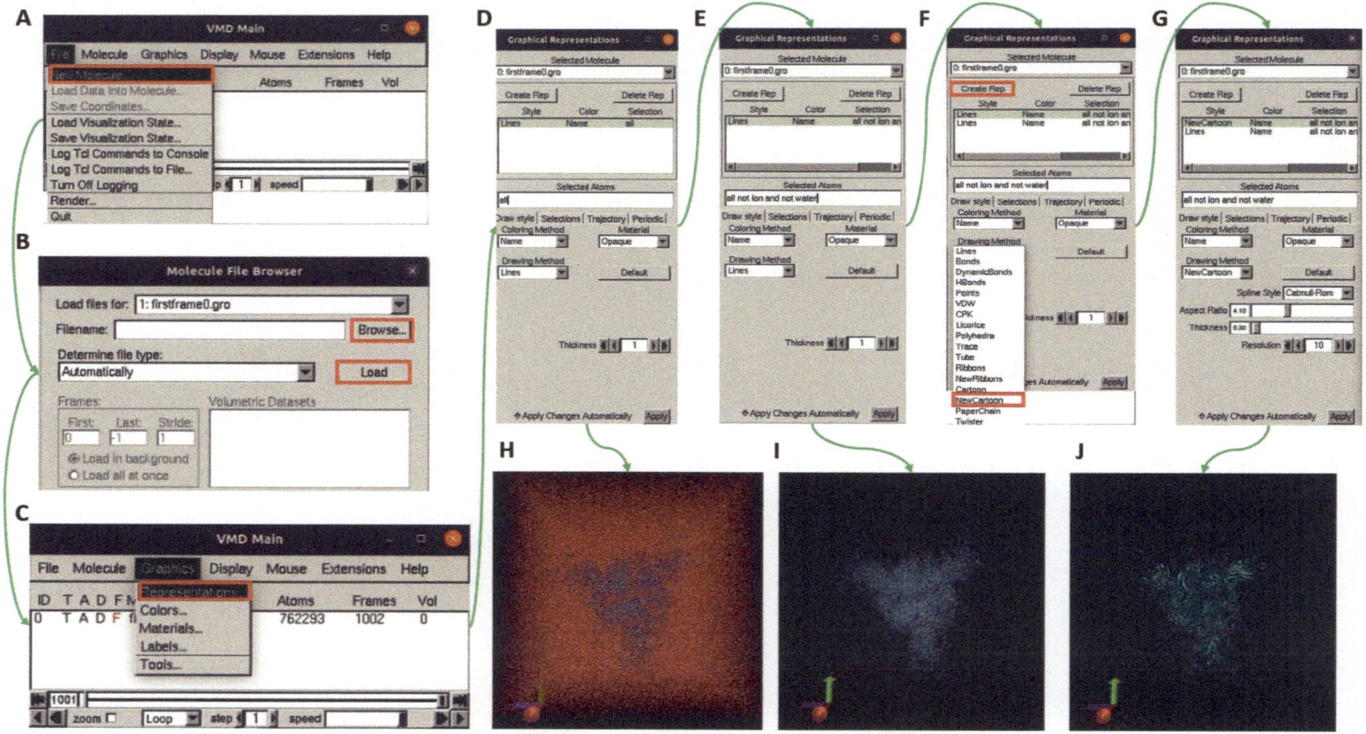

Fig. 6 Analysis and visualization workflow using the VMD software. (A and B) Load the molecule and its trajectory into VMD. In VMD main, go to File - > New molecule (A). Load the first frame of the production simulation, browsing for the .gro file, then load the .xtc file into the molecule (B). (C–G) Define the representation of the molecule. Open the Graphical Representation panel by selecting Graphics → Representations… (C). In this panel (D), change the "selected atoms" panel to represent only the glycoprotein, by excluding the water atoms and the ions from the representation (E). Create a new representation, select only the protein atoms and change the "Drawing method" to "New Cartoon" (F). The final setup in the Graphical Representation panel is showed in (G). (H) All atom representation. (I) glycoprotein representation. (J) Glycoprotein representation using NewCartoon drawing method for the protein.

This will generate 3 output files:

hbonds-details.dat, referring to each residue that interacted with a glycan during the simulation and that interaction occupancy, Fig. 7.

RMSD.txt, which outputs a comparison of a frame of the simulation with the first frame, Fig. 8A;

RMSF.txt, which represents the average fluctuations of each aminoacid in the model during the simulations, Fig. 8B.

4. SASA/AbASA calculations and analysis

To perform the SASA and AbASA measurements, the `gmx sasa` command was used within the GROMACS software (Eisenhaber, Lijnzaad, Argos, Sander, & Scharf, 1995):

```
-f: the recentered trajectory file
-s: the protein coordinates file
-n: the generated index with the whole glycoprotein group (.ndx)
-o: the output file name for total exposed area (.xvg)
-or: the output file name for exposed area for each aminoacid residue (.xvg)
-probe: radius of the solvent in nm (0.14 for water; 0.7 for Ab)
```

The selected probe radius for Ab was based on the estimated size of the hypervariable loop performed by Grant, Montgomery, Ito, & Woods, 2020. The `gmx sasa` command performs a double cubic lattice method for surface area calculation and uses as input the group of interest to have its area measured and which group to be in the results. For the first two iterations of this command, the surface selection should be the whole glycoprotein group defined in the index file, and for the output the group selection that implies only the protein. This will generate a resarea.xvg file with the average GlySASA (for the probe radius of 0.14) and GlyAbASA (for the probe radius of 0.7) for each amino acid residue considering the presence of the oligosaccharides. Following this, a second calculation with the same inputs should be performed, using only the protein for both surface calculation and output writing SASA (for the probe radius of 0.14) and AbASA (for the probe radius of 0.7). This step is performed to evaluate the effect of site-specific glycosylation on solvent and antibody accessibility. Based on the

A

Donor	Acceptor	Occupancy
BGLCN1-Main	ILE1132-Main	93.11%
AFUC13-Side	GLU132-Side	66.17%
BGLCN1-Main	PRO579-Main	52.79%
BGLCN1-Side	ILE235-Main	51.30%
LYS129-Side	BGLCN2-Main	49.10%
BGLCN1-Main	ASN125-Side	48.50%
BGLCN1-Main	ASN710-Side	46.71%
BGLCN2-Main	TYR160-Side	42.22%
BGAL12-Side	GLU484-Side	42.12%
AMAN5-Side	GLU702-Side	40.12%
TYR351-Side	BGLCN2-Main	39.32%
BGLCN1-Main	THR1100-Side	38.22%
AMAN6-Side	GLU169-Side	37.03%
TYR449-Main	ANE5A7-Side	34.53%
GLN895-Side	BGLCN1-Main	34.33%
ALA163-Main	AMAN4-Side	34.23%
AMAN4-Side	ASN164-Side	33.73%
BGLCN5-Side	LEU492-Main	32.24%
TYR1110-Side	AFUC10-Side	29.54%
AMAN5-Side	SER161-Main	29.34%
BMAN3-Side	GLU583-Side	26.65%
THR236-Side	BGLCN1-Side	26.55%
SER494-Main	BGLCN5-Side	25.15%
BGLCN5-Side	GLU918-Side	24.35%
AMAN6-Side	ASP88-Side	22.75%
BGLCN1-Main	GLU281-Side	22.36%
AMAN5-Side	ASP88-Side	22.26%
AFUC9-Side	ASN556-Main	21.66%
BGLCN1-Main	GLN580-Main	21.36%
AMAN4-Side	GLU702-Main	21.06%
TYR160-Side	BGLCN1-Side	20.86%
ASN17-Main	AFUC10-Side	20.56%
BGLCN1-Main	HSD655-Main	20.36%

B

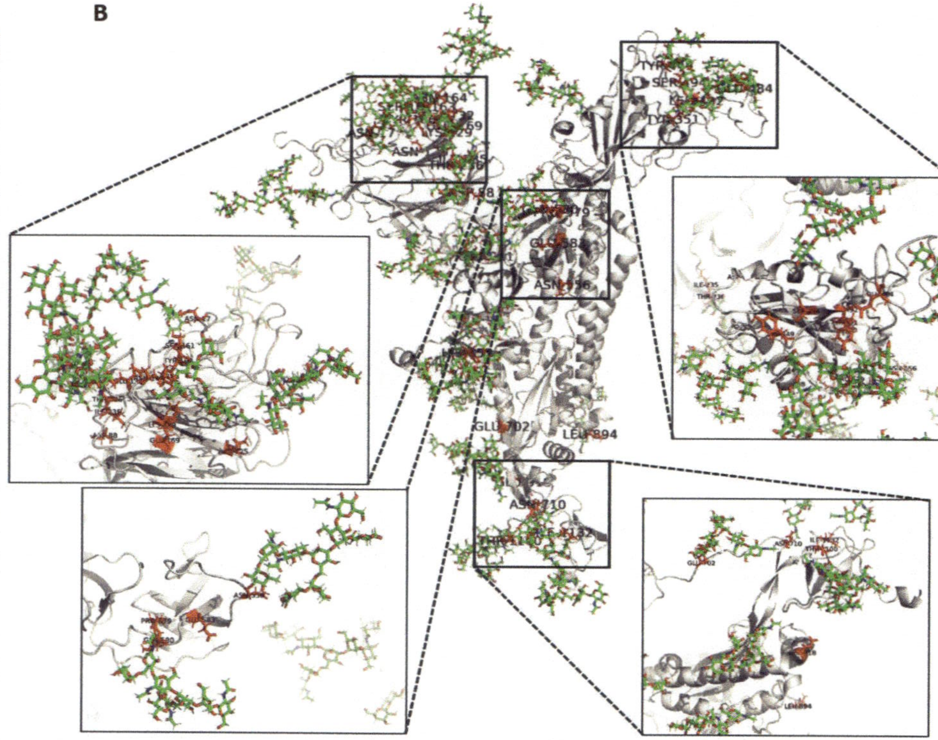

Fig. 7 Analysis of H-bridges content. (A) Relation of hydrogen bonds with above 20% occupancy and involving a glycan chain with a protein chain found in the simulation. BGLCN: *N*-acetyl-D-glucosamine; AFUC: L-fucose; A/BMAN: D-Mannose. (B) Graphical representation of a monomer of the SPIKE protein with its glycans. Aminoacid residues that presented a high occupancy H-bridge with glycans where highlighted in red and labeled using Pymol.

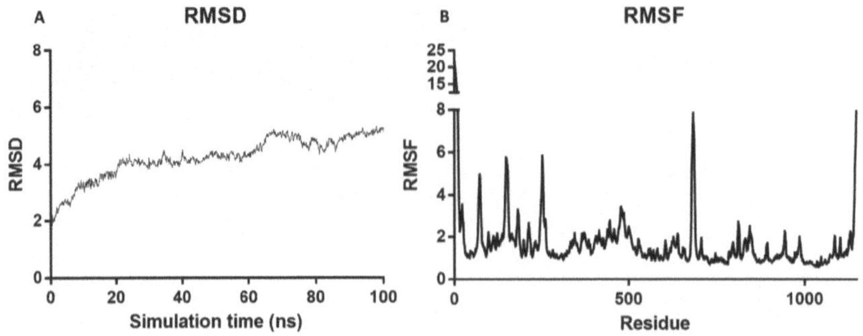

Fig. 8 Calculation of the root mean square deviation (A) during the simulation and root mean square fluctuation (B) per residue.

ratios of GlyAbASA/AbASA for each residue, the glycan-dependent occlusion for the whole protein can be determined for each residue. Also, using the ratios of SASA/Max.SASA for each residue, being Max.SASA defined as the maximum available area per residue calculated by Tien, Meyer, Sydykova, Spielman, and Wilke (2013), it is possible to discern if a residue is buried into the structure (SASA/Max.SASA < 0.15). Comparing the GlyAbASA/AbASA in a specific region against the whole glycoprotein and removing from the analysis residues that are buried based on the SASA/Max.SASA < 0.15, it is possible to calculate which residues are statistically occluded (Fig. 9). Other analysis can be derived from the output trajectories using this methodology, such as salt bridges, that could indicate protein stabilization, or radius of gyration, that is related to protein compactness, but they were not the scope of this project.

There is extensive literature discussing the relevance of the simulation length and sampling, which must be decided based on the hypothesis (Henzler-Wildman & Kern, 2007). Convergence in the carbohydrate structure conformations has been discussed as one of the main issues and, phenomena such as ring puckering were shown to require microsecond length simulations and cannot be addressed by nanoscecond length simulations (Sattelle, Hansen, Gardiner, & Almond, 2010). That being said, there are papers in the literature that successfully reported interesting findings by studying glycoprotein molecular dynamics trajectory files with the nanosecond timescale (Yokoyama et al., 2017; Bernardi, Kirschner, & Faller, 2017). In the latter timescale, it is not possible to explore all the glycan conformations that could be populated in the system; however, glycan-protein interactions can still be assessed by this methodology.

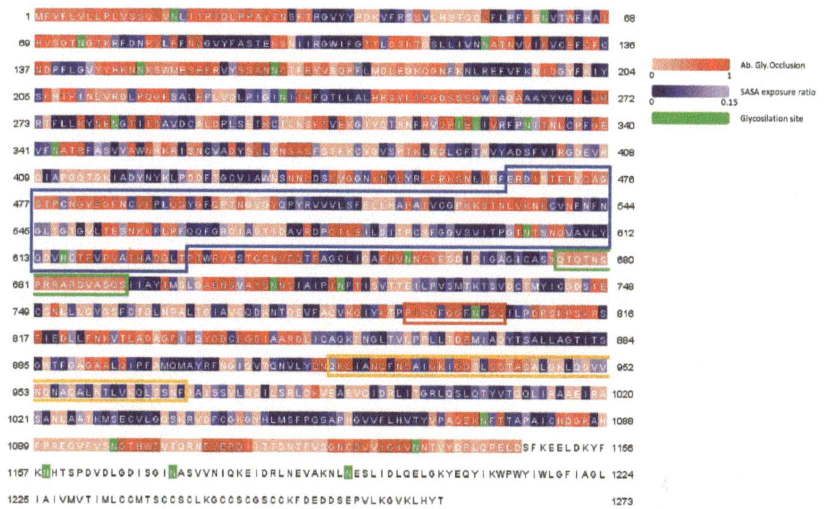

Fig. 9 Representation of Ab accessibility for each aminoacid residue. Residues in red indicate antibody occlusion dependent on the glycan chains. Residues in blue indicate solvent exposure in the protein without considering the glycan chains. Residues in green indicate asparagine with N-linked glycosylation. Blue box: RBD (465–630). Green box: Furin cleavage loop (675–692). Red box: Modulated region (792–804) with a significant shift toward higher glycan occlusion ($P = 0.0003$). Orange box: Heptad repeat 1 (920–970).

5. Results and discussion

The applied methodology was already performed in other MD simulations (Barros et al., 2021; Liguori, Croce, Marrink, & Thallmair, 2020; Mehdipour & Hummer, 2021; Turoňová et al., 2020). The analysis of the simulation energy files (Fig. 4) showed that the simulation correctly followed the defined parameters and the plot in JalView 2.11.1.7 (Waterhouse, Procter, Martin, Clamp, & Barton, 2009) allowed to evaluate and quantify the differential glycan shield protection in solvent-exposed residues (Fig. 9), such as the modulated regions in 792–804. We were also able to identify a large array of protein-glycan hydrogen bonds and assess their interactions inside the structure (Figs. 7 and 9). This methodology can be applied to other glycoproteins to evaluate a large array of effects that the glycosylation might impose to the structure and accessibility of the protein in question, such as evaluating glycan promoted structure stability, glycan occlusion from solvent or antibody and ligand interaction.

6. Conclusions

The growing interest in protein glycosylation and structure to function relationship calls for reliable analytical and computational tools. MD simulation of glycoproteins can help in elucidating the role of oligosaccharides on protein structure, antibody binding, receptor interaction and function. In infectious diseases, glycoproteins can serve as chemotherapeutic targets and biomarkers giving a deeper understanding on the biological mechanisms of host-pathogen interaction. In this step-by-step guide, we provide a detailed view of the different computational platforms that can be used to perform MD simulation of glycoproteins focusing on the SARS-CoV-2 SPIKE glycoprotein. We believe that this guide will serve as a platform to start MD simulations of glycoproteins in several infectious diseases and help in improving the tools currently available.

Acknowledgments

We are grateful for the financial support provided by the São Paulo Research Foundation (FAPESP, grants processes no 2018/18257-1 (GP), 2018/15549-1 and 2020/04923-0 (GP), 2021/00140-3 (JMDS), 2015/26722-8 (CW), 2021/14751-4 (SNM)); by the Conselho Nacional de Desenvolvimento Científico e Tecnológico ("Bolsa de Produtividade" (307854/2018-3 to GP); by the Coordenação de Aperfeiçoamento de Pessoal de Nível Superior (CAPES bolsa PNPD 88887.372048/2019-00 to LRF and to JVPC).

References

Aebi, M. (2013). N-linked protein glycosylation in the ER. *Biochimica et Biophysica Acta, 1833*, 2430–2437.

Alocci, D., Mariethoz, J., Gastaldello, A., Gasteiger, E., Karlsson, N. G., Kolarich, D., et al. (2019). GlyConnect: Glycoproteomics goes visual, interactive, and analytical. *Journal of Proteome Research, 18*, 664–677. https://doi.org/10.1021/acs.jproteome.8b00766.

Baloch, S., Baloch, M. A., Zheng, T., & Pei, X. (2020). The coronavirus disease 2019 (COVID-19) pandemic. *The Tohoku Journal of Experimental Medicine, 250*, 271–278. https://doi.org/10.1620/tjem.250.271.

Banerjee, N., & Mukhopadhyay, S. (2016). Viral glycoproteins: biological role and application in diagnosis. *Virusdisease, 27*, 1–11. https://doi.org/10.1007/s13337-015-0293-5.

Barros, E. P., Casalino, L., Gaieb, Z., Dommer, A. C., Wang, Y., Fallon, L., et al. (2021). The flexibility of ACE2 in the context of SARS-CoV-2 infection. *Biophysical Journal, 120*, 1072–1084. https://doi.org/10.1016/J.BPJ.2020.10.036.

Bauer, P., Hess, B., & Lindahl, E. (2022a). *GROMACS 2022 source code*. https://doi.org/10.5281/ZENODO.6103835.

Bauer, P., Hess, B., & Lindahl, E. (2022b). *GROMACS 2022 manual*. https://doi.org/10.5281/ZENODO.6103568.

Bernardi, A., Kirschner, K. N., & Faller, R. (2017). Structural analysis of human glycoprotein butyrylcholinesterase using atomistic molecular dynamics: The importance of glycosylation site ASN241. *PLoS One, 12*, e0187994. https://doi.org/10.1371/JOURNAL.PONE.0187994.

Campbell, M. P. (2017). A review of software applications and databases for the interpretation of glycopeptide data. *Trends in Glycoscience and Glycotechnology, 29*, E51–E62. https://doi.org/10.4052/tigg.1601.1E.

Casalino, L., Gaieb, Z., Goldsmith, J. A., Hjorth, C. K., Dommer, A. C., Harbison, A. M., et al. (2020). Beyond shielding: The roles of glycans in the SARS-CoV-2 spike protein. *ACS Central Science, 6*, 1722–1734. https://doi.org/10.1021/acscentsci.0c01056.

Casarotto, P. C., Girych, M., Fred, S. M., Kovaleva, V., Moliner, R., Enkavi, G., et al. (2021). Antidepressant drugs act by directly binding to TRKB neurotrophin receptors. *Cell, 184*, 1299–1313. e19 https://doi.org/10.1016/J.CELL.2021.01.034.

Chang, V. T., Crispin, M., Aricescu, A. R., Harvey, D. J., Nettleship, J. E., Fennelly, J. A., et al. (2007). Glycoprotein structural genomics: Solving the glycosylation problem. *Structure, 15*, 267–273. https://doi.org/10.1016/j.str.2007.01.011.

Cheong, Y., Shim, G., Kang, D., & Kim, Y. (1999). Carbohydrate binding specificity of pea lectin studied by NMR spectroscopy and molecular dynamics simulations. *Journal of Molecular Structure, 475*, 219–232. https://doi.org/10.1016/S0022-2860(98)00511-0.

Chernykh, A., Kawahara, R., & Thaysen-Andersen, M. (2021). Towards structure-focused glycoproteomics. *Biochemical Society Transactions, 49*, 161–186. https://doi.org/10.1042/BST20200222.

Cook, J. D., & Lee, J. E. (2013). The secret life of viral entry glycoproteins: Moonlighting in immune evasion. *PLoS Pathogens, 9*, e1003258. https://doi.org/10.1371/journal.ppat.1003258.

Dalziel, M., Crispin, M., Scanlan, C. N., Zitzmann, N., & Dwek, R. A. (2014). Emerging principles for the therapeutic exploitation of glycosylation. *Science (New York, N.Y.), 343*. https://doi.org/10.1126/SCIENCE.1235681.

Davis, S. J., & Crispin, M. (2010). Solutions to the glycosylation problem for low- and high-throughput structural glycoproteomics. In R. Owens, & J. Nettleship (Eds.), *Functional and structural proteomics of glycoproteins* (pp. 127–158). Dordrecht: Springer Netherlands. https://doi.org/10.1007/978-90-481-9355-4_6.

Duan, L., Zheng, Q., Zhang, H., Niu, Y., Lou, Y., & Wang, H. (2020). The SARS-CoV-2 spike glycoprotein biosynthesis, structure, function, and antigenicity: Implications for the design of spike-based vaccine immunogens. *Frontiers in Immunology, 11*, 2593. https://doi.org/10.3389/FIMMU.2020.576622/BIBTEX.

Eisenhaber, F., Lijnzaad, P., Argos, P., Sander, C., & Scharf, M. (1995). The double cubic lattice method: Efficient approaches to numerical integration of surface area and volume and to dot surface contouring of molecular assemblies. *Journal of Computational Chemistry, 16*, 273–284. https://doi.org/10.1002/JCC.540160303.

Evans, D. J., & Holian, B. L. (1998). The Nose–Hoover thermostat. *The Journal of Chemical Physics, 83*, 4069. https://doi.org/10.1063/1.449071.

Grant, O. C., Montgomery, D., Ito, K., & Woods, R. J. (2020). Analysis of the SARS-CoV-2 spike protein glycan shield reveals implications for immune recognition. *Scientific Reports, 2020*, 1–11. https://doi.org/10.1038/s41598-020-71748-7. 10:1 10.

Hacisuleyman, E., Hale, C., Saito, Y., Blachere, N. E., Bergh, M., Conlon, E. G., et al. (2021). Vaccine breakthrough infections with SARS-CoV-2 variants. *The New England Journal of Medicine, 384*, 2212–2218. https://doi.org/10.1056/NEJMoa2105000.

Hanisch, F. G. (2001). O-glycosylation of the mucin type. *Biological Chemistry, 382*, 143–149. https://doi.org/10.1515/BC.2001.022/MACHINEREADABLECITATION/RIS.

Hargett, A. A., Marcella, A. M., Yu, H., Li, C., Orwenyo, J., Battistel, M. D., et al. (2021). Glycosylation states on intact proteins determined by NMR spectroscopy. *Molecules, 26*, 4308. https://doi.org/10.3390/molecules26144308.

Hastie, K. M., Zandonatti, M. A., Kleinfelter, L. M., Heinrich, M. L., Rowland, M. M., Chandran, K., et al. (2017). Structural basis for antibody-mediated neutralization of Lassa virus. *Science*, *356*, 923–928. https://doi.org/10.1126/SCIENCE.AAM7260.

Henzler-Wildman, K., & Kern, D. (2007). Dynamic personalities of proteins. *Nature*, *450*(2007), 964–972. 4507172 https://doi.org/10.1038/nature06522.

Hess, B., Bekker, H., Berendsen, H. J. C., & Fraaije, J. G. E. M. (1997). LINCS: A linear constraint solver for molecular simulations. *Journal of Computational Chemistry*, *18*, 14631472. https://doi.org/10.1002/(SICI)1096-987X(199709)18:12.

Hollingsworth, S. A., & Dror, R. O. (2018). Molecular dynamics simulation for all. *Neuron*, *99*, 1129–1143. https://doi.org/10.1016/J.NEURON.2018.08.011.

Huang, J., Rauscher, S., Nawrocki, G., Ran, T., Feig, M., De Groot, B. L., et al. (2016). CHARMM36m: An improved force field for folded and intrinsically disordered proteins. *Nature Methods*, *14*(1), 71–73. https://doi.org/10.1038/nmeth.4067.

Humphrey, W., Dalke, A., & Schulten, K. (1996). VMD: Visual molecular dynamics. *Journal of Molecular Graphics*, *14*, 33–38. https://doi.org/10.1016/0263-7855(96)00018-5.

Hünenberger, P. H. (2005). Thermostat algorithms for molecular dynamics simulations. *Advances in Polymer Science*, *173*, 105–147. https://doi.org/10.1007/B99427.

Jo, S., Cheng, X., Lee, J., Kim, S., Park, S. J., Patel, D. S., et al. (2017). CHARMM-GUI 10 years for biomolecular modeling and simulation. *Journal of Computational Chemistry*, *38*, 1114–1124. https://doi.org/10.1002/JCC.24660.

Jo, S., Kim, T., Iyer, V. G., & Im, W. (2008). CHARMM-GUI: A web-based graphical user interface for CHARMM. *Journal of Computational Chemistry*, *29*, 1859–1865.

Jo, S., Song, K. C., Desaire, H., MacKerell, A. D., & Im, W. (2011). Glycan reader: Automated sugar identification and simulation preparation for carbohydrates and glycoproteins. *Journal of Computational Chemistry*, *32*, 3135–3141. https://doi.org/10.1002/JCC.21886.

Kaji, H., Saito, H., Yamauchi, Y., Shinkawa, T., Taoka, M., Hirabayashi, J., et al. (2003). Lectin affinity capture, isotope-coded tagging and mass spectrometry to identify N-linked glycoproteins. *Nature Biotechnology*, *21*, 667–672. https://doi.org/10.1038/nbt829.

Kang, P., Mechref, Y., & Novotny, M. V. (2008). High-throughput solid-phase permethylation of glycans prior to mass spectrometry. *Rapid Communications in Mass Spectrometry*, *22*, 721–734. https://doi.org/10.1002/rcm.3395.

Kawahara, R., Chernykh, A., Alagesan, K., Bern, M., Cao, W., Chalkley, R. J., et al. (2021). Community evaluation of glycoproteomics informatics solutions reveals high-performance search strategies for serum glycopeptide analysis. *Nature Methods*, *18*, 1304–1316. https://doi.org/10.1038/s41592-021-01309-x.

Kirschner, K. N., Yongye, A. B., Tschampel, S. M., González-Outeiriño, J., Daniels, C. R., Foley, B. L., et al. (2008). GLYCAM06: A generalizable biomolecular force field. *Carbohydrates. Journal of Computational Chemistry*, *29*, 622–655. https://doi.org/10.1002/JCC.20820.

Kwon, Y. D., Pancera, M., Priyamvada, A., Georgiev, I. S., Crooks, E. T., et al., & Kwong, P. D. (2015). Crystal structure, conformational fixation and entry-related interactions of mature ligand-free HIV-1. *Nature Structural & Molecular Biology*, *22*, 522–531.

Larsen, M. R., Jensen, S. S., Jakobsen, L. A., & Heegaard, N. H. H. (2007). Exploring the Sialiome using titanium dioxide chromatography and mass spectrometry. *Molecular & Cellular Proteomics*, *6*, 1778–1787. https://doi.org/10.1074/mcp.M700086-MCP200.

Lasky, L. A. (1991). Lectin cell adhesion molecules (LEC-CAMs): A new family of cell adhesion proteins involved with inflammation. *Journal of Cellular Biochemistry*, *45*, 139–146. https://doi.org/10.1002/JCB.240450204.

Lee, P., Ohshima, N., Stanfield, R., Yu, W., Okuno, Y., Kurosawa, Y., et al. (2014). Receptor mimicry by antibody F045–092 facilitates universal binding to the H3 subtype of influenza virus. *Nature Communications*, *5*, 3614.

Lemieux, R. U., & Koto, S. (1974). The conformational properties of glycosidic linkages. *Tetrahedron*, *30*, 1933–1944. https://doi.org/10.1016/S0040-4020(01)97324-7.

Liguori, N., Croce, R., Marrink, S. J., & Thallmair, S. (2020). Molecular dynamics simulations in photosynthesis. *Photosynthesis Research*, *144*(2), 273–295. https://doi.org/10.1007/S11120-020-00741-Y.

Lippi, G., Mattiuzzi, C., & Henry, B. M. (2022). Updated picture of SARS-CoV-2 variants and mutations. *Diagnosis*, *9*, 11–17. https://doi.org/10.1515/DX-2021-0149/PDF.

Lis, H., & Sharon, N. (1993). Protein glycosylation. Structural and functional aspects. *European Journal of Biochemistry*, *218*, 1–27. https://doi.org/10.1111/j.1432-1033.1993.tb18347.x.

Liu, H., Wei, P., Zhang, Q., Chen, Z., Aviszus, K., Downing, W., et al. (2021). 501Y.V2 and 501Y.V3 variants of SARS-CoV-2 lose binding to bamlanivimab in vitro. *bioRxiv*, *13*. https://doi.org/10.1080/19420862.2021.1919285.

Liu, H., Zhang, Q., Wei, P., Chen, Z., Aviszus, K., Yang, J., & Zhang, G. (2021). The basis of a more contagious 501Y.V1 variant of SARS-COV-2. *Cell Res*, *31*, 720–722. https://doi.org/10.1101/2021.02.02.428884.

Lupala, C. S., Ye, Y., Chen, H., Su, X. D., & Liu, H. (2022). Mutations on RBD of SARS-CoV-2 omicron variant result in stronger binding to human ACE2 receptor. *Biochemical and Biophysical Research Communications*, *590*, 34–41. https://doi.org/10.1016/J.BBRC.2021.12.079.

Macedo-da-Silva, J., Santiago, V. F., Rosa-Fernandes, L., Marinho, C. R. F., & Palmisano, G. (2021). Protein glycosylation in extracellular vesicles: Structural characterization and biological functions. *Molecular Immunology*, *135*, 226–246. https://doi.org/10.1016/j.molimm.2021.04.017.

Machiels, B., Lété, C., Guillaume, A., Mast, J., Stevenson, P. G., Vanderplasschen, A., et al. (2011). Antibody evasion by a Gammaherpesvirus O-glycan shield. *PLoS Pathogens*, *7*, e1002387. https://doi.org/10.1371/JOURNAL.PPAT.1002387.

Mehdipour, A. R., & Hummer, G. (2021). Dual nature of human ACE2 glycosylation in binding to SARS-CoV-2 spike. *Proceedings of the National Academy of Sciences of the United States of America*, *118*, 2022.

Meng, B., Kemp, S. A., Papa, G., Datir, R., Ferreira, I. A. T. M., Marelli, S., et al. (2021). Recurrent emergence of SARS-CoV-2 spike deletion H69/V70 and its role in the Alpha variant B.1.1.7. *Cell Reports*, *35*, 109292. https://doi.org/10.1016/J.CELREP.2021.109292.

Mohan, G. S., Li, W., Ye, L., Compans, R. W., & Yang, C. (2012). Antigenic subversion: A novel mechanism of host immune evasion by Ebola virus. *PLoS Pathogens*, *8*, e1003065. https://doi.org/10.1371/JOURNAL.PPAT.1003065.

Morelle, W., & Michalski, J.-C. (2007). Analysis of protein glycosylation by mass spectrometry. *Nature Protocols*, *2*, 1585–1602. https://doi.org/10.1038/nprot.2007.227.

Mule, S. N., Rosa-Fernandes, L., Coutinho, J. V. P., Gomes, V. D. M., Macedo-da-Silva, J., Santiago, V. F., et al. (2021). Systems-wide analysis of glycoprotein conformational changes by limited deglycosylation assay. *Journal of Proteomics*, *248*, 104355. https://doi.org/10.1016/j.jprot.2021.104355.

Nagae, M., & Yamaguchi, Y. (2012). Function and 3D structure of the N-glycans on glycoproteins. *International Journal of Molecular Sciences*, *13*, 8398–8429. https://doi.org/10.3390/ijms13078398.

Naveca, F. G., Nascimento, V., de Souza, V. C., de Corado, A. L., Nascimento, F., Silva, G., et al. (2021). COVID-19 in Amazonas, Brazil, was driven by the persistence of endemic lineages and P.1 emergence. *Nature Medicine*, *27*, 1230–1238. https://doi.org/10.1038/s41591-021-01378-7.

Oliveira, T., Thaysen-Andersen, M., Packer, N. H., & Kolarich, D. (2021). The Hitchhiker's guide to glycoproteomics. *Biochemical Society Transactions*, *49*, 1643–1662. https://doi.org/10.1042/BST20200879.

Ongay, S., Boichenko, A., Govorukhina, N., & Bischoff, R. (2012). Glycopeptide enrichment and separation for protein glycosylation analysis: Sample preparation. *Journal of Separation Science, 35*, 2341–2372. https://doi.org/10.1002/jssc.201200434.

Ou, X., Liu, Y., Lei, X., Li, P., Mi, D., Ren, L., et al. (2020). Characterization of spike glycoprotein of SARS-CoV-2 on virus entry and its immune cross-reactivity with SARS-CoV. *Nature Communications, 11*(1), 1–12. https://doi.org/10.1038/s41467-020-15562-9.

Palmisano, G., Larsen, M. R., Packer, N. H., & Thaysen-Andersen, M. (2013). Structural analysis of glycoprotein sialylation—Part II: LC-MS based detection. *RSC Advances, 3*, 22706. https://doi.org/10.1039/c3ra42969e.

Palmisano, G., Lendal, S. E., Engholm-Keller, K., Leth-Larsen, R., Parker, B. L., & Larsen, M. R. (2010). Selective enrichment of sialic acid–containing glycopeptides using titanium dioxide chromatography with analysis by HILIC and mass spectrometry. *Nature Protocols, 5*, 1974–1982. https://doi.org/10.1038/nprot.2010.167.

Palmisano, G., Melo-Braga, M. N., Engholm-Keller, K., Parker, B. L., & Larsen, M. R. (2012). Chemical deamidation: A common pitfall in large-scale N-linked glycoproteomic mass spectrometry-based analyses. *Journal of Proteome Research, 11*, 1949–1957. https://doi.org/10.1021/pr2011268.

Park, S. J., Lee, J., Patel, D. S., Ma, H., Lee, H. S., Jo, S., et al. (2017). Glycan reader is improved to recognize most sugar types and chemical modifications in the protein data bank. *Bioinformatics, 33*, 3051–3057. https://doi.org/10.1093/BIOINFORMATICS/BTX358.

Park, S.-J., Lee, J., Qi, Y., Kern, N. R., Lee, H. S., Jo, S., et al. (2019). CHARMM-GUI *glycan modeler* for modeling and simulation of carbohydrates and glycoconjugates. *Glycobiology, 29*, 320–331. https://doi.org/10.1093/glycob/cwz003.

Parker, B. L., Gupta, P., Cordwell, S. J., Larsen, M. R., & Palmisano, G. (2011). Purification and identification of *O*-Glc *N* Ac-modified peptides using phosphate-based alkyne CLICK chemistry in combination with titanium dioxide chromatography and mass spectrometry. *Journal of Proteome Research, 10*, 1449–1458. https://doi.org/10.1021/pr100565j.

Parrinello, M., & Rahman, A. (1980). Crystal structure and pair potentials: A molecular-dynamics study. *Physical Review Letters, 45*, 1196. https://doi.org/10.1103/PhysRevLett.45.1196.

Pascarella, S., Ciccozzi, M., Zella, D., Bianchi, M., Benedetti, F., Benvenuto, D., et al. (2021). SARS-CoV-2 B.1.617 Indian variants: Are electrostatic potential changes responsible for a higher transmission rate? *Journal of Medical Virology, 93*, 6551–6556. https://doi.org/10.1002/JMV.27210.

Pasing, Y., Sickmann, A., & Lewandrowski, U. (2012). N-glycoproteomics: Mass spectrometry-based glycosylation site annotation. *Biological Chemistry, 393*, 249–258. https://doi.org/10.1515/HSZ-2011-0245/MACHINEREADABLECITATION/RIS.

Pearlman, D. A., Case, D. A., Caldwell, J. W., Ross, W. S., Cheatham, T. E., DeBolt, S., et al. (1995). AMBER, a package of computer programs for applying molecular mechanics, normal mode analysis, molecular dynamics and free energy calculations to simulate the structural and energetic properties of molecules. *Computer Physics Communications, 91*, 1–41. https://doi.org/10.1016/0010-4655(95)00041-D.

Peters, T., Meyer, B., Stuike-Prill, R., Somorjai, R., & Brisson, J. R. (1993). A Monte Carlo method for conformational analysis of saccharides. *Carbohydrate Research, 238*, 49–73. https://doi.org/10.1016/0008-6215(93)87005-D.

Planas, D., Veyer, D., Baidaliuk, A., Staropoli, I., Guivel-Benhassine, F., Rajah, M. M., et al. (2021). Reduced sensitivity of SARS-CoV-2 variant Delta to antibody neutralization. *Nature, 596*, 276–280. https://doi.org/10.1038/S41586-021-03777-9.

Reiding, K. R., Blank, D., Kuijper, D. M., Deelder, A. M., & Wuhrer, M. (2014). High-throughput profiling of protein N-glycosylation by MALDI-TOF-MS employing linkage-specific sialic acid esterification. *Analytical Chemistry*, *86*, 5784–5793. https://doi.org/10.1021/ac500335t.

Reily, C., Stewart, T. J., Renfrow, M. B., & Novak, J. (2019). Glycosylation in health and disease. *Nature Reviews. Nephrology*, *15*, 346–366. https://doi.org/10.1038/s41581-019-0129-4.

Rodriguez, R., Chinea, G., Lopez, N., Pons, T., & Vriend, G. (1998). Homology modeling, model and software evaluation: Three related resources. *Bioinformatics*, *14*, 523–528. https://doi.org/10.1093/BIOINFORMATICS/14.6.523.

Sattelle, B. M., Hansen, S. U., Gardiner, J., & Almond, A. (2010). Free energy landscapes of iduronic acid and related monosaccharides. *Journal of the American Chemical Society*, *132*, 13132–13134. https://doi.org/10.1021/JA1054143/SUPPL_FILE/JA1054143_SI_001.PDF.

Schindler, B., Barnes, L., Renois, G., Gray, C., Chambert, S., Fort, S., et al. (2017). Anomeric memory of the glycosidic bond upon fragmentation and its consequences for carbohydrate sequencing. *Nature Communications*, *8*, 973. https://doi.org/10.1038/s41467-017-01179-y.

Schjoldager, K. T., Narimatsu, Y., Joshi, H. J., & Clausen, H. (2020). Global view of human protein glycosylation pathways and functions. *Nature Reviews. Molecular Cell Biology*, *21*, 729–749. https://doi.org/10.1038/s41580-020-00294-x.

Stanca-Kaposta, E. C., Gamblin, D. P., Cocinero, E. J., Frey, J., Kroemer, R. T., Fairbanks, A. J., et al. (2008). Solvent interactions and conformational choice in a core N-glycan segment: Gas phase conformation of the central, branching trimannose unit and its singly hydrated complex. *Journal of the American Chemical Society*, *130*, 10691–10696. https://doi.org/10.1021/JA801892H/SUPPL_FILE/JA801892H-FILE010.PDF.

Struwe, W. B., & Robinson, C. V. (2019). Relating glycoprotein structural heterogeneity to function—Insights from native mass spectrometry. *Current Opinion in Structural Biology*, *58*, 241–248. https://doi.org/10.1016/j.sbi.2019.05.019.

Sugrue, R. J. (2007). Viruses and glycosylation. In R. J. Sugrue (Ed.), *Glycovirology protocols. Methods in molecular biology* (pp. 1–13). Totowa, NJ: Humana Press. https://doi.org/10.1007/978-1-59745-393-6_1.

Tachibana, Y., Fletcher, G. L., Fujitani, N., Tsuda, S., Monde, K., & Nishimura, S. I. (2004). Antifreeze glycoproteins: Elucidation of the structural motifs that are essential for antifreeze activity. *Angewandte Chemie, International Edition*, *43*, 856–862. https://doi.org/10.1002/ANIE.200353110.

Thaysen-Andersen, M., & Packer, N. H. (2014). Advances in LC–MS/MS-based glycoproteomics: Getting closer to system-wide site-specific mapping of the N- and O-glycoproteome. *Biochimica et Biophysica Acta (BBA) - Proteins and Proteomics*, *1844*, 1437–1452. https://doi.org/10.1016/j.bbapap.2014.05.002.

Thaysen-Andersen, M., Packer, N. H., & Schulz, B. L. (2016). Maturing Glycoproteomics technologies provide unique structural insights into the N-glycoproteome and its regulation in health and disease. *Molecular & Cellular Proteomics*, *15*, 1773–1790. https://doi.org/10.1074/mcp.O115.057638.

Tian, D., Sun, Y. H., Zhou, J. M., & Ye, Q. (2022). The global epidemic of SARS-CoV-2 variants and their mutational immune escape. *Journal of Medical Virology*, *94*, 847–857. https://doi.org/10.1002/JMV.27376.

Tien, M. Z., Meyer, A. G., Sydykova, D. K., Spielman, S. J., & Wilke, C. O. (2013). Maximum allowed solvent accessibilites of residues in proteins. *PLoS One*, *8*, e80635. https://doi.org/10.1371/JOURNAL.PONE.0080635.

Turoňová, B., Sikora, M., Schürmann, C., Hagen, W. J. H., Welsch, S., Blanc, F. E. C., et al. (2020). In situ structural analysis of SARS-CoV-2 spike reveals flexibility mediated by

three hinges. *Science, 370*, 203–208. https://doi.org/10.1126/SCIENCE.ABD5223/SUPPL_FILE/ABD5223S1.MOV.
Unione, L., Ardá, A., Jiménez-Barbero, J., & Millet, O. (2021). NMR of glycoproteins: Profiling, structure, conformation and interactions. *Current Opinion in Structural Biology, 68*, 9–17. https://doi.org/10.1016/j.sbi.2020.09.009.
Valverde, P., Quintana, J. I., Santos, J. I., Ardá, A., & Jiménez-Barbero, J. (2019). Novel NMR avenues to explore the conformation and interactions of glycans. *ACS Omega, 4*, 13618–13630. https://doi.org/10.1021/acsomega.9b01901.
Van Der Spoel, D., Lindahl, E., Hess, B., Groenhof, G., Mark, A. E., & Berendsen, H. J. C. (2005). GROMACS: Fast, flexible, and free. *Journal of Computational Chemistry, 26*, 1701–1718. https://doi.org/10.1002/JCC.20291.
Vanommeslaeghe, K., Hatcher, E., Acharya, C., Kundu, S., Zhong, S., Shim, J., et al. (2010). CHARMM general force field: A force field for drug-like molecules compatible with the CHARMM all-atom additive biological force fields. *Journal of Computational Chemistry, 31*, 671–690. https://doi.org/10.1002/JCC.21367.
Vigerust, D. J., & Shepherd, V. L. (2007). Virus glycosylation: Role in virulence and immune interactions. *Trends in Microbiology, 15*, 211–218. https://doi.org/10.1016/j.tim.2007.03.003.
Volz, E., Hill, V., McCrone, J. T., Price, A., Jorgensen, D., O'Toole, Á., et al. (2021). Evaluating the effects of SARS-CoV-2 spike mutation D614G on transmissibility and pathogenicity. *Cell, 184*, 64–75. e11 https://doi.org/10.1016/J.CELL.2020.11.020.
Walls, A. C., Park, Y. J., Tortorici, M. A., Wall, A., McGuire, A. T., & Veesler, D. (2020). Structure, function, and antigenicity of the SARS-CoV-2 spike glycoprotein. *Cell, 181*, 281–292. e6 https://doi.org/10.1016/j.cell.2020.02.058.
Walls, A. C., Tortorici, M. A., Frenz, B., Snijder, J., Li, W., Rey, F. A., et al. (2016). Glycan shield and epitope masking of a coronavirus spike protein observed by cryo-electron microscopy. *Nature Structural and Molecular Biology, 23*, 899–905. https://doi.org/10.1038/nsmb.3293.
Walton, E. B., & VanVliet, K. J. (2006). Equilibration of experimentally determined protein structures for molecular dynamics simulation. *Physical Review. E, Statistical, Nonlinear, and Soft Matter Physics, 74*. https://doi.org/10.1103/PHYSREVE.74.061901.
Wang, J., Fan, Q., Satoh, T., Arii, J., Lanier, L. L., Spear, P. G., et al. (2009). Binding of herpes simplex virus glycoprotein B (gB) to paired immunoglobulin-like type 2 receptor α depends on specific sialylated O-linked glycans on gB. *Journal of Virology, 83*, 13042–13045. https://doi.org/10.1128/JVI.00792-09/ASSET/82927C0C-8656-48E9-8BDF-0E30EA5E71B7/ASSETS/GRAPHIC/ZJV0240926790004.JPEG.
Watanabe, Y., Allen, J. D., Wrapp, D., McLellan, J. S., & Crispin, M. (2020). Site-specific glycan analysis of the SARS-CoV-2 spike. *Science*, eabb9983. https://doi.org/10.1126/science.abb9983.
Watanabe, Y., Berndsen, Z. T., Raghwani, J., Seabright, G. E., Allen, J. D., Pybus, O. G., et al. (2020). Vulnerabilities in coronavirus glycan shields despite extensive glycosylation. *Nature Communications, 11*, 1–10. https://doi.org/10.1038/s41467-020-16567-0.
Watanabe, Y., Bowden, T. A., Wilson, I. A., & Crispin, M. (2019). Exploitation of glycosylation in enveloped virus pathobiology. *Biochimica et Biophysica Acta (BBA) - General Subjects, 1863*, 1480–1497. https://doi.org/10.1016/J.BBAGEN.2019.05.012.
Waterhouse, A. M., Procter, J. B., Martin, D. M. A., Clamp, M., & Barton, G. J. (2009). Jalview version 2—A multiple sequence alignment editor and analysis workbench. *Bioinformatics, 25*, 1189–1191. https://doi.org/10.1093/BIOINFORMATICS/BTP033.
WHO Coronavirus (COVID-19) Dashboard. (n.d.). WHO Coronavirus (COVID-19) Dashboard with Vaccination Data. [WWW Document]. https://covid19.who.int/. (Accessed 28 February 2022).

Wilkinson, H., & Saldova, R. (2020). Current methods for the characterization of O-glycans. *Journal of Proteome Research*, *19*, 3890–3905. https://doi.org/10.1021/acs.jproteome.0c00435.

Wolter, N., Jassat, W., Walaza, S., Welch, R., Moultrie, H., Groome, M., et al. (2022). Early assessment of the clinical severity of the SARS-CoV-2 omicron variant in South Africa: A data linkage study. *The Lancet*, *399*, 437–446. https://doi.org/10.1016/S0140-6736(22)00017-4.

Woo, H., Park, S. J., Choi, Y. K., Park, T., Tanveer, M., Cao, Y., et al. (2020). Developing a fully glycosylated full-length SARS-COV-2 spike protein model in a viral membrane. *Journal of Physical Chemistry B*, *124*, 7128–7137. https://doi.org/10.1021/acs.jpcb.0c04553.

Woods, R. J. (2018). Predicting the structures of glycans, glycoproteins, and their complexes. *Chemical Reviews*, *118*, 8005–8024. https://doi.org/10.1021/ACS.CHEMREV.8B00032/ASSET/IMAGES/ACS.CHEMREV.8B00032.SOCIAL.JPEG_V03.

Wormald, M. R., & Dwek, R. A. (1999). Glycoproteins: Glycan presentation and protein-fold stability. *Structure*, *7*, R155–R160. https://doi.org/10.1016/S0969-2126(99)80095-1.

Wu, D., & Robinson, C. V. (2022). Understanding glycoprotein structural heterogeneity and interactions: Insights from native mass spectrometry. *Current Opinion in Structural Biology*, *74*, 102351. https://doi.org/10.1016/j.sbi.2022.102351.

Yang, Z., & Hancock, W. S. (2004). Approach to the comprehensive analysis of glycoproteins isolated from human serum using a multi-lectin affinity column. *Journal of Chromatography. A*, *1053*, 79–88.

Yokoyama, M., Fujisaki, S., Shirakura, M., Watanabe, S., Odagiri, T., Ito, K., et al. (2017). Molecular Dynamics Simulation of the Influenza A (H3N2) hemagglutinin trimer reveals the structural basis for adaptive evolution of the recent epidemic clade 3c.2a. *Frontiers in Microbiology*, *8*, 584. https://doi.org/10.3389/fmicb.2017.00584.

Zhao, Y., Ren, J., Harlos, K., Jones, D., Zeltina, A., et al., & Stuart, D. I. (2016). Toremifene interacts with and destabilizes the Ebola virus glycoprotein. *Nature*, *535*, 169–172.

CHAPTER TEN

Exploring COVID-19 pathogenesis on command-line: A bioinformatics pipeline for handling and integrating omics data

Janaina Macedo-da-Silva[a], João Victor Paccini Coutinho[a], Livia Rosa-Fernandes[a], Suely Kazue Nagahashi Marie[b], and Giuseppe Palmisano[a,c,*]

[a]GlycoProteomics Laboratory, Department of Parasitology, ICB, University of São Paulo, São Paulo, Brazil
[b]Cellular and Molecular Biology Laboratory (LIM 15), Neurology Department, Faculdade de Medicina FMUSP, Universidade de Sao Paulo, Sao Paulo, Brazil
[c]School of Natural Sciences, Macquarie University, Sydney, NSW, Australia
*Corresponding author: e-mail addresses: palmisano.gp@gmail.com; palmisano.gp@usp.br

Contents

Advances in Protein Chemistry and Structural Biology, Volume 131
ISSN 1876-1623
https://doi.org/10.1016/bs.apcsb.2022.04.002

Abstract

Severe acute respiratory syndrome coronavirus 2 (SARS-CoV-2) was first identified in late 2019 in Wuhan, China, and has proven to be highly pathogenic, making it a global public health threat. The immediate need to understand the mechanisms and impact of the virus made omics techniques stand out, as they can offer a holistic and comprehensive view of thousands of molecules in a single experiment. Mastering bioinformatics tools to process, analyze, integrate, and interpret omics data is a powerful knowledge to enrich results. We present a robust and open access computational pipeline for extracting information from quantitative proteomics and transcriptomics public data. We present the entire pipeline from raw data to differentially expressed genes. We explore processes and pathways related to mapped transcripts and proteins. A pipeline is presented to integrate and compare proteomics and transcriptomics data using also packages available in the Bioconductor and providing the codes used. Cholesterol metabolism, immune system activity, ECM, and proteasomal degradation pathways increased in infected patients. Leukocyte activation profile was overrepresented in both proteomics and transcriptomics data. Finally, we found a panel of proteins and transcripts regulated in the same direction in the lung transcriptome and plasma proteome that distinguish healthy and infected individuals. This panel of markers was confirmed in another cohort of patients, thus validating the robustness and functionality of the tools presented.

1. Introduction

Severe acute respiratory syndrome coronavirus 2 (SARS-CoV-2) was first identified in late 2019 in Wuhan, China, and has proven to be highly pathogenic, making it a global public health threat (Mahalmani et al., 2020). There are a variety of manifestations, from asymptomatic to severe cases, that include symptoms such as muscle damage, prolonged tiredness, shortness of breath, loss of taste/smell, and severe pneumonia (Dixon et al., 2021; Nalbandian et al., 2021). Cumulative reports also associate post-acute effects of the infection, including pulmonary, hematological, cardiovascular, neuronal, renal, endocrine, and gastrointestinal sequelae (Hayes, Ingram, & Sculthorpe, 2021; Nalbandian et al., 2021). The quest to understand the infectious mechanisms of SARS-CoV-2, diagnostic alternatives, and treatments have been approached by several techniques (Das, Ahmed, Akhtar, Begum, & Banu, 2021). Among them, omics sciences generate large amounts of information to enable holistic understanding of cell, tissue or organism function and reaction against a disease. These are high-performance techniques able to explore an organism at the level of genes (genomics), proteins (proteomics), metabolites (metabolomics), and lipids (lipidomics). Integration of this vast information provides a comprehensive and powerful tool for exploring the infection mechanism of SARS-CoV-2 (Overmyer et al., 2021; Wu et al., 2021).

In the last 10 years, with the advancement of technologies and processing capacity, proteomics and transcriptomics were readily implemented in several laboratories (Martens & Vizcaíno, 2017), while metabolomics and lipidomics were integrated along the years (Wenk, 2005). The more sophisticated data becomes, the greater the demand for new computational methods to deal with them. Currently, numerous tools can assess protein-protein interaction, calculate correlation between expression of different genes, identify enriched pathways and biological processes, and thus, drive data to find the most relevant points (Mangul et al., 2019; Mishra et al., 2021).

2. Applications

Omics sciences combined with bioinformatics technologies, in particular, contributed to the identification of therapeutic targets (Bojkova et al., 2020; Li et al., 2021); to the elucidation of virus pathophysiological mechanisms (Desai et al., 2020; Rosa-Fernandes et al., 2019; Wu et al., 2020); pathogen–host interactions (Terracciano et al., 2021); detection of patients predisposed to manifestations of severe symptoms (Lazari et al., 2021), selection of prognostic markers (Rosa-Fernandes et al., 2020; Terracciano et al., 2021) and to determine differential gene/protein/metabolite expression profiles (Desai et al., 2020; Leng et al., 2020; Wu et al., 2020). In addition to these fundamental applications, omics tools are also helpful in stratifying clinical manifestations in viral infections (Macedo-da-Silva et al., 2020). Furthermore, integrating data from different biological matrices may offer a comprehensive overview of disease pathogenesis, helping in prioritizing biomarkers and therapeutic targets (Wu et al., 2021; Zhu et al., 2022). Storing, analyzing, and interpreting these data can be a big challenge for researchers. However, it is a worthwhile task to overcome as omics approach enable answering multiple biological questions.

Notably the generated raw data in previous studies are cumulatively stored and accessible in public repositories, such as PRIDE (Vizcaíno et al., 2013) and SRA (Kodama et al., 2012), allowing reanalysis by different research groups. Therefore, mastering bioinformatics tools to process, analyze, integrate, and interpret these omic data is a powerful knowledge to build and validate hypotheses. The objective of this study is to present bioinformatics tools that can be used to reanalyze transcriptomic and proteomic data deposited in public databases and integrate them to increase the understanding of the pathophysiology of SARS-CoV-2.

3. Methodologies

3.1 Selected datasets

Four datasets, including proteomics and transcriptomics, were selected from public repositories. Shu et al. (PXD019106) evaluated plasma samples from patients infected with SARS-CoV-2, subdivided into fatal (FT, $n=5$), severe (SV, $n=7$), and moderate (MD, $n=10$) groups by proteomics. In addition, the study included samples from healthy patients (HE, $n=8$) who tested negative on throat swab and serological tests (Shu et al., 2020). Zhong et al. (S-BSST719) determined the plasma proteome profile of 50 individuals with mild and moderate disease. Collections were performed in the first 24 h (D0) after the positive PCR test and after 14 days (D14) with negative PCR result (Zhong et al., 2021). Kwan et al. (PRJNA692253) conducted RNA sequencing analyses of whole blood samples from 64 patients, of whom 45 tested positive for COVID-19 (INF) and 19 healthy participants not exposed to the virus (CTRL) (Kwan et al., 2021). Wu et al. (PRJNA646224) explored the lung and colon tissue transcriptome of patients who died due to COVID-19 (INF, $n=8$) and healthy counterparts were obtained from cancer patients who underwent surgical resection or biopsy (CTRL) (Wu et al., 2020).

3.2 Hardware features

The analyzes presented here were performed on an Ubuntu 20.04.3 LTS system (https://www.ubuntu.com/), with a total memory of 62GB, 40 CPUs, 2 threads per core, 10 cores per socket, and 2 sockets. The local storage of raw files required a large amount of space. However, the size of the files varied substantially as it depended on the quality of the data and equipment used.

3.3 Data download

In this protocol, we did not address the processing of raw files from proteomic approaches. The protein quantification tables provided in the selected manuscripts were downloaded and analyzed. The detailed pipelines for processing proteomic raw data are available in Kong, Leprevost, Avtonomov, Mellacheruvu, and Nesvizhskii (2017), Carvalho et al. (2016), Guangcan et al. (2021), and Tyanova, Temu, and Cox (2016).

For transcriptomic analysis, the fastq raw files were downloaded through the SRA Explorer platform (https://sra-explorer.info/) with the access numbers PRJNA692253 and PRJNA646224. Information regarding each file, such as SRA Accession, instrument, and total bases (Mb) were accessed. The SRA repository stores raw high-throughput sequencing data from different search fields. SRA files consist of raw data and require quality scores per basis for the submitted data. Tools like the sratoolkit download sequencing data from this database. Fastq files are raw data that have a definition line (defline) that contains a read identifier and nucleotide base information, all in text form.

After selection, all files from PRJNA692253 were downloaded by clicking on "Add to collection." A total of 64 SRA files were added. The platform offers the option of downloading .fastq files directly, but also .sra files and metadata. In the "FastQ Downloads" tab, the corresponding URL for all files could be downloaded. Kwan et al. used paired-end sequencing, resulting in 128 fastq files. Accession number PRJNA646224 provides the possibility to download colon and lung transcriptome data. Here, we downloaded only data referring to the lung (CTRL and INF), resulting in a total of 19 fastq files since Wu et al. used single-end sequencing. After creating a directory named "fastq_files" in the terminal console, files were downloaded using the "wget" command.

```
# Creating a directory named "fastq_files"
mkdir fastq_files

cd fastq_files


# Download the files

wget ftp://ftp.sra.ebi.ac.uk/vol1/fastq/SRR134/015/SRR13440815/SRR13440815_1.fastq.gz

wget ftp://ftp.sra.ebi.ac.uk/vol1/fastq/SRR134/015/SRR13440815/SRR13440815_2.fastq.gz
```

3.4 FastQC

FastQC is a highly recommended software for checking the quality of raw data coming from high-throughput sequencing pipelines. To download the software *via* the terminal console, "wget" command must be used.

The download is also possible by using directly the web browser from the Brabraham Institut website (https://www.bioinformatics.babraham.ac.uk/projects/fastqc/). After unzipping the file, turned the tool executable and added FastQC to the path. A directory to allocate FastQC outputs files was created and then the software was executed.

```
# FastQC software download
cd ..
wget https://www.bioinformatics.babraham.ac.uk/projects/fastqc/fastqc_v0.11.9.zip
unzip fastqc_v0.11.9.zip
chmod +x FastQC


# Creating a directory named "fastqc_output"
mkdir fastqc_output
# Run FastQC for all files in the "fastq_files" directory
fastqc -o fastqc_output fastq/*fastq.gz
```

At the end of the process, two output files will be created for each fastq file. By opening the generated .html file, the user can access the sample quality control report. The first module presents the results of basic statistics, including Total Sequences, Sequences flagged as poor quality, Sequence length, and %GC. Below is a BoxWhisker chart, which on the y-axis indicates the quality scores and on the x-axis the position in the read. The background indicates the quality of the sequences: green (very good), orange (fair), and red (poor). The "Per sequence quality scores" graph shows the average quality score on the x-axis and the number of sequences with that average on the y-axis. Generally, most samples' readings are expected to have a high-quality score. The parameter "Per base sequence content" reveals the proportion of each position of a base. If the difference between A and T or G and C bases is greater than 10%, the software will report a warning. If the difference is greater than 20%, this module will fail. In "Per sequence GC content," the GC content measurement is reported. A normal distribution profile is expected. In "Overrepresented Sequences" frequent sequences are reported, indicating that it has some biological relevance or that the library is not diverse or contaminated.

3.5 Trim Galore!

Trim Galore! is a platform that performs sample quality control by efficiently removing poor-quality parts of the readings and trimming the adapter.

This tool requires the previous installation of cutadapt and FastQC tools. The step of removing adapter sequences may or not be adopted in a transcriptome data analysis pipeline. In the selected datasets, Wu et al. chose to apply Trim Galore! while Kwan et al. started the sequence alignment right after running FastQC. In this pipeline, we applied the tool only to the dataset from Wu et al. study, based on the obtained FastQC report. The command used followed the structure for single-end sequencing data.

```
# Running Trim Galore
Command: trim_galore [options] <filename(s)>
trim_galore -j 2 -e 0.1 -q 20 --stringency 1 -O SRR12816734 SRR12816734.fastq.gz

# Running in a bash mode
#!/usr/bin/bash
SAMPLES="SRR12816734 SRR12816735"
for SAMPLE in $SAMPLES; do
trim_galore -j 2 -e 0.1 -q 20 --stringency 1 -O Trimmed {SAMPLE}.fastq.gz
done
```

In the above command, the "-j" option indicates the number of cores to be used for trimming; "-e" corresponds to the maximum allowed error rate; "-q" allows the trimming of low-quality ends from reads in addition to adapter removal; "--stringency" points out the overlap with adapter sequence required to trim a sequence; the "-O" option creates a directory to save the output files. The command "-a" is used to add the adapter sequence. The Trim Galore! can automatically detect whether the Illumina universal, Nextera transposase, or Illumina small RNA adapter sequence was used. However, it is possible to add the adapter sequence manually. When the command is missing, the software auto-detects the adapter.

3.6 Reference genome

Before starting the sequence alignment, it is necessary to download the reference genome. As indicated in the original manuscripts, Wu et al. used the h19 genome and Kwan et al. used the GRCh38 genome. The downloaded file will be allocated in the created directory. Then the HISAT2 sequence alignment software will be downloaded, unzipped, and added to the path. The genome will be indexed using the "hisat2-build" script, allowing HISAT2 to perform the read alignment faster in the next step. At the end of the indexing, eight .ht2 files will be created.

```
# Creating a directory and downloading the reference genome

mkdir reference_genome
cd reference_genome
#Kwan et. al
wget ftp://ftp.ensembl.org/pub/release-
86/fasta/homo_sapiens/dna/Homo_sapiens.GRCh38.dna_sm.primary_assembly.fa.gz
gunzip Homo_sapiens.GRCh38.dna_sm.primary_assembly.fa.gz
#or
#Wu et. al
wget https://hgdownload.soe.ucsc.edu/goldenPath/hg19/bigZips/hg19.fa.gz
gunzip hg19.fa.gz
cd ..

# Downloading the HISAT2 software and building the reference genome

mkdir hisat2
cd hisat2
wget https://cloud.biohpc.swmed.edu/index.php/s/oTtGWbWjaxsQ2Ho/download
unzip hisat2-2.2.1-Linux_x86_64.zip
cd ..
mkdir index
cd index
Command: hisat2-build [options]* <reference_in> <ht2_index_base>
hisat2-build -p 4 /home/reference_genome/hg19.fa hg19_index
#or
hisat2-build -p 4 /home/reference_genome/Homo_sapiens.GRCh38.dna_sm.primary_assembly.fa GRCh38_index
```

3.7 Alignment of raw readings

Sequence alignment was performed with HISAT2 software (Kim, Paggi, Park, Bennett, & Salzberg, 2019). Other programs are available for free, such as STAR (Dobin et al., 2013), Bowtie2 (Langmead & Salzberg, 2012), and TopHat2 (Kim et al., 2013). A comparison between the methods was recently reported and can be accessed at Schaarschmidt, Fischer, Zuther, and Hincha (2020) and Musich, Cadle-Davidson, and Osier (2021).

```
# For single-end alignment
Command: hisat2  -x genome_index -U input_file.fq -S output_name.sam --no-spliced-alignment
# For paired-end alignment
Command: hisat2 -x reference_genome -1 input_file_1.fq -2 input_file_2.fq -S output_name.sam

cd ..
mkdir aligned
#!/usr/bin/bash
SAMPLES="SRR12816730_trimmed SRR12816729_trimmed SRR12816728_trimmed SRR12816727_trimmed SRR12816726_trimmed
SRR12816725_trimmed SRR12816735_trimmed SRR12816732_trimmed SRR12816724_trimmed SRR12816723_trimmed
SRR12816722_trimmed SRR12816721_trimmed SRR12816736_trimmed SRR12816720_trimmed SRR12816719_trimmed
SRR12816718_trimmed SRR12816731_trimmed SRR12816733_trimmed SRR12816734_trimmed"
for SAMPLE in $SAMPLES; do
hisat2 -p 6  -x /home/index/hg19_index -U /home/Trimmed/${SAMPLE}.fq.gz -S /home/aligned/${SAMPLE}.sam
done
#or
hisat2 -p 8 -x /home/index/GRCh38_index -1 /home/fastq/{SAMPLE}_1.fastq.gz -2 /home/fastq/{SAMPLE}_2.fastq.gz -S {SAMPLE}.sam
done
```

The "-p" command indicates the number of threads to perform the alignment; "-x" is used to enter the path of previously generated reference files. It is important not to add the file extension (e.g., .ht2). The "-U" option is used to show the path of unpaired files. To perform paired-end sequencing alignment, the synthesis used is "-1" and "-2" to assign files _1 and _2. The "-S" command designates the .sam file that will be generated. SAM files store aligned sequences and take up much space, so these files can be converted to compressed version .BAM by the samtools software.

3.8 Samtools to convert SAM to BAM files

Samtools is a set of programs for interacting with high-throughput sequencing data. It is helpful for converting SAM, BAM and CRAM files. One of the most used commands is the "samtools view," which takes .BAM/.SAM files as input and converts them to .SAM/.BAM, respectively. The "-S" and "-b" commands are used. The alignment of fastq files occurs in random order with the position in the reference genome. Therefore, in "samtools sort," the BAM files sorting is performed. The "-o" command indicates the output file.

```
# Running samtools

samtools view -@ 4 -S -b SRR12816736_trimmed.sam > SRR12816736_trimmed.bam
samtools sort -@ 4 SRR12816736_trimmed.bam -o SRR12816736_trimmed.sorted.bam
```

3.9 FeatureCounts

FeatureCounts is a platform that performs reading counting, also called reading summarization, by gene (Liao, Smyth, & Shi, 2014). The htseq-count tool (Anders, Pyl, & Huber, 2015), and featureCounts, are widely used for this purpose. Both share the same file input format (BAM or SAM) and need an annotation file that includes the chromosomal coordinates of features. Among the mandatory arguments required to run the software are "-a" and "-o," which specifies the name/path of an annotation file and the name of the output file including read counts. Among the optional arguments, we used the "-T" command to indicate the number of threads to be used in the analysis; the "-t" option specifies the feature type in GTF annotation, and "-g" specifies the attribute type in the GTF annotation.

```
# Running featureCounts

featureCounts -T 4 -t exon -g gene_id -a Homo_sapiens.GRCh38.86.gtf -o SRR12816736.txt SRR12816736_trimmed.sorted.bam
```

3.10 Exploring data with R packages

After obtaining the read counts for each sample and downloading the quantitative proteomics tables, a set of R packages were used to handle the data and identify the relevant biological findings. The packages used are: limma, Glimma, edgeR, Homo.sapiens, metboAnalystR, Enhancedvolcano, combiroc, pROC, clusterProfiler, flashClust, uwot, NbClust, richplot, ggplot2, WCGNA, gprofiler2, ggvenn, DOSE, wordcloud, FactoMineR, factoextra, ggstatsplot, among others. The codes used to generate the data and figures will be available in Supplementary File 1 in the online version at https://doi.org/10.1016/bs.apcsb.2022.04.002.

The gene expression count files for each sample mapped by the transcriptomics pipeline were gathered in a single directory and, with the "readDGE" function available in the edgeR package, they were merged into a single object DGEList class. The raw counts were transformed into counts per million (CPM) so that the size of the libraries was taken into account and then scaled to log (log CPM). Most samples genes that did not have sufficiently large counts were removed with the "filterByExpr" function. The resulting matrix was normalized by a trimmed mean of M-values (TMM) (Law et al., 2016).

The differentially expressed genes and proteins were determined by applying the standard analysis pipeline of the limma package, with *p*-value correction by the Benjamini-Hochberg method (q-value). The Homo.sapiens package performed the transformation of ENSMBL ID into Gene symbol. Interactive MAplots were plotted with the Glimma package. The table resulting from the analysis of differentially regulated genes/proteins was used as input to visualize of volcano plots by Enhancedvolcano package. The x and y values were the columns logFC and adjusted *p*-value, respectively. Principal component analysis (PCA) was conducted by applying the FactoMineR package using the "PCA" function. The "fviz_pca_ind()" function, available in the factoextra package, was applied to visualize the result. Samples were identified as outliers using the MetaboAnalystR package, with the "PlotRF.Outlier" function. The input data used were filtered and normalized matrices. Venn diagrams were built with the ggvenn package. The gprofiler2 package was used to perform gene ontology (GO). The "gost" function was applied to access functional enrichments of gene lists by applying a threshold of q-value < 0.05 and "gostplot" to visualize the results, with the capped and interactive options set to TRUE. Only ontologies with 5 or more proteins/genes were considered. The wordcloud package was used to build the word cloud based on the name frequency of ontologies identified by gprofiler2. Bubble plots were built with the ggplot package. The x-axis shows the gene count in each ontology shown on the y-axis. The size of the bubbles indicates the -log of the q-value and the color indicates the ontology category. To access pathways related to differentially expressed genes/proteins, the clusterProfiler and enrichplot packages were used. The "GSA" function with the options minGSSize set to 5 and maxGSSize to 100. The p-value correction method was indicated as "BH" (Benjamini-Hochberg). The regulated pathways network was built with the "cnetplot" function.

Data integration was performed by applying the "merge" function to select Gene symbols common between lung transcriptomics and plasma proteomics normalized tables. The "boxplot" function was applied to verify data distribution. Then the "removeBatchEffect" function, available in limma, was applied. The data were then normalized by the "scale" function (z-score). The "umap" function, available in the uwot package, was used to produce a low-dimensional embedding that summarizes the overall structure of high-dimensional data. The parameters adopted were min_dist=0.01, n_neighbors=20, n_epochs=10,000, verbose=2 and spread=2. The NbClust package was applied to determine the best number of clusters. Then the data were clustered by the "eclust" function. The cluster number was set to k=8. The "silhouette" function, available in the cluster package, was applied to assess the clustering quality. Genes and proteins from the same cluster, which have logFC > |1| (same direction) and q-value < 0.05 were selected and applied to construct ROC curves based on data from other datasets. The WGCNA package was applied to the expression data from Wu et al. and Kwan et al. The intersection of the ENSEMBL IDs mapped in the two datasets was performed by the "intersect" function. Correlation and connectivity between two datasets were determined using the softPower parameter set to 5. The number of modules was set to 4 and minClusterSize to 30. The "combiroc" package was used to determine the best marker combinations, and the "pROC" package was applied to plot and build the ROC curves.

4. Results

Given the increasing availability of public omics data, we present a robust and open access pipeline for extracting insights from quantitative biomolecular data (Fig. 1). Initially, raw files generated by RNA sequencing were downloaded and processed. The proteomics and transcriptomics data were then analyzed to find differently regulated features between the groups. In addition, outlier samples, biological processes, cellular components, molecular functions, and pathways of interest were identified. Transcriptomics data from different matrices were correlated, and proteomics

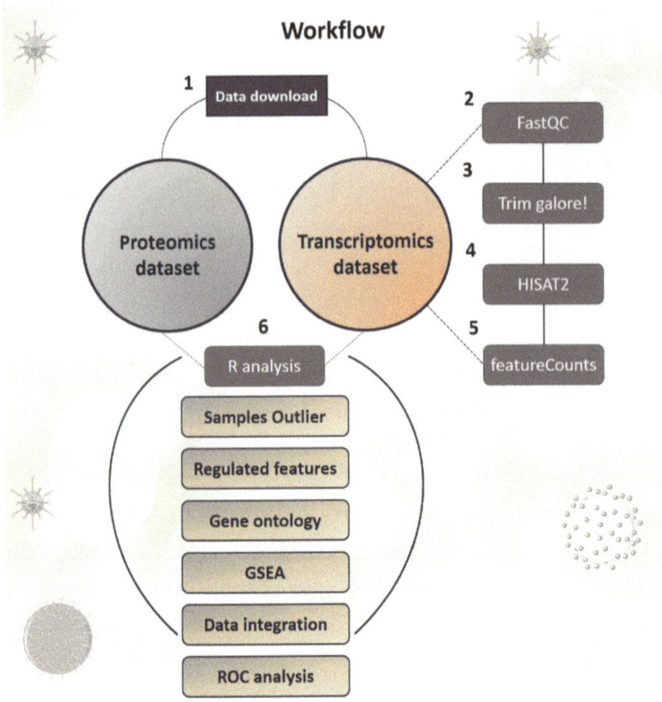

Fig. 1 Computational workflow and tools adopted to explore quantitative proteomics and transcriptomics data from plasma, whole blood, and lung of healthy individuals and patients infected with SARS-CoV-2.

and transcriptomics data from different biological samples were fitted and integrated. Finally, we evaluated through ROC curves the ability to separate samples from infected and healthy individuals based on the panel of relevant proteins and transcripts. Proteins and transcripts with better resolution between the clinical conditions were prioritized and discussed.

4.1 Plasma proteome presents different patterns according to the patient's COVID-19 grade

After filtering the original data provided by the authors, the PCA plot demonstrated that the plasma proteome of infected patients presented a different pattern than that of healthy patients (HE). The groups SV (severe) and FT (fatal) showed a very similar pattern, which differed from the MD (moderate) group (Fig. 2A). The protein profile of plasma from the same patient collected on day 0 (D0) of infection (PCR positive) and day

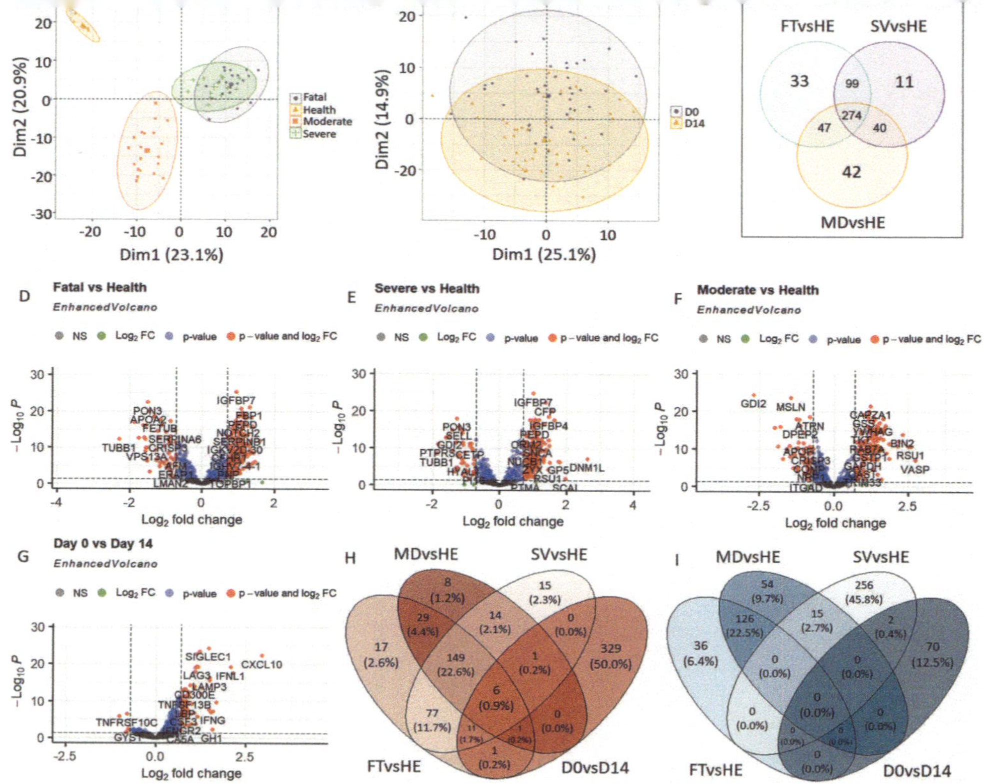

Fig. 2 Exploring the plasma proteome alterations resulting from SARS-CoV-2 infection. (A) Principal component analysis (PCA) indicating the separation between the groups that developed fatal (FT), severe (SV), moderate (MD) COVID-19 disease, and healthy donors (HE); (B) PCA indicating the separation between the same patients, with samples collected on day 0 of infection (D0) and day 14 (D14); (C) Venn diagram showing differently regulated proteins between FTvsHE, SVsvHE and MDvsHE comparisons; (D–G) Volcanos plot of differentially expressed proteins among FTvcHE, SVvsHE, MDvsHE, and D0vsD14 conditions; (H) Venn diagram showing upregulated proteins in the SASRS-CoV-2 infected group in both evaluated studies and (I) Venn diagram showing downregulated proteins in the SASRS-CoV-2 infected group in both evaluated studies.

14 (D14) (PCR negative) showed similarity in the majority of cases, however few presented distinct recovery pattern (Fig. 2B).

A total of 453 proteins were differentially regulated in the comparison between FT and HE (Fig. 2D); 424 between SV and HE (Fig. 2E), and 403 between MD and HE (Fig. 2F). Among patients in groups D0 and D14, 421 regulated proteins were identified (Fig. 2G). Comparing the proteins upregulated and downregulated in all comparisons (Fig. 2H and I), 6 proteins (C1QA, CD93, FLT4, INHBC, ENPP2, and NID1) are identified upregulated in the infected groups, indicating a possible role in the SARS-CoV-2 pathology (Fig. 3).

The boxplot plots (Fig. 3) showed that in the three degrees of infectious disease severity (FT, SV, MD) the expression level of the six proteins were higher than the healthy group (HE). The intra-patient comparisons (Fig. 3B) showed that after 14 days (D14) of disease onset (negative PCR), the levels of these proteins were lower than at the time of diagnosis (D0).

4.2 Plasma proteome analysis indicates dysregulation of immune system and cholesterol metabolism

Enriched GO terms were associated to the dysregulation of immune system, proteasome, cytoskeleton, cell adhesion, and lipoprotein metabolism (Fig. 4A). Upregulated proteins in the infected groups (FT, SV, MD, and D0) are related to the activation of the immune system, especially with the response mediated by leukocytes. Cytoskeleton dynamics and proteasomal degradation were also upregulated (Fig. 4B). Processes related to lipid metabolism and cell adhesion and migration are downregulated (Fig. 4C). Pathway analysis was performed using differentially regulated proteins and confirmed findings from GO analysis (Fig. 4D and E). In fact, downregulated proteins in the infected groups (FT, SV, MD, and D0) are related to cholesterol and lipoprotein metabolism. The ECM-proteoglycans pathway and cell adhesion molecules were also associated with down-regulated features. On the other hand, the pathways of infection, immune system, apoptosis, and extracellular matrix (ECM) organization are upregulated.

4.3 Lung and whole blood transcriptome reinforce findings in the plasma proteome

PCA analysis based on blood transcriptome from CTRL and INF patients with different degrees of symptoms did not show a clear separation (Fig. 5A). On the other hand, analysis of lung tissue from patients with fatal

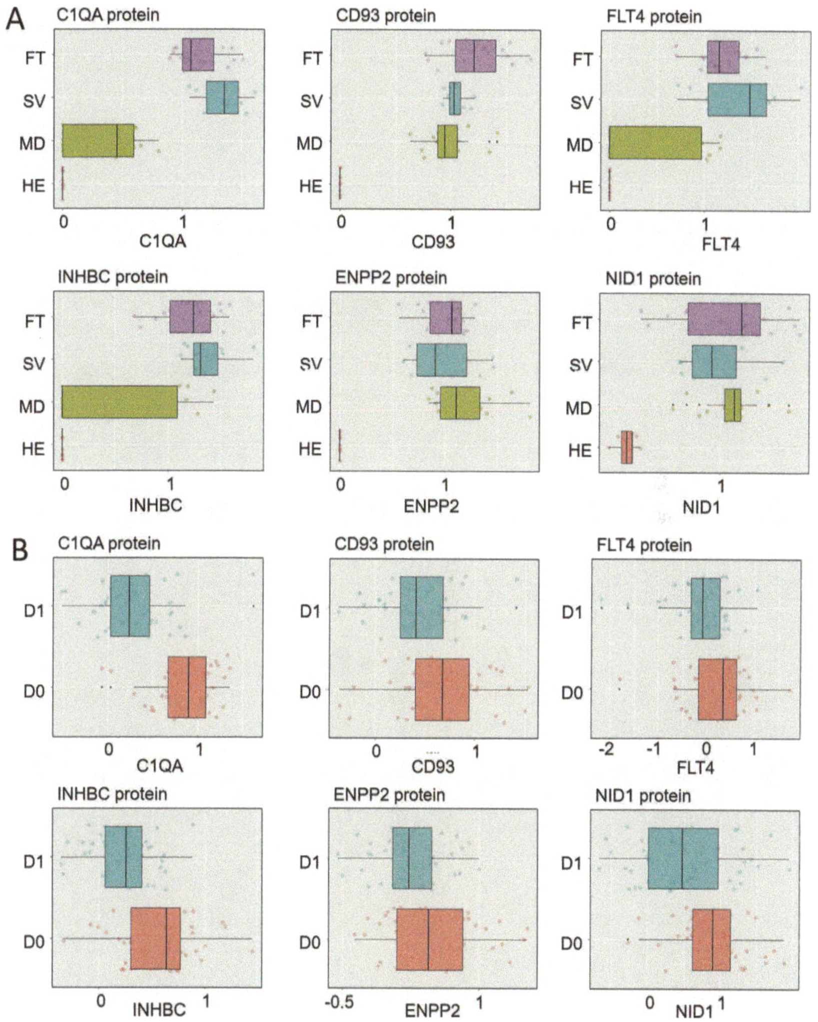

Fig. 3 Boxplot of upregulated proteins in SARS-CoV-2 infection. (A) Upregulated proteins in the groups that developed fatal (FT), severe (SV), and moderate (MD) COVID-19; (B) Upregulated proteins at the onset of COVID-19 (D0).

COVID-19 (INF) and healthy counterparts (CTRL) showed separation between groups (Fig. 5B). A total of 1039 genes were identified differentially expressed in the blood transcriptome (Fig. 5C**)** and 1034 in the lung tissue analysis (Fig. 5D) (logFc |1| and q-value < 0.05). A higher number of enriched GO terms (q value < 0.05) were found in the lung tissue compared to the blood transcriptome giving more information about the COVID-19 pathophysiology. Immune response mediated by leukocytes, apoptotic

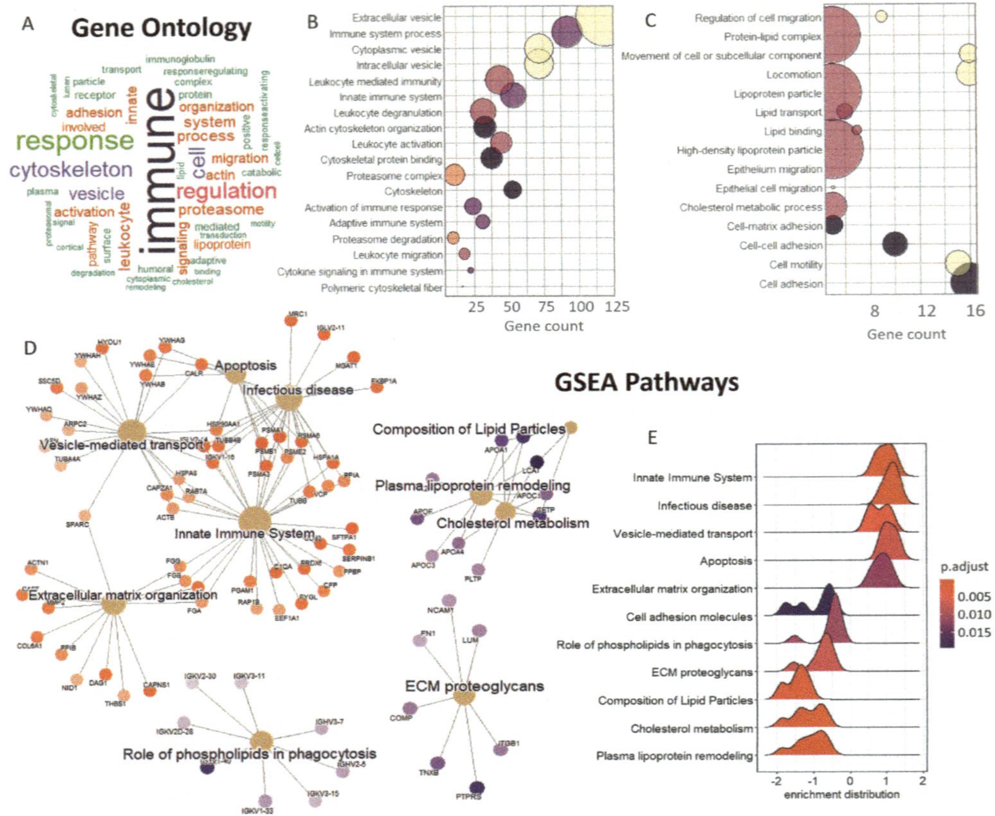

Fig. 4 Gene ontology and GSEA analysis of differentially regulated plasma proteins. (A) Word cloud generated from differentially regulated identified terms (gene count > 5 and q-value < 0.05); (B–C) Biological processes, cellular components, and molecular functions of upregulated and downregulated proteins in infected groups, respectively; (D) Network of proteins that participate in the main enriched pathways. Red and blue dots indicate upregulated and downregulated proteins, respectively; (E) Pathways associated with differentially regulated proteins.

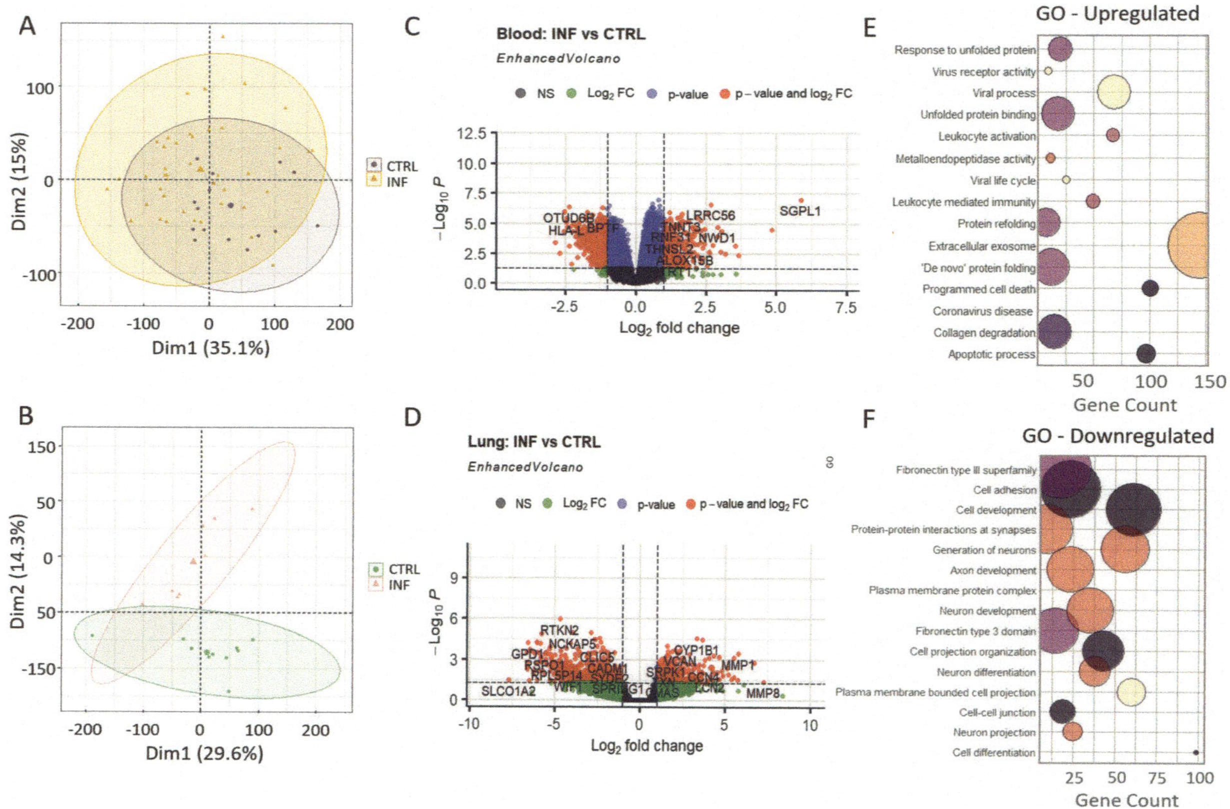

Fig. 5 Alteration of blood and lung transcriptome due to SARS-CoV-2 infection. (A) Principal component analysis (PCA) based on blood transcriptome from healthy (CTRL) and infected patients (INF); (B) PCA showing lung transcriptome-based separation of groups that developed fatal COVID-19 (INF) and non-infected individuals (CTRL); (C) Volcano plot indicating differentially expressed genes identified in blood; (D) Volcano plot indicating differentially expressed genes identified in the lung; Gene ontology (GO) analysis of upregulated (E) and downregulated (F) genes in the INF (blood and lung) groups.

processes, collagen degradation, and viral life cycle were among the upregulated enriched GO terms corroborating the findings of plasma proteomics. On the other hand, there was an enrichment associated with misfolded proteins that were not identified in plasma (Fig. 5E). Ontologies related to downregulated genes showed greater differences with respect to proteomic findings, although processes such as cell adhesion are common to both. Terms related to cell development and differentiation, especially of the brain components, are highlighted (Fig. 5F). Processes related to lipid and cholesterol metabolism were not identified.

4.4 The integration of omics data identifies clusters of proteins and genes related to key processes in infection

Weighted Gene Coexpression Network Analysis (WGCNA) is a popular method applied to identify correlation patterns between genes in samples. Integration of the 5000 common genes between lung and blood transcriptome did not show high modulus preservation between studies (Fig. 6A). However, the green (Fig. 6B) and turquoise (Fig. 6C) modules have $5 < Z < 10$, indicating moderate preservation. By selecting and working with the 400 key genes (labeled on the basis of kME) differentially expressed in the assessed groups of the turquoise and green modules, we identified enriched biological processes and pathways. The green module has genes related to leukocyte activation and proteasome activity, which was also identified in the turquoise module (Fig. 6D). However, it was also possible to verify processes linked to the virus, including proteins associated with spike viral glycoprotein maturation and deregulation of host-glycosylation (Fig. 6E), a key process during infection.

The clustering of 421 features common to lung transcriptome and plasma proteome data from healthy (CTRL and HE) and fatal (INF and FT) COVID-19 patients resulted in 8 clusters (Fig. 7A and B). The silhouette plot indicates genes possibly associated with the wrong cluster ($n = 27$) (Fig. 7C). These genes were disregarded in the enrichment analyses. Clusters 3, 4, 6, 7, and 8 show the highest correlation between the LogFC of the proteome and transcriptome (Fig. 7D). Considering only genes regulated in the same direction in both matrices, we identified enriched biological pathways and processes. Cluster 1 highlights terms associated with ECM; cluster 2, terms linked to the immune system and enzymatic activity; cluster 3, cytoskeletal activities; cluster 4, lipid metabolism; cluster 5, immune system and leukocyte activation; cluster 6,

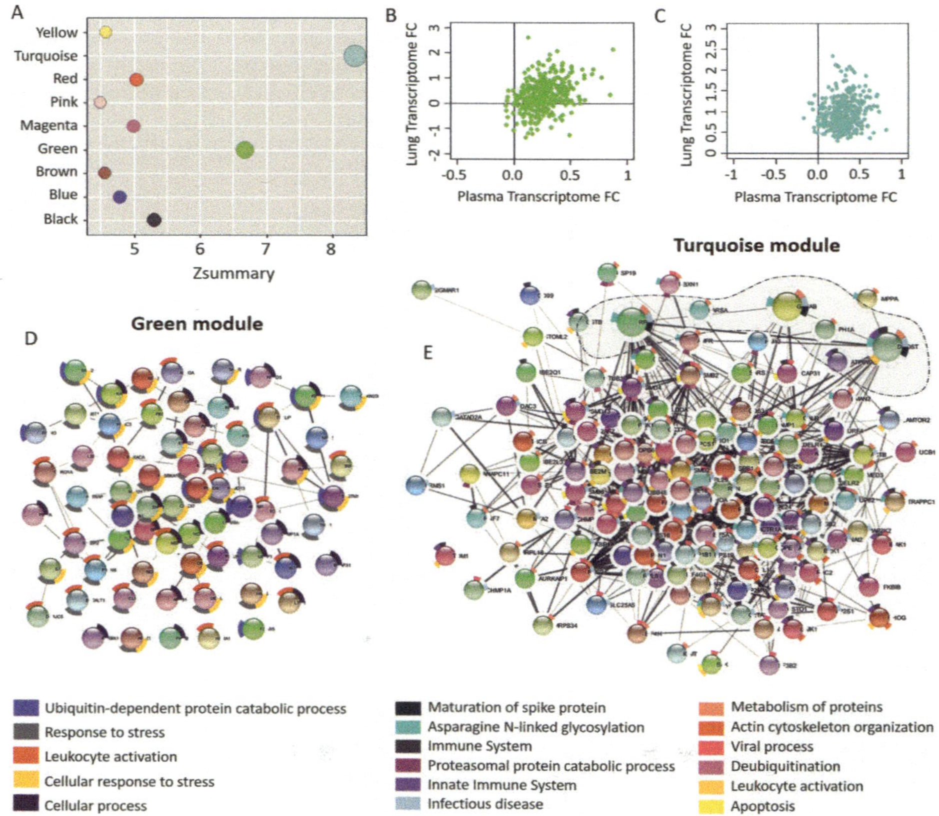

Fig. 6 Weighted gene co-expression network analysis (WGCNA). (A) *Z*-score for modules preserved between lung transcriptome and whole blood; (B–C) Scatter plot indicating the LogFC of the 400 genes comprising the green and turquoise modules ($Z > 5$), respectively, selected by the topGenesKME function; (D–E) Network indicating differentially expressed proteins from the turquoise and green modules. The donut plot indicates enriched processes and pathways (q-value < 0.05).

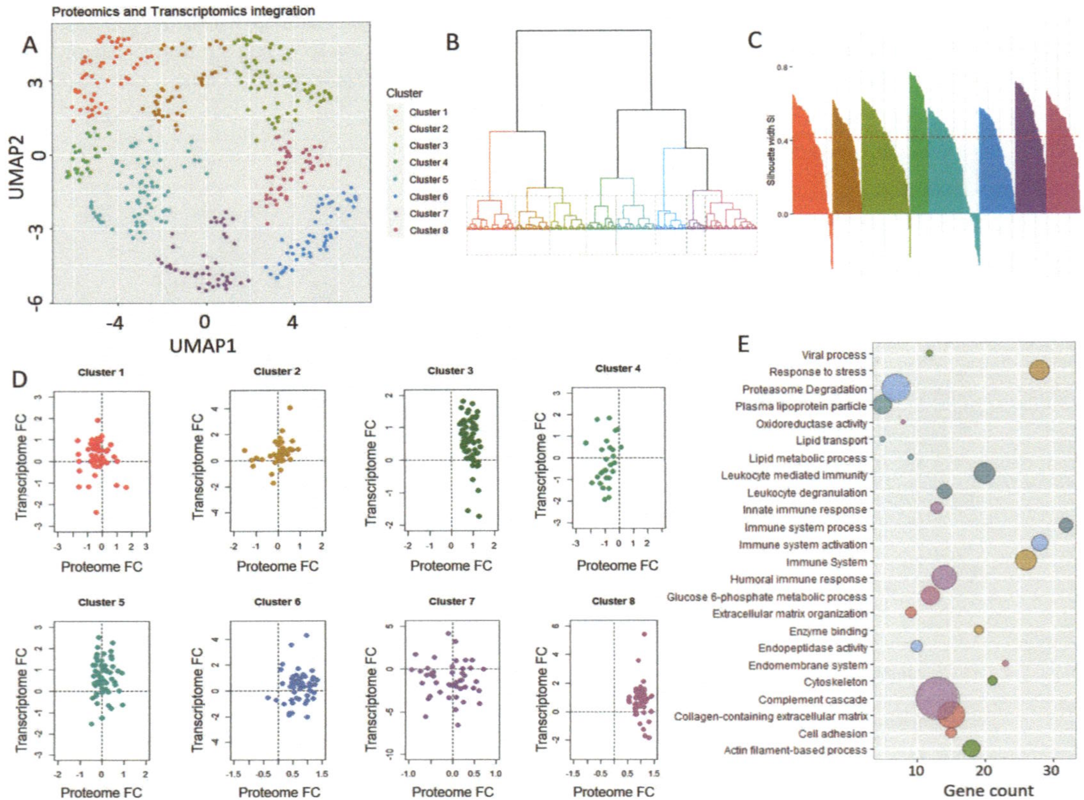

Fig. 7 Integration of lung transcriptomics and plasma proteomics data. (A) UMAP plot indicating the clustering of 421 genes/proteins identified in both evaluated datasets; (B) Dendrogram indicating the separation of 8 clusters identified; (C) Silhouette graph showing the quality of clustering applied. Negative values indicate genes that were possibly wrongly clustered. These genes were disregarded; (D) Scatter plot representing the LogFC of the proteome (x) and transcriptome (y) analysis; (E) Gene ontology and enriched pathways associated with identified clusters. The size of the bubbles indicates the -log-q-value and the color represents the associated cluster.

proteasomal degradation, immune system, and enzyme activity; cluster 7, complement system and cluster 8, glucose metabolism (Fig. 7E).

The transcripts and proteins differentially regulated with a logFC > |1| in the same direction (upregulation or downregulation) in both the proteome and the transcriptome were selected for application of the ROC curve (Fig. 8). The targets with the highest specificity and sensitivity were

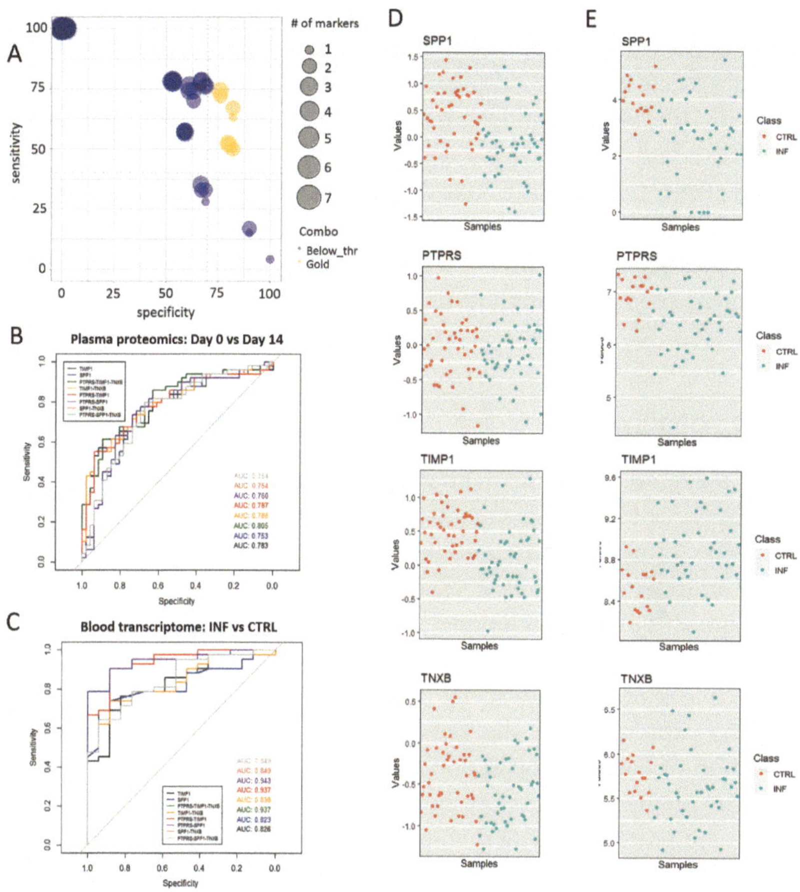

Fig. 8 ROC curves based on proteins involved in SARS-CoV-2 pathology. (A) Bubble chart indicating combinations between proteins/genes that present better results (golden combinations); (B) ROC curves of gold combinations indicating the potential for protein separation in distinguishing between groups D0 (PCR positive, day 01) and D14 (PCR negative, day 14); (C) ROC curves of gold combinations indicating the gene splitting potential in distinguishing between infected (INF) and healthy (CTRL) groups from whole blood transcriptomics analysis; (D–E) Boxplots indicating the separation of patients based on the selected protein/gene.

selected based in the integrated expression dataset (plasma proteome and lung transcriptome) (Fig. 8A). Additionally, the predictive power of these targets to stratify infected and healthy conditions was tested in another cohort (plasma proteome of D0 and D14 and whole blood transcriptome from INF and CTRL).

Three biomolecules gave the best AUC: TIMP1, a natural inhibitor of the matrix metalloproteinases involved in extracellular matrix remodeling, cell proliferation and regulated in response to cytokines; TNXB, tenascin XB, an extracellular matrix glycoprotein involved in wound healing, and PTPRS, protein tyrosine phosphatase receptor type S, the cell surface receptor that binds to glycosaminoglycans and involved in cell-cell interaction and cell growth and differentiation (Fig. 8B). The combination of the three showed an AUC greater than 0.8. At the gene level, the same combination resulted in AUC higher than 0.9 (Fig. 8C). Other combinations that showed AUC values close to 1 were PTPRS and TIMP1; and PTPRS and SPP1, a sialoprotein related to lymphocytic activation (Fig. 8D and E).

5. Discussion

To explore and understand the mechanisms of SARS-CoV-2 infection, the virus responsible for the death of more than 5 million people worldwide (https://covid19.who.int/), we present here an optimized pipeline for omics data analysis (Fig. 1), as well as extract meaningful information from high-throughput techniques. Dealing with codes in command line may not be a common expertise; however, being able to access and interpret public data is of great advantage for performing *in silico* validations and designing experiments in the wet lab to test hypotheses. We present the application of computational tools to analyze raw transcriptomics data and funnel the findings to access key disease processes. In addition, we integrated quantitative proteomics data to improve our understanding of COVID-19.

The rapid need to understand the mechanisms and impacts of SARS-CoV-2 makes omics strategies stand out as they can offer a holistic and comprehensive view of thousands of molecules (Colinge & Bennett, 2007; Haas, Muralidharan, Krogan, Kaake, & Hüttenhain, 2021). The ability to study alterations of a large number of proteins and transcripts in a single experiment is attractive (Al-Amrani, Al-Jabri, Al-Zaabi, Alshekaili, & Al-Khabori, 2021). Furthermore, accessing the modulation of molecules based on a clinical status can answer many questions about a given disease (Amiri-Dashatan, Koushki, Abbaszadeh, Rostami-Nejad, & Rezaei-Tavirani, 2018). However, to access these countless possibilities offered by omics, it is necessary to conduct information processing to reduce noisy data and focus only on what is relevant. Computational and bioinformatic analyses complement the sample preparation and data acquisition steps allowing a comprehensive visualization of dysregulated pathways (Cannataro, 2007).

The mechanisms altered by SARS-CoV-2 during infection can be accessed through the application of shotgun proteomics (Badua, Baldo, & Medina, 2021; Rais, Fu, & Drabovich, 2021). In addition, this approach can provide the quantification of viral proteins in complex clinical samples (Bojkova et al., 2020; Lachén-Montes, Corrales, Fernández-Irigoyen, & Santamaría, 2020; Lazari et al., 2021). Reports of sequelae resulting from viral infection are increasingly frequent. Chen et al. (2021) applied a proteomic approach monitoring the health status of patients who developed COVID-19. The study showed changes in proteins related to cholesterol metabolism and heart disease. Our *in silico* analyses resulting from the application of the deft pipeline detailed here showed an alteration in processes linked to lipoproteins and cholesterol. These results are in agreement with changes in the HDL proteome associated with COVID-19 complications (Souza Junior et al., 2021). The data obtained in this manuscript show that the plasma proteome reflects the immunological status of infected patients, since pathways and proteins were identified as increased in these groups. The intra-patient comparison at different time points showed there is an activation of the immune system, especially of pathways linked to leukocytes in the first 24h. However, these responses are diminished after 14days. Other studies also reported the ability of the plasma and serum proteome to reflect the action of the immune system in the infection (Arthur et al., 2021; Kumar, 2021; Villar et al., 2021; Zhong et al., 2021). Among the immune system proteins that stood out were C1QA, which is crucial in the activation of the classical pathway of the complement system (Ghebrehiwet, Hosszu, Valentino, & Peerschke, 2012; Kouser et al., 2015; Nayak, Pednekar, Reid, & Kishore, 2012), and CD93, involved in the regulation of phagocytosis of apoptotic cells *in vivo*. Transcriptomics approaches have also been extensively applied to study SARS-CoV-2 infection (Butler et al., 2021; Chakraborty, Sharma, Bhattacharya, Zayed, & Lee, 2021; Islam et al., 2021; O'Donnell et al., 2021; Sun et al., 2020; Wong et al., 2021). We identified regulated pathways linked to misfolded proteins in the lung of infected patients, as previously reported (Rosa-Fernandes et al., 2021). Moreover, we identified an increase in the enzymes involved ECM remodeling, extravasation of intracellular contents and activation of the immune system (Leeming et al., 2021; Leng et al., 2020).

Leukocyte activation was recurrent in our analyses of the integrated proteomic and transcriptomic data (Alon et al., 2021; Coradi & Vieira, 2021). Regarding our data integration approach, we saw that there is little correlation between transcripts mapped in lung and whole blood transcriptome. However, looking at moderately conserved transcripts, we found processes linked to leukocyte activation and host glycosylation. Our integrated data also mapped transcripts and proteins linked to proteasome activity in the infected groups. In fact, other groups have identified the importance of the proteasome in COVID-19 from the application of other techniques (Chatterjee et al., 2020; Longhitano et al., 2020; Wang et al., 2021).

6. Conclusions

We conducted integrated omic data analysis with focus on open tools to analyze and interpret this data type. We presented the entire pipeline for handling from raw files to differentially expressed features using the FastQC, Trim Galore!, HISAT2, samtools, featureCounts and R softwares. We also looked at useful tools to explore processes and pathways related to mapped transcripts and proteins. An important point was establishing a pipeline to integrate proteomics and transcriptomics data, since adjusting these data to make them comparable can be a big challenge. We also consistently reported applying various packages available in the bioconductor, providing the codes used. Regarding the biological findings, we evidenced increase in the cholesterol metabolism, immune system activity, ECM and proteasome degradation increased in infected patients. We also noticed a leukocyte activation profile in both proteomics and transcriptomics data. Finally, we identified a panel of proteins and transcripts that are regulated in the same direction in the lung transcriptome and plasma proteome that have a great ability to distinguish between healthy and infected groups. This panel of markers was applied to another cohort of patients and showed good results, corroborating the robustness and usefulness of the computational tools presented in this manuscript.

Acknowledgments

We are grateful for the financial support provided by the São Paulo Research Foundation (FAPESP, grants processes n° 2018/18257-1 (G.P.), 2018/15549-1 (G.P.), 2021/00140-3

(J.M.D.S.)); by the Conselho Nacional de Desenvolvimento Científico e Tecnológico ("Bolsa de Produtividade" (S.K.N.M. and G.P.)); by the Coordenação de Aperfeiçoamento de Pessoal de Nível Superior (process n° 88882.463201/2019-01 (L.R.F.)).

Data and code availability

The codes used in this manuscript were made available in the Supplementary File 1 in the online version at https://doi.org/10.1016/bs.apcsb.2022.04.002. The input and output data are available in Supplementary File 2 in the online version at https://doi.org/10.1016/bs.apcsb.2022.04.002.

References

Al-Amrani, S., Al-Jabri, Z., Al-Zaabi, A., Alshekaili, J., & Al-Khabori, M. (2021). Proteomics: Concepts and applications in human medicine. *World Journal of Biological Chemistry*, *12*, 57–69. https://doi.org/10.4331/wjbc.v12.i5.57.

Alon, R., Sportiello, M., Kozlovski, S., Kumar, A., Reilly, E. C., Zarbock, A., et al. (2021). Leukocyte trafficking to the lungs and beyond: Lessons from influenza for COVID-19. *Nature Reviews. Immunology*, *21*, 49–64. https://doi.org/10.1038/s41577-020-00470-2.

Amiri-Dashatan, N., Koushki, M., Abbaszadeh, H.-A., Rostami-Nejad, M., & Rezaei-Tavirani, M. (2018). Proteomics applications in health: Biomarker and drug discovery and food industry. *Iranian Journal of Pharmaceutical Research*, *17*, 1523–1536.

Anders, S., Pyl, P. T., & Huber, W. (2015). HTSeq—A Python framework to work with high-throughput sequencing data. *Bioinformatics*, *31*, 166–169. https://doi.org/10.1093/bioinformatics/btu638.

Arthur, L., Esaulova, E., Mogilenko, D. A., Tsurinov, P., Burdess, S., Laha, A., et al. (2021). Cellular and plasma proteomic determinants of COVID-19 and non-COVID-19 pulmonary diseases relative to healthy aging. *Nature Aging*, *1*, 535–549. https://doi.org/10.1038/s43587-021-00067-x.

Badua, C. L. D. C., Baldo, K. A. T., & Medina, P. M. B. (2021). Genomic and proteomic mutation landscapes of SARS-CoV-2. *Journal of Medical Virology*, *93*, 1702–1721. https://doi.org/10.1002/jmv.26548.

Bojkova, D., Klann, K., Koch, B., Widera, M., Krause, D., Ciesek, S., et al. (2020). Proteomics of SARS-CoV-2-infected host cells reveals therapy targets. *Nature*, *583*, 469–472. https://doi.org/10.1038/s41586-020-2332-7.

Butler, D., Mozsary, C., Meydan, C., Foox, J., Rosiene, J., Shaiber, A., et al. (2021). Shotgun transcriptome, spatial omics, and isothermal profiling of SARS-CoV-2 infection reveals unique host responses, viral diversification, and drug interactions. *Nature Communications*, *12*, 1660. https://doi.org/10.1038/s41467-021-21361-7.

Cannataro, M. (2007). Computational proteomics: Management and analysis of proteomics data. *Briefings in Bioinformatics*, *9*, 97–101. https://doi.org/10.1093/bib/bbn011.

Carvalho, P. C., Lima, D. B., Leprevost, F. V., Santos, M. D. M., Fischer, J. S. G., Aquino, P. F., et al. (2016). Integrated analysis of shotgun proteomic data with PatternLab for proteomics 4.0. *Nature Protocols*, *11*, 102–117. https://doi.org/10.1038/nprot.2015.133.

Chakraborty, C., Sharma, A. R., Bhattacharya, M., Zayed, H., & Lee, S.-S. (2021). Understanding gene expression and transcriptome profiling of COVID-19: An initiative towards the mapping of protective immunity genes against SARS-CoV-2 infection. *Frontiers in Immunology*, *12*, 724936. https://doi.org/10.3389/fimmu.2021.724936.

Chatterjee, P., Ponnapati, M., Kramme, C., Plesa, A. M., Church, G. M., & Jacobson, J. M. (2020). Targeted intracellular degradation of SARS-CoV-2 via computationally optimized peptide fusions. *Communications Biology*, *3*, 715. https://doi.org/10.1038/s42003-020-01470-7.

Chen, Y., Yao, H., Zhang, N., Wu, J., Gao, S., Guo, J., et al. (2021). Proteomic analysis identifies prolonged disturbances in pathways related to cholesterol metabolism and myocardium function in the COVID-19 recovery stage. *Journal of Proteome Research, 20*, 3463–3474. https://doi.org/10.1021/acs.jproteome.1c00054.

Colinge, J., & Bennett, K. L. (2007). Introduction to computational proteomics. *PLoS Computational Biology, 3*, e114. https://doi.org/10.1371/journal.pcbi.0030114.

Coradi, C., & Vieira, S. L. V. (2021). Alterações leucocitárias em pacientes com COVID-19 observadas em extensão de sangue periférico. *Research, Society and Development, 10*, e400101119838. https://doi.org/10.33448/rsd-v10i11.19838.

Das, A., Ahmed, R., Akhtar, S., Begum, K., & Banu, S. (2021). An overview of basic molecular biology of SARS-CoV-2 and current COVID-19 prevention strategies. *Gene Reports, 23*, 101122. https://doi.org/10.1016/j.genrep.2021.101122.

Desai, N., Neyaz, A., Szabolcs, A., Shih, A. R., Chen, J. H., Thapar, V., et al. (2020). Temporal and spatial heterogeneity of host response to SARS-CoV-2 pulmonary infection. *Nature Communications, 11*, 6319. https://doi.org/10.1038/s41467-020-20139-7.

Dixon, B. E., Wools-Kaloustian, K. K., Fadel, W. F., Duszynski, T. J., Yiannoutsos, C., Halverson, P. K., et al. (2021). Symptoms and symptom clusters associated with SARS-CoV-2 infection in community-based populations: Results from a statewide epidemiological study. *PLoS One, 16*, e0241875. https://doi.org/10.1371/journal.pone.0241875.

Dobin, A., Davis, C. A., Schlesinger, F., Drenkow, J., Zaleski, C., Jha, S., et al. (2013). STAR: Ultrafast universal RNA-seq aligner. *Bioinformatics, 29*, 15–21. https://doi.org/10.1093/bioinformatics/bts635.

Ghebrehiwet, B., Hosszu, K. K., Valentino, A., & Peerschke, E. I. B. (2012). The C1q family of proteins: Insights into the emerging non-traditional functions. *Frontiers in Immunology, 3*. https://doi.org/10.3389/fimmu.2012.00052.

Guangcan, S., Yong, C., Zhenlin, C., Chao, L., Shangtong, L., Hao, C., et al. (2021). How to use open-pFind in deep proteomics data analysis?—A protocol for rigorous identification and quantitation of peptides and proteins from mass spectrometry data. *Biophysics Reports*, 7, 207–226. https://doi.org/10.52601/bpr.2021.210004.

Haas, P., Muralidharan, M., Krogan, N. J., Kaake, R. M., & Hüttenhain, R. (2021). Proteomic approaches to study SARS-CoV-2 biology and COVID-19 pathology. *Journal of Proteome Research, 20*, 1133–1152. https://doi.org/10.1021/acs.jproteome.0c00764.

Hayes, L. D., Ingram, J., & Sculthorpe, N. F. (2021). More than 100 persistent symptoms of SARS-CoV-2 (long COVID): A scoping review. *Frontiers in Medicine, 8*, 750378. https://doi.org/10.3389/fmed.2021.750378.

Islam, A. B. M. M. K., Khan, M. A.-A.-K., Ahmed, R., Hossain, M. S., Kabir, S. M. T., Islam, M. S., et al. (2021). Transcriptome of nasopharyngeal samples from COVID-19 patients and a comparative analysis with other SARS-CoV-2 infection models reveal disparate host responses against SARS-CoV-2. *Journal of Translational Medicine, 19*, 32. https://doi.org/10.1186/s12967-020-02695-0.

Kim, D., Paggi, J. M., Park, C., Bennett, C., & Salzberg, S. L. (2019). Graph-based genome alignment and genotyping with HISAT2 and HISAT-genotype. *Nature Biotechnology, 37*, 907–915. https://doi.org/10.1038/s41587-019-0201-4.

Kim, D., Pertea, G., Trapnell, C., Pimentel, H., Kelley, R., & Salzberg, S. L. (2013). TopHat2: Accurate alignment of transcriptomes in the presence of insertions, deletions and gene fusions. *Genome Biology, 14*, R36. https://doi.org/10.1186/gb-2013-14-4-r36.

Kodama, Y., Shumway, M., Leinonen, R., & International Nucleotide Sequence Database Collaboration. (2012). The sequence read archive: Explosive growth of sequencing data. *Nucleic Acids Research, 40*, D54–D56. https://doi.org/10.1093/nar/gkr854.

Kong, A. T., Leprevost, F. V., Avtonomov, D. M., Mellacheruvu, D., & Nesvizhskii, A. I. (2017). MSFragger: Ultrafast and comprehensive peptide identification in mass spectrometry-based proteomics. *Nature Methods*, *14*, 513–520. https://doi.org/10.1038/nmeth.4256.

Kouser, L., Madhukaran, S. P., Shastri, A., Saraon, A., Ferluga, J., Al-Mozaini, M., et al. (2015). Emerging and novel functions of complement protein C1q. *Frontiers in Immunology*, *6*. https://doi.org/10.3389/fimmu.2015.00317.

Kumar, V. (2021). Can proteomics-based approaches further help COVID-19 prevention and therapy? *Expert Review of Proteomics*, *18*, 241–245. https://doi.org/10.1080/14789450.2021.1924684.

Kwan, P. K. W., Cross, G. B., Naftalin, C. M., Ahidjo, B. A., Mok, C. K., Fanusi, F., et al. (2021). A blood RNA transcriptome signature for COVID-19. *BMC Medical Genomics*, *14*, 155. https://doi.org/10.1186/s12920-021-01006-w.

Lachén-Montes, M., Corrales, F. J., Fernández-Irigoyen, J., & Santamaría, E. (2020). Proteomics insights into the molecular basis of SARS-CoV-2 infection: What we can learn from the human olfactory axis. *Frontiers in Microbiology*, *11*, 2101. https://doi.org/10.3389/fmicb.2020.02101.

Langmead, B., & Salzberg, S. L. (2012). Fast gapped-read alignment with Bowtie 2. *Nature Methods*, *9*, 357–359. https://doi.org/10.1038/nmeth.1923.

Law, C. W., Alhamdoosh, M., Su, S., Dong, X., Tian, L., Smyth, G. K., et al. (2016). RNA-seq analysis is easy as 1-2-3 with limma, Glimma and edgeR. *F1000 Research*, *5*. https://doi.org/10.12688/f1000research.9005.3. ISCB Comm J-1408.

Lazari, L. C., Ghilardi, F. D. R., Rosa-Fernandes, L., Assis, D. M., Nicolau, J. C., Santiago, V. F., et al. (2021). Prognostic accuracy of MALDI-TOF mass spectrometric analysis of plasma in COVID-19. *Life Science Alliance*, *4*, e202000946. https://doi.org/10.26508/lsa.202000946.

Leeming, D. J., Genovese, F., Sand, J. M. B., Rasmussen, D. G. K., Christiansen, C., Jenkins, G., et al. (2021). Can biomarkers of extracellular matrix remodelling and wound healing be used to identify high risk patients infected with SARS-CoV-2?: Lessons learned from pulmonary fibrosis. *Respiratory Research*, *22*, 38. https://doi.org/10.1186/s12931-020-01590-y.

Leng, L., Cao, R., Ma, J., Mou, D., Zhu, Y., Li, W., et al. (2020). Pathological features of COVID-19-associated lung injury: A preliminary proteomics report based on clinical samples. *Signal Transduction and Targeted Therapy*, *5*, 240. https://doi.org/10.1038/s41392-020-00355-9.

Li, Y., Hou, G., Zhou, H., Wang, Y., Tun, H. M., Zhu, A., et al. (2021). Multi-platform omics analysis reveals molecular signature for COVID-19 pathogenesis, prognosis and drug target discovery. *Signal Transduction and Targeted Therapy*, *6*, 155. https://doi.org/10.1038/s41392-021-00508-4.

Liao, Y., Smyth, G. K., & Shi, W. (2014). featureCounts: An efficient general purpose program for assigning sequence reads to genomic features. *Bioinformatics*, *30*, 923–930. https://doi.org/10.1093/bioinformatics/btt656.

Longhitano, L., Tibullo, D., Giallongo, C., Lazzarino, G., Tartaglia, N., Galimberti, S., et al. (2020). Proteasome inhibitors as a possible therapy for SARS-CoV-2. *International Journal of Molecular Sciences*, *21*, E3622. https://doi.org/10.3390/ijms21103622.

Macedo-da-Silva, J., Rosa-Fernandes, L., Barbosa, R. H., Angeli, C. B., Carvalho, F. R., de Oliveira Vianna, R. A., et al. (2020). Serum proteomics reveals alterations in protease activity, axon guidance, and visual phototransduction pathways in infants with in utero exposure to zika virus without congenital zika syndrome. *Frontiers in Cellular and Infection Microbiology*, *10*, 577819. https://doi.org/10.3389/fcimb.2020.577819.

Mahalmani, V. M., Mahendru, D., Semwal, A., Kaur, S., Kaur, H., Sarma, P., et al. (2020). COVID-19 pandemic: A review based on current evidence. *Indian Journal of Pharmacology, 52*, 117–129. https://doi.org/10.4103/ijp.IJP_310_20.

Mangul, S., Martin, L. S., Hill, B. L., Lam, A. K.-M., Distler, M. G., Zelikovsky, A., et al. (2019). Systematic benchmarking of omics computational tools. *Nature Communications, 10*, 1393. https://doi.org/10.1038/s41467-019-09406-4.

Martens, L., & Vizcaíno, J. A. (2017). A golden age for working with public proteomics data. *Trends in Biochemical Sciences, 42*, 333–341. https://doi.org/10.1016/j.tibs.2017.01.001.

Mishra, S., Shah, M. I., Udhaya Kumar, S., Thirumal Kumar, D., Gopalakrishnan, C., Al-Subaie, A. M., et al. (2021). Network analysis of transcriptomics data for the prediction and prioritization of membrane-associated biomarkers for idiopathic pulmonary fibrosis (IPF) by bioinformatics approach. In *Advances in protein chemistry and structural biology* (pp. 241–273). Elsevier. https://doi.org/10.1016/bs.apcsb.2020.10.003.

Musich, R., Cadle-Davidson, L., & Osier, M. V. (2021). Comparison of short-read sequence aligners indicates strengths and weaknesses for biologists to consider. *Frontiers in Plant Science, 12*, 657240. https://doi.org/10.3389/fpls.2021.657240.

Nalbandian, A., Sehgal, K., Gupta, A., Madhavan, M. V., McGroder, C., Stevens, J. S., et al. (2021). Post-acute COVID-19 syndrome. *Nature Medicine, 27*, 601–615. https://doi.org/10.1038/s41591-021-01283-z.

Nayak, A., Pednekar, L., Reid, K. B., & Kishore, U. (2012). Complement and non-complement activating functions of C1q: A prototypical innate immune molecule. *Innate Immunity, 18*, 350–363. https://doi.org/10.1177/1753425910396252.

O'Donnell, K. L., Pinski, A. N., Clancy, C. S., Gourdine, T., Shifflett, K., Fletcher, P., et al. (2021). Pathogenic and transcriptomic differences of emerging SARS-CoV-2 variants in the Syrian golden hamster model. *eBioMedicine, 73*, 103675. https://doi.org/10.1016/j.ebiom.2021.103675.

Overmyer, K. A., Shishkova, E., Miller, I. J., Balnis, J., Bernstein, M. N., Peters-Clarke, T. M., et al. (2021). Large-scale multi-omic analysis of COVID-19 severity. *Cell Systems, 12*, 23–40.e7. https://doi.org/10.1016/j.cels.2020.10.003.

Rais, Y., Fu, Z., & Drabovich, A. P. (2021). Mass spectrometry-based proteomics in basic and translational research of SARS-CoV-2 coronavirus and its emerging mutants. *Clinical Proteomics, 18*, 19. https://doi.org/10.1186/s12014-021-09325-x.

Rosa-Fernandes, L., Barbosa, R. H., Dos Santos, M. L. B., Angeli, C. B., Silva, T. P., Melo, R. C. N., et al. (2020). Cellular imprinting proteomics assay: A novel method for detection of neural and ocular disorders applied to congenital zika virus syndrome. *Journal of Proteome Research, 19*, 4496–4515. https://doi.org/10.1021/acs.jproteome.0c00320.

Rosa-Fernandes, L., Cugola, F. R., Russo, F. B., Kawahara, R., de Melo Freire, C. C., Leite, P. E. C., et al. (2019). Zika virus impairs neurogenesis and synaptogenesis pathways in human neural stem cells and neurons. *Frontiers in Cellular Neuroscience, 13*, 64. https://doi.org/10.3389/fncel.2019.00064.

Rosa-Fernandes, L., Lazari, L. C., da Silva, J. M., de Morais Gomes, V., Machado, R. R. G., dos Santos, A. F., et al. (2021). SARS-CoV-2 activates ER stress and Unfolded protein response (preprint). *Biochemistry*. https://doi.org/10.1101/2021.06.21.449284.

Schaarschmidt, S., Fischer, A., Zuther, E., & Hincha, D. K. (2020). Evaluation of seven different RNA-seq alignment tools based on experimental data from the model plant *Arabidopsis thaliana*. *International Journal of Molecular Sciences, 21*, E1720. https://doi.org/10.3390/ijms21051720.

Shu, T., Ning, W., Wu, D., Xu, J., Han, Q., Huang, M., et al. (2020). Plasma proteomics identify biomarkers and pathogenesis of COVID-19. *Immunity, 53*, 1108–1122.e5. https://doi.org/10.1016/j.immuni.2020.10.008.

Souza Junior, D. R., Silva, A. R. M., Rosa-Fernandes, L., Reis, L. R., Alexandria, G., Bhosale, S. D., et al. (2021). HDL proteome remodeling associates with COVID-19 severity. *Journal of Clinical Lipidology*, *15*, 796–804. https://doi.org/10.1016/j.jacl.2021.10.005.

Sun, J., Ye, F., Wu, A., Yang, R., Pan, M., Sheng, J., et al. (2020). Comparative transcriptome analysis reveals the intensive early stage responses of host cells to SARS-CoV-2 infection. *Frontiers in Microbiology*, *11*, 593857. https://doi.org/10.3389/fmicb.2020.593857.

Terracciano, R., Preianò, M., Fregola, A., Pelaia, C., Montalcini, T., & Savino, R. (2021). Mapping the SARS-CoV-2-host protein-protein interactome by affinity purification mass spectrometry and proximity-dependent biotin labeling: A rational and straightforward route to discover host-directed anti-SARS-CoV-2 therapeutics. *International Journal of Molecular Sciences*, *22*, E532. https://doi.org/10.3390/ijms22020532.

Tyanova, S., Temu, T., & Cox, J. (2016). The MaxQuant computational platform for mass spectrometry-based shotgun proteomics. *Nature Protocols*, *11*, 2301–2319. https://doi.org/10.1038/nprot.2016.136.

Villar, M., Urra, J. M., Rodríguez-del-Río, F. J., Artigas-Jerónimo, S., Jiménez-Collados, N., Ferreras-Colino, E., et al. (2021). Characterization by quantitative serum proteomics of immune-related prognostic biomarkers for COVID-19 symptomatology. *Frontiers in Immunology*, *12*, 730710. https://doi.org/10.3389/fimmu.2021.730710.

Vizcaíno, J. A., Côté, R. G., Csordas, A., Dianes, J. A., Fabregat, A., Foster, J. M., et al. (2013). The PRoteomics IDEntifications (PRIDE) database and associated tools: Status in 2013. *Nucleic Acids Research*, *41*, D1063–D1069. https://doi.org/10.1093/nar/gks1262.

Wang, Z., Zhao, Y., Wang, Q., Xing, Y., Feng, L., Kong, J., et al. (2021). Identification of proteasome and caspase inhibitors targeting SARS-CoV-2 Mpro. *Signal Transduction and Targeted Therapy*, *6*, 214. https://doi.org/10.1038/s41392-021-00639-8.

Wenk, M. R. (2005). The emerging field of lipidomics. *Nature Reviews. Drug Discovery*, *4*, 594–610. https://doi.org/10.1038/nrd1776.

Wong, H. S.-C., Guo, C.-L., Lin, G.-H., Lee, K.-Y., Okada, Y., & Chang, W.-C. (2021). Transcriptome network analyses in human coronavirus infections suggest a rational use of immunomodulatory drugs for COVID-19 therapy. *Genomics*, *113*, 564–575. https://doi.org/10.1016/j.ygeno.2020.12.041.

Wu, P., Chen, D., Ding, W., Wu, P., Hou, H., Bai, Y., et al. (2021). The trans-omics landscape of COVID-19. *Nature Communications*, *12*, 4543. https://doi.org/10.1038/s41467-021-24482-1.

Wu, M., Chen, Y., Xia, H., Wang, C., Tan, C. Y., Cai, X., et al. (2020). Transcriptional and proteomic insights into the host response in fatal COVID-19 cases. *Proceedings of the National Academy of Sciences of the United States of America*, *117*, 28336–28343. https://doi.org/10.1073/pnas.2018030117.

Zhong, W., Altay, O., Arif, M., Edfors, F., Doganay, L., Mardinoglu, A., et al. (2021). Next generation plasma proteome profiling of COVID-19 patients with mild to moderate symptoms. *eBioMedicine*, *74*, 103723. https://doi.org/10.1016/j.ebiom.2021.103723.

Zhu, Z., Zhang, S., Wang, P., Chen, X., Bi, J., Cheng, L., et al. (2022). A comprehensive review of the analysis and integration of omics data for SARS-CoV-2 and COVID-19. *Briefings in Bioinformatics*, *23*, bbab446. https://doi.org/10.1093/bib/bbab446.

FSC
www.fsc.org
MIX
Papier | Fördert gute Waldnutzung
FSC® C083411

Kolibri 360 GmbH
Ferdinand-Jühlke-Straße 7
99095 Erfurt, Deutschland
produktsicherheit@kolibri360.de